Concepts and Applications
of Finite Element Analysis

Board of Advisors, Engineering

Concepts and Applications
of Finite Element Analysis

*A Treatment of the Finite Element Method as used
for the Analysis of Displacement, Strain, and Stress.*

Robert D. Cook
University of Wisconsin

John Wiley & Sons, Inc.

New York   London   Sydney   Toronto

Library of Congress Cataloging in Publication Data:

Cook, Robert Davis.
Concepts and applications of finite element analysis.

Includes bibliographical references.
1. Finite element method.   I.   Title
TA646.C66   624'.171   74-7053
ISBN 0-471-16915-3

Preface

Since about 1956 the finite element method has developed enormously. It began as a numerical method of stress analysis and is still most widely used for this purpose. Also it has become useful in many other areas including heat conduction, seepage flow, fluid dynamics, and electric and magnetic fields. Mathematicians now recognize it as a proper area of study.

This book is intended for students of stress analysis. It is assumed that they are familiar with mechanics of materials, computer programming in Fortran, and matrix algebra. Learning will be easier if they are familiar with numerical analysis techniques and basic theory of elasticity, but introductory material regarding these topics is provided. Therefore, an undergraduate education is sufficient for an understanding of most of the contents.

Let me describe in more detail the background required for a profitable study of this book. I assume that the student intends to use the finite element method for practical stress analysis and wants to understand why the method behaves as it does, but is not greatly concerned with mathematical proofs and derivations. He should be well versed in the undergraduate courses of statics, dynamics, and mechanics of materials. As an approximation he should understand enough computer programming to be able to use subroutines, auxilliary storage such as disc files, and COMMON and EQUIVALENCE statements. The need for programming ability exists because computer implementation is mandatory but cannot be done entirely by persons who are unfamiliar with the method. The student must be familiar with basic matrix operations: for example, matrix multiplication, transposition, and differentiation must be mastered; and the meaning of an inverse must be clear but not necessarily the calculation procedure by which it is obtained. Although the use of matrix notation is not absolutely necessary, a description of the method would be very cumbersome without

2 /8

it. In addition, matrix formulation leads easily to computer implementation, a fact that is familiar to those who have studied matrix methods for analysis of framed structures.

Knowledge of theory of elasticity, plate and shell theory, energy methods, and numerical analysis is also desirable. However, only the more elementary notions from these subjects are really necessary. I will summarize these in Chapter One and, where necessary, in later chapters. Rarely, an advanced concept is used. Here the reader may accept the result as plausible or he may study the subject further on his own. This resembles the situation that prevails in elementary mechanics of materials: the student must ignore certain effects and accept many explanations as plausible, or he must study the theory of elasticity to obtain clarification.

I feel that this volume is related to finite element analysis in the same way that a book on strength of materials is related to stress analysis in general. In other words, the book emphasizes topics that are simple, useful, and have physical appeal; it is not primarily concerned with the mathematical foundations of the subject or its extensions to the most complicated problems.

The subject is so large that even an annotated bibliography on finite element theory and practice would be a book in itself. Many references are provided, but no attempt is made to list all of them. Here are some of the topics that the book does and does not discuss:

- Our subject is analysis, not design. Optimization and treatment of structural modifications are not described.
- The orientation is toward general methods and continua instead of special methods for framed structures. However, truss and beam elements are used as helpful examples.
- Elements discussed in detail are based on assumed displacement fields, work well in practice, are not restricted to special shape, and have no strain or curvature quantities as nodal degrees of freedom. Nodal freedoms are displacements or displacements and rotations. Many elements, including some good ones, are not discussed, but these often differ from the simpler elements because of algebraic detail, not because of principle.
- Linear static analysis is emphasized for both practical and pedagogical reasons. Dynamic and nonlinear problems are treated less thoroughly.
- Some items of interest and importance are presented as homework problems. The casual reader, therefore, may wish to examine the problems without working them.
- Fortran coding is given if it is available, helpful, relatively self-contained, and not very long.

To keep the book from being overly long, detailed discussion of elements is limited to those based on assumed displacement fields. Other approaches are less popular but are occasionally valuable. For example, hybrid elements (Section 17.1) are particularly good for plate bending. This is especially true because of the discovery that element $Q19$ (reference 9 of Chapter Twelve), which was formerly considered excellent, is not very accurate if its length is greater than about 1.5 times its width.

The currently available textbooks on finite elements emphasize basic theory and give many examples of how the method works when a functioning computer program is appled to practical problems. In between basic theory and the production computer run lies the conversion of theory into practice. This book adds a "how to go about it" component to the available textbook literature. I encourage the student to consult other books to benefit from different approaches and other points of view.

In practice, where a working computer program is essential to success in finite element work, the analyst must either do programming himself or work closely with others who do programming. To compel study of this vital area, I usually assign a term project in which a program is to be written or modified, and the work verified by means of test cases. The project also has the benefit of encouraging clear understanding, since fuzzy thinking is not apt to coax correct answers from a computer.

I thank the students whose questions prompted my study of certain topics, and who made valuable and clever suggestions. V. N. Shah contributed the scheme of Figures 6.1.3 and 6.1.4. H. Takemoto did the programming and numerical work for Section 10.5. J. A. Stricklin and W. E. Haisler at Texas A.&M. University generously supplied technical reports and other information.

Robert D. Cook

Madison, Wisconsin, 1973

Contents

Notation

{ }	A column vector, even when coefficients are written horizontally to save space; for example $$\begin{Bmatrix} u \\ v \end{Bmatrix} \equiv \{u \ v\}$$
[ ]	Row vector, or rectangular or square matrix.
⌈ ⌋	Diagonal matrix.
$[\,]^T$, $\{\,\}^T$	Transpose of a matrix or column vector.
$[\,]^{-1}$	Inverse of a square matrix.
$[\,]^{-T}$	Inverse of the transpose of a square matrix, $$[\,]^{-T} \equiv ([\,]^{-1})^T \equiv ([\,]^T)^{-1}$$
det [ ]	Determinant of a square matrix.
· (dot)	Differentiation with respect to time; for example, $$\dot{u} = du/dt, \qquad \ddot{u} = d^2u/dt^2$$
, (comma)	Indicates partial differentiation when used before a subscript; for example, $$w_{,x} = \partial w/\partial x \qquad w_{,xy} = \partial^2 w/\partial x\, \partial y$$
– (bar)	Denotes amplitude of motion when used in Chapter Twelve; for example, $$w = \bar{w} \sin \omega t$$
δQ	Virtual (infinitesimal) change in a quantity Q.

ΔQ	Small but finite change in a quantity Q.
d.o.f.	Degrees of freedom.
E, ν	Elastic modulus and Poisson's ratio of an isotropic material.
λ	An eigenvalue; or a loading parameter whose critical (buckling) value is the eigenvalue λ_{cr}.
ρ	Mass density.
T	Temperature above an arbitrary zero.
t	Thickness of a member, or time (Chapter Twelve).
ω	Natural frequency of vibration, rad/sec.
U, U_o	Strain energy, strain energy per unit volume.
Π_p	Total potential energy.
$\{f\}$	$\{f\} = \{u \ v \ w\}$, displacements of a point in the coordinate directions. *Note.* Following common usage, w has the same direction as θ in Chapter Seven, and v has the same direction as θ in Chapter Eight, where θ is the circumferential coordinate.
$\{F\}$	$\{F\} = \{F_x \ F_y \ F_z\}$, body forces per unit volume.
$\{\Phi\}$	$\{\Phi_x \ \Phi_y \ \Phi_z\}$, surface tractions per unit area.
$\{d\}, \{D\}$	Nodal degrees of freedom for element and structure, respectively.
$\{r\}, \{R\}$	Generalized forces applied to nodes, corresponding to $\{d\}$ and $\{D\}$, respectively.
$\{\bar{r}\}$	Forces applied to element by nodes; $\{\bar{r}\} = - \{r\}$.
$\{P\}$	Concentrated loads applied to structure nodes. $\{R\} = \{P\}$ if body forces, thermal strains, etc., are all zero.
$\{\sigma\}, \{\epsilon\}$	Engineering stresses and strains; see Sections 1.2 and 1.3.
$\{\sigma_o\}, \{\epsilon_o\}$	Initial stresses and strains; see Sections 1.2 and 1.3.
$[E]$	Elastic stress-strain relation; $\{\sigma\} = [E]\{\epsilon\}$ for $\{\epsilon_o\} = 0$.
$[N]$	Shape function matrix, $\{f\} = [N]\{d\}$.
$[B]$	Strain-displacement matrix, $\{\epsilon\} = [B]\{d\}$.
$[DA]$	Matrix relating $\{d\}$ to generalized coordinates (amplitudes) $\{a\}$ in assumed displacement fields, $\{d\} = [DA]\{a\}$ (see Section 4.2).
$[k], [K]$	Element and structure stiffness matrices, respectively.
$[k_\sigma], [K_\sigma]$	Element and structure initial-stress stiffness matrices, respectively.
$[J], [J^*]$	Jacobian matrix and its inverse, $[J^*] \equiv [J]^{-1}$.
$[m], [M]$	Element and structure mass matrices, respectively.
$[T]$	Coordinate transformation matrix, used with and without subscripts (see Chapter Eleven).
$U_L, [K_L]$	Strain energy and structure stiffness matrix arising from linear strain-displacement relations. Subscript L used in Chapter Fourteen to distinguish these quantities from nonlinear forms U_{NL} and $[K_{NL}]$.

$[I]$ A unit matrix (1's on-diagonal an 0's elsewhere).

$\left\{\dfrac{\partial \Pi_p}{\partial a}\right\}$ Represents the vector

$$\left\{\frac{\partial \Pi_p}{\partial a_1}\ \frac{\partial \Pi_p}{\partial a_2} \cdots \frac{\partial \Pi_p}{\partial a_n}\right\}$$

where Π_p is a scalar function of parameters a_1 through a_n.

Infrequently used notation and modifications of the above (e.g., by addition of subscripts) are defined where they are used.

CHAPTER ONE

Background and Introductory Material

1.1 The Finite Element Method

Suppose that the structure of Fig. 1.1.1a must be analyzed to find its stresses and displacements. Although the structure is a cantilever beam of rectangular cross section, beam theory does not apply because the beam is too short. The methods of theory of elasticity serve to formulate this problem in terms of partial differential equations. The solution of these equations would provide an exact solution of the stress analysis problem. But the loading and support conditions are such that a solution is very difficult indeed.

Engineers cannot devote long hours to the solution of the partial differential equations of elasticity when each new problem comes along. They are quite content with good approximate solutions that can be gotten at reasonable effort. Approximate solutions often involve replacing the continuum with a substitute structure having a finite number of degrees of freedom. One such method involves use of a lattice of elastic bars (1)[1] (see Fig. 1.1.1b). If elastic properties of the bars can be appropriately specified, displacements of the framework will closely approximate displacements of the original structure. When displacements are known, strains and stresses may be computed by means of relations described in Section 1.2. The d.o.f. (degrees of freedom) involved are the displacements of the points where the bars are connected to one another. This "lattice analogy" seeks to capitalize on well-established methods for analysis of framed structures.

The finite element method uses a substitute structure whose parts are, in a sense, pieces of the actual structure. Thus in Fig. 1.1.1c each rectangular

---

[1] Numbers in parentheses indicate references listed at the back of the book.

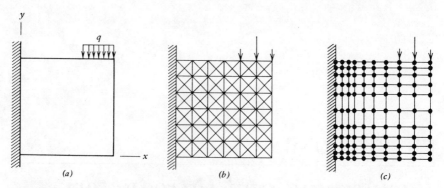

Figure 1.1.1. (*a*) Flat plate under in-plane loading. (*b*) A framework model. (*c*) A finite element model.

area is a flat sheet of material. The grid lines merely outline the separate areas, and are not framework members. We might intuitively expect that as the subdivision is made finer, the substitute structure models the original structure more closely. This is true if one follows the rules of formulation that are discussed in Chapter Four. Our substitute structure is a *finite element structure*, and each separate area is a *finite element*. Points where the elements are connected to one another are called *nodes* and are indicated by dots in Fig. 1.1.1*c*.

A finite element is *not* simply a piece of the actual structure. Suppose, as perhaps suggested by Fig. 1.1.1*c*, that many small rectangular plates were to be connected together by pins at their corners. As load is applied, we would expect distortions to be larger near the corners than elsewhere, with element sides becoming curved and gaps appearing between some elements. Clearly such behavior is not representative of the actual structure. Accordingly, for the description of element behavior to be consistent with our interest in modeling structure behavior, certain element distortions must be excluded, while others must be permitted. This consideration forces us to adopt a somewhat theoretical point of view if element properties are to be properly formulated; physical insight, while helpful and perhaps essential, is not alone sufficient.

For each element, one may write equations of the form

$$k_{11}\,d_1 + k_{12}\,d_2 + \cdots + k_{1n}\,d_n = \bar{r}_1$$
$$k_{21}\,d_1 + k_{22}\,d_2 + \cdots + k_{2n}\,d_n = \bar{r}_2$$
$$\vdots \qquad\qquad \vdots \qquad \vdots$$
$$k_{n1}\,d_1 + k_{n2}\,d_2 + \cdots + k_{nn}\,d_n = \bar{r}_n$$

$$(1.1.1)$$

Here n is the number of d.o.f. per element, each d_i is a node point displacement, and $\bar{r}_i$ is a node point force corresponding to d_i. The coefficients k_{ij} are called *stiffness coefficients*. A single d.o.f. example is that of the linear spring; a force $\bar{r} = kd$ is required to stretch the spring d units. If gathered into matrix form, Eqs. 1.1.1 are

$$[k]\{d\} = \{\bar{r}\} \tag{1.1.2}$$

where $[k]$ is called the *element stiffness matrix*, $\{d\}$ the *element nodal displacement vector*, and $\{\bar{r}\}$ the *vector of element nodal loads*.

An example may help to explain the meaning of $[k]$ (a more detailed explanation appears in Chapter Two). Consider a beam (Fig. 1.1.2a). There is a node at each end, and element d.o.f. consist of two displacements and two rotations. Equation 1.1.2 becomes

$$\underset{4\times4}{[k]}\{v_1 \quad \theta_1 \quad v_2 \quad \theta_2\} = \{\bar{r}\} \tag{1.1.3}$$

For the displacement configuration of Fig. 1.1.2b, we have $\{d\} = \{1 \quad 0 \quad 0 \quad 0\}$, so that $\{\bar{r}\}$ is equal to the first column of $[k]$. Thus, in general, the ith column of $[k]$ represents the forces that must be applied to the element to preserve static equilibrium when $d_i = 1$ and all other d's are zero. This description, however, does not tell us how to form $[k]$ in the first place. The proper formulation of various element types occupies much of Chapters Four through Ten.

A finite element structure, such as Fig. 1.1.1c, is built by assembly of the component elements. (Assembly is discussed in Chapter Two.) Equations analogous to Eq. 1.1.2 are produced:

$$[K]\{D\} = \{R\} \tag{1.1.4}$$

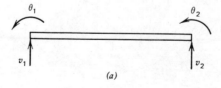

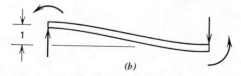

Figure 1.1.2.   (a) Beam element. (b) Forces required for $v_1 = 1$, $v_2 = \theta_1 = \theta_2 = 0$.

where $[K]$, $\{D\}$, and $\{R\}$ are named as below Eq. 1.1.2, except that "structure" replaces "element." The meaning of $[K]$ is directly analogous to the meaning of $[k]$ as described in the foregoing paragraph.

When the structural Eqs. 1.1.4 have been formulated, all quantities are known but the displacements $\{D\}$, which may then be found by a standard solution technique for linear algebraic equations. When $\{D\}$ is known, displacements are known at all points in the finite element structure. This is because each element $\{d\}$ is contained in the structure $\{D\}$, and because element properties are formulated using a simple polynomial interpolation scheme that gives displacements at any point within an element in terms of

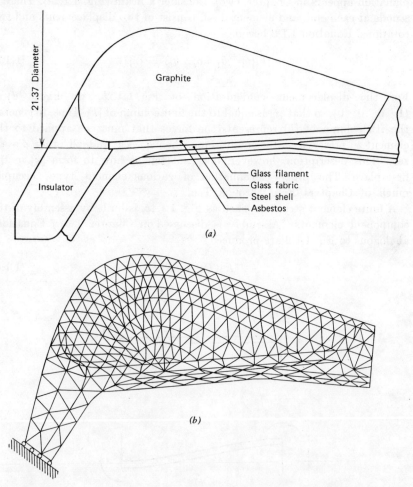

Figure 1.1.3. (a) Axisymmetric rocket nozzle. (b) Finite element mesh. (From AIAA Journal paper.)

the nodal displacements of the element. This method of element formulation is known as the *assumed displacement field method*. There are other ways to formulate elements, and some are mentioned in Section 17.1. However, to date the assumed displacement field method has been most popular and most successful, and we will confine our attention to this approach.

The finite element method is most widely used for analysis of stress and displacement in bodies under static load. But problems of buckling, vibration, and dynamic response may also be solved, and the method has been extended to handle the nonlinear problems of large displacement, plasticity, and creep.

Two additional stress analysis problems may help illustrate the power of the method. Figure 1.1.3a shows a rocket nozzle problem (1 of Chapter Seven).[2] The nozzle is axisymmetric, and the figure shows one half of a cross section through the axis of symmetry. The nozzle is composed of several different materials, and is loaded by a combination of temperature gradient and internal pressure. Figure 1.1.3b shows the division into finite elements. Each element is a toroidal ring of triangular cross section. Figure 1.1.4 shows the finite element mesh used on one quarter of a cylinder-to-cylinder intersection (4.12 of Chapter Eighteen). Note that the mesh is refined near the juncture where stresses are high and change considerably over short distances. The elements are flat triangles and are formulated in such a way as to model thin shell behavior.

The foregoing examples suggest some advantages of the finite element method.

- Its ability to use elements of various types, sizes, and shapes, and to model a structure of arbitrary geometry.

- Its ability to accommodate arbitrary support conditions and arbitrary loading, including thermal loading.

- Its ability to model composite structures involving different structural components, such as stiffening members on a shell and combinations of plates, bars, solids, etc.

- The finite element structure closely resembles the actual structure instead of being a quite different abstraction that is hard to visualize.

Other approximate methods—for example, the finite difference method—lack these attributes or accommodate them less readily. Of course, there are also disadvantages. These include:

- A specific numerical result is obtained for a specific problem. A general closed form solution, which would permit one to examine system response to changes in various parameters, is not produced.

[2] This notation means Ref. 1 in Chapter Seven.

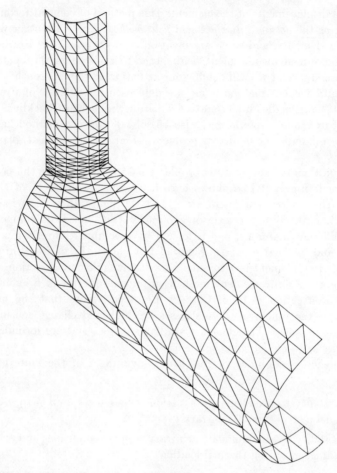

Figure 1.1.4.   Finite element mesh for cylinder-to-cylinder intersection.

- Experience and judgment are needed in order to construct a good finite element model.
- A large computer and a reliable computer program are essential.
- Input and output data may be large and tedious to prepare and interpret.

These drawbacks, however, are not unique to the finite element method.

1.2   *Theory of Elasticity*

A sound understanding of finite elements demands familiarity with the elementary concepts of equilibrium, compatibility, and strain-displacement

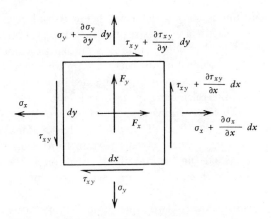

Figure 1.2.1. Stresses on plane differential element.

relations. We will discuss these concepts with particular reference to plane problems in cartesian coordinates, leaving the three-dimensional case and cylindrical coordinates largely to elasticity textbooks (3).

Figure 1.2.1 shows stresses acting on a differential (not finite!) element. Let t be the element thickness, and let F_x and F_y be body forces per unit of volume. That is, forces $F_x\,dV$ and $F_y\,dV$ are exerted on each unit of volume dV, these forces being caused perhaps by gravity, acceleration, or a magnetic field. Stresses are generally functions of the coordinates; thus, for example, $\partial\sigma_x/\partial x$ is the rate of change of σ_x with respect to x, and the change of σ_x in a distance dx is $(\partial\sigma_x/\partial x)\,dx$. Equilibrium of forces in the x direction requires that

$$-\sigma_x t\,dy - \tau_{xy}t\,dx + \left(\sigma_x + \frac{\partial\sigma_x}{\partial x}dx\right)t\,dy$$

$$+ \left(\tau_{xy} + \frac{\partial\tau_{xy}}{\partial y}dy\right)t\,dx + F_x t\,dx\,dy = 0 \quad (1.2.1)$$

A similar summation of forces may be written in the y direction. The two equations reduce to

$$\frac{\partial\sigma_x}{\partial x} + \frac{\partial\tau_{xy}}{\partial y} + F_x = 0 \qquad\qquad \sigma_{x,x} + \tau_{xy,y} + F_x = 0$$

$$\qquad\qquad\qquad\qquad , \qquad \text{or} \qquad\qquad\qquad\qquad (1.2.2)$$

$$\frac{\partial\tau_{xy}}{\partial x} + \frac{\partial\sigma_y}{\partial y} + F_y = 0 \qquad\qquad \tau_{xy,x} + \sigma_{y,y} + F_y = 0$$

The latter notation, where a comma denotes partial differentiation, will be used throughout most of this book. Equations 1.2.2 are known as the *differential equations of equilibrium.*

When a continuous body is deformed below its failure point, no cracks or gaps appear in the body, and no part overlaps another. Stated more elegantly, the displacement field is both continuous and single-valued. This common-place observation is known as the *compatibility condition*. It assumes special importance in the finite element method, since in a finite element model of a continuous body, interelement gaps or overlapping could conceivably occur.

As we will see below, strains ϵ_x, ϵ_y, and γ_{xy} are derivable from only two displacement field quantities. This implies that a definite relation exists among ϵ_x, ϵ_y, and γ_{xy} if these strains pertain to a compatible deformation. This relation is called the "compatibility equation." We will not have to use it, as we will subsequently work with assumptions on displacement instead of strain.

Boundary conditions are of two types. Boundary conditions on displacement require certain displacements to prevail at certain points; for example, in Fig. 1.1.1a no displacement is permitted along the vertical wall. Similarly, there may be boundary conditions on stress. In Fig. 1.1.1a, $\sigma_y = \tau_{xy} = 0$ is prescribed along the bottom and part of the top, $\sigma_x = \tau_{xy} = 0$ along the right side, and $\tau_{xy} = 0$, $\sigma_y = -q$ along another part of the top. Where displacements are prescribed, stresses are unknown and are free to assume values dictated by the solution. Similarly, where stresses are prescribed, displacements are unknown.

Let surface tractions Φ_x and Φ_y act on the boundary of an object (Fig. 1.2.2). Tractions have dimensions of force per unit area. For equilibrium to prevail, applied tractions and internal stresses must satisfy the following

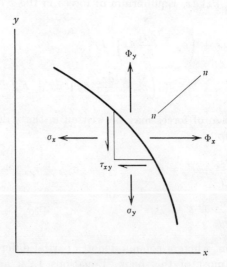

Figure 1.2.2. Surface tractions at a boundary.

equations at the boundary.

$$l\,\sigma_x + m\,\tau_{xy} = \Phi_x$$
$$l\,\tau_{xy} + m\,\sigma_y = \Phi_y$$

$$(1.2.3)$$

where l and m are direction cosines of the boundary-normal line n–n.

When a stress field or a displacement field has been found that satisfies the differential equations of equilibrium and the compatibility conditions at every point, and satisfies boundary conditions at every boundary point, then a solution has been found. If no nonlinearities are present, then no other solution is possible; that is, the solution is unique. How do these observations relate to finite element analysis? A finite element solution seeks to satisfy the necessary conditions in an approximate way. If elements are based on assumed displacement fields, the compatibility condition is satisfied within elements if each assumed field is continuous. Suitably chosen fields will also satisfy interelement compatibility and permit all displacement boundary conditions to be met. The equilibrium equations and boundary conditions on stress (Eqs. 1.2.2 and 1.2.3) are satisfied only approximately. The approximation improves as more elements are used, and, were it not for computational problems, the conditions on stress would be exactly satisfied in the limit of an infinitely fine mesh.

A finite element method based on assumed stress fields would employ only fields that satisfy the equilibrium conditions *a priori*. Increasingly fine meshes would then yield better and better approximations to the compatibility conditions.

Next we must describe the relation between displacement and strain. A general two-dimensional displacement field causes lines 0–1–2 in Fig. 1.2.3 to move to configuration 0'–1'–2'. The motion is a combination of rigid body motion and distortion. Let u and v represent the x and y direction displacements of the arbitrary point 0. As is the case with stress, u and v are functions of the coordinates. Accordingly the displacements of points 1 and 2 are not merely u and v, but u and v plus increments. We will assume that these increments are infinitesimals. Now let us consider the normal strain in the x direction, ϵ_x. Using the definition of strain as change in length divided by original length, we find

$$\epsilon_x = \frac{[dx + (u + u_{,x}\,dx) - u] - dx}{dx} = u_{,x}$$

$$(1.2.4)$$

A similar analysis yields the y-direction normal strain as

$$\epsilon_y = v_{,y}$$

$$(1.2.5)$$

The shear strain γ_{xy} is defined as

$$\gamma_{xy} = \phi_1 + \phi_2 = \frac{(u + u_{,y}\,dy) - u}{dy} + \frac{(v + v_{,x}\,dx) - v}{dx} = u_{,y} + v_{,x} \quad (1.2.6)$$

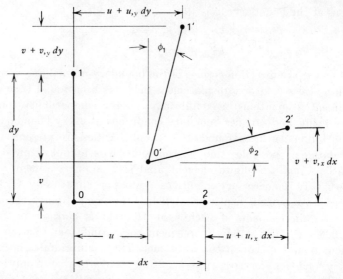

Figure 1.2.3. Displacement and distortion of an infinitesimal element. Commas denote partial differentiation.

where the computation is based on small-displacement assumptions; for example, $\phi_2 \approx \tan \phi_2$ and $dx \approx [dx + (u + u_{,x} dx) - u]$. Equations 1.2.4, 1.2.5, and 1.2.6 are the *strain-displacement relations*. In the three-dimensional case similar expressions are obtained and may be written in matrix form as

$$
\begin{Bmatrix} \epsilon_x \\ \epsilon_y \\ \epsilon_z \\ \gamma_{xy} \\ \gamma_{yz} \\ \gamma_{zx} \end{Bmatrix} = \begin{bmatrix} \dfrac{\partial}{\partial x} & 0 & 0 \\ 0 & \dfrac{\partial}{\partial y} & 0 \\ 0 & 0 & \dfrac{\partial}{\partial z} \\ \dfrac{\partial}{\partial y} & \dfrac{\partial}{\partial x} & 0 \\ 0 & \dfrac{\partial}{\partial z} & \dfrac{\partial}{\partial y} \\ \dfrac{\partial}{\partial z} & 0 & \dfrac{\partial}{\partial x} \end{bmatrix} \begin{Bmatrix} u \\ v \\ w \end{Bmatrix} \tag{1.2.7}
$$

where u, v, w are respectively the x, y, z components of displacement.

Finally, for future reference, an alternate form of the differential equations of equilibrium is presented. Again we consider the two-dimensional case.

Hooke's law for an isotropic material is

$$\epsilon_x = \frac{(\sigma_x - \nu\sigma_y)}{E}$$

$$\epsilon_y = \frac{(\sigma_y - \nu\sigma_x)}{E} \qquad (1.2.8)$$

$$\gamma_{xy} = \frac{\tau_{xy}}{G} = 2(1 + \nu)\frac{\tau_{xy}}{E}$$

where E = elastic modulus, G = shear modulus, and ν = Poisson's ratio. Equations 1.2.8 may be solved for stress and substituted into Eqs. 1.2.2, then the strain-displacement relations substituted into the result. Thus, for the case $F_x = F_y = 0$,

$$u_{,xx} + u_{,yy} = \frac{1 + \nu}{2}(u_{,yy} - v_{,xy})$$

$$\qquad (1.2.9)$$

$$v_{,xx} + v_{,yy} = \frac{1 + \nu}{2}(v_{,xx} - u_{,xy})$$

where $u_{,xx} = \partial^2 u/\partial x^2$, etc. Equations 1.2.9 are the equilibrium equations expressed in terms of displacements.

1.3 Stress-Strain Relations

In the general three-dimensional case of an elastic material, the strains $\{\epsilon\}$ are related to stresses and initial strains by the equation

$$\begin{Bmatrix} \epsilon_x \\ \epsilon_y \\ \epsilon_z \\ \gamma_{xy} \\ \gamma_{yz} \\ \gamma_{zx} \end{Bmatrix} = [C]_{6\times6} \begin{Bmatrix} \sigma_x \\ \sigma_y \\ \sigma_z \\ \tau_{xy} \\ \tau_{yz} \\ \tau_{zx} \end{Bmatrix} + \begin{Bmatrix} \epsilon_{xo} \\ \epsilon_{yo} \\ \epsilon_{zo} \\ \gamma_{xyo} \\ \gamma_{yzo} \\ \gamma_{zxo} \end{Bmatrix} \qquad (1.3.1)$$

where $[C]$ is a matrix of elastic coefficients. Equation 1.3.1 is "Hooke's law"; for real materials it is an approximation valid for small strains. For the familiar case of heating of an isotropic material, the initial strain vector contains the terms

$$\epsilon_{xo} = \epsilon_{yo} = \epsilon_{zo} = \alpha T$$

$$\gamma_{xzo} = \gamma_{yzo} = \gamma_{zxo} = 0 \qquad (1.3.2)$$

where α = coefficient of thermal expansion and T = temperature change. Thus, in the absence of stress, the strains $\{\epsilon\}$ are merely those of free thermal expansion. Also, for isotropy, we have the familiar terms $C_{11} = 1/E$, $C_{12} = C_{13} = -\nu/E$, $C_{14} = C_{15} = C_{16} = 0$, etc., where E = elastic modulus and ν = Poisson's ratio.

Let us abbreviate Eq. 1.3.1 as

$$\{\epsilon\} = [C]\{\sigma\} + \{\epsilon_o\} \tag{1.3.3}$$

The inverse of Eq. 1.3.3 is

$$\{\sigma\} = [E](\{\epsilon\} - \{\epsilon_o\}) = [E]\{\epsilon\} + \{\sigma_o\} \tag{1.3.4}$$

where $[E] = [C]^{-1}$ and the vector of initial stresses, $\{\sigma_o\}$, is given by $\{\sigma_o\} = -[E]\{\epsilon_o\}$. Thus if all mechanical strain $\{\epsilon\}$ were prohibited while a body is heated, the stresses $\{\sigma_o\}$ would prevail. Equation 1.3.4, with either $\{\epsilon_o\}$ or $\{\sigma_o\}$, is the form of Hooke's law we will use throughout this book.

Matrix $[E]$ is symmetric. This is easily shown if the reader will grant in advance the validity of Eq. 3.5.6. By differentiation,

$$
\begin{aligned}
\frac{\partial^2 U_o}{\partial \epsilon_x \, \partial \epsilon_y} &= \frac{\partial^2 U_o}{\partial \epsilon_y \, \partial \epsilon_x} = E_{12} = E_{21}, \\
\frac{\partial^2 U_o}{\partial \epsilon_x \, \partial \epsilon_z} &= \frac{\partial^2 U_o}{\partial \epsilon_z \, \partial \epsilon_x} = E_{13} = E_{31}, \text{ etc.}
\end{aligned}
\tag{1.3.5}
$$

Thus in the most general case of elastic properties, $[E]$ contains 21 independent coefficients.

If the material is orthotropic and axes x, y, z coincide with principal directions of the material, then matrix $[C]$ has the form (2):

$$
[C] =
\begin{bmatrix}
\dfrac{1}{E_x} & -\dfrac{\nu_{yx}}{E_y} & -\dfrac{\nu_{zx}}{E_z} & 0 & 0 & 0 \\[2mm]
-\dfrac{\nu_{xy}}{E_x} & \dfrac{1}{E_y} & -\dfrac{\nu_{zy}}{E_z} & 0 & 0 & 0 \\[2mm]
-\dfrac{\nu_{xz}}{E_x} & -\dfrac{\nu_{yz}}{E_y} & \dfrac{1}{E_z} & 0 & 0 & 0 \\[2mm]
0 & 0 & 0 & \dfrac{1}{G_{xy}} & 0 & 0 \\[2mm]
0 & 0 & 0 & 0 & \dfrac{1}{G_{yz}} & 0 \\[2mm]
0 & 0 & 0 & 0 & 0 & \dfrac{1}{G_{zx}}
\end{bmatrix}
\tag{1.3.6}
$$

Each Poisson ratio ν_{ij} characterizes strain in the j direction produced by stress in the i direction $(i, j = x, y, z)$. For example, uniaxial stress σ_z produces strains $\epsilon_z = \sigma_z/E_z$, $\epsilon_x = -\nu_{zx}\epsilon_z$, $\epsilon_y = -\nu_{zy}\epsilon_z$. Because of the symmetry requirement,

$$E_x\nu_{yx} = E_y\nu_{xy}$$
$$E_y\nu_{zy} = E_z\nu_{yz} \qquad (1.3.7)$$
$$E_z\nu_{xz} = E_x\nu_{zx}$$

If a temperature change T is applied, $\gamma_{xyo} = \gamma_{yzo} = \gamma_{zzo} = 0$ and $\epsilon_{xo} = \alpha_x T$, $\epsilon_{yo} = \alpha_y T$, $\epsilon_{zo} = \alpha_z T$. If the material is orthotropic but x, y, z are not principal directions, a coordinate transformation must be used (Section 11.2).

If the material is isotropic, material properties are the same in every direction, and Eq. 1.3.4 becomes

$$\{\sigma\} = \frac{E}{(1 + \nu)(1 - 2\nu)}\begin{bmatrix} 1 - \nu & \nu & \nu & 0 & 0 & 0 \\ \nu & 1 - \nu & \nu & 0 & 0 & 0 \\ \nu & \nu & 1 - \nu & 0 & 0 & 0 \\ 0 & 0 & 0 & \dfrac{1 - 2\nu}{2} & 0 & 0 \\ 0 & 0 & 0 & 0 & \dfrac{1 - 2\nu}{2} & 0 \\ 0 & 0 & 0 & 0 & 0 & \dfrac{1 - 2\nu}{2} \end{bmatrix}\{\epsilon\}$$

$$+ \frac{E\alpha T}{1 - 2\nu}\{-1 \quad -1 \quad -1 \quad 0 \quad 0 \quad 0\} \qquad (1.3.8)$$

where thermal strains αT have been presumed. The last three diagonal terms are the shear moduli $G = E/2(1 + \nu)$.

For the (two-dimensional) case of plane stress, we have by definition $\sigma_z = \tau_{yz} = \tau_{zx} = \gamma_{yz} = \gamma_{zx} = 0$, and strain ϵ_z, while in general not zero, is ignored. Thus Eq. 1.3.4 specializes to

$$\begin{Bmatrix} \sigma_x \\ \sigma_y \\ \tau_{xy} \end{Bmatrix} = [E]_{3 \times 3}\begin{Bmatrix} \epsilon_x \\ \epsilon_y \\ \gamma_{xy} \end{Bmatrix} + \begin{Bmatrix} \sigma_{xo} \\ \sigma_{yo} \\ \tau_{xyo} \end{Bmatrix} \qquad (1.3.9)$$

For the (two-dimensional) case of plane strain, we have by definition $\epsilon_z = \tau_{yz} = \tau_{zx} = \gamma_{yz} = \gamma_{zx} = 0$, and stress σ_z, while in general not zero, need not enter into computations. The symbolic statement Eq. 1.3.9 applies to plane strain as well as plane stress. The appropriate form of $[E]$ is obtained as follows: (a) for plane stress, discard rows and columns 3, 5, and 6 from

$[C]$ in Eq. 1.3.1, and invert the remaining 3 by 3 matrix; (b) for plane strain, discard rows and columns 3, 5, and 6 from $[E] = [C]^{-1}$. (The reader should ask himself *why* discarding these terms is proper.) Note that in order for plane stress or plane strain to prevail in the xy plane, the xy plane must be a plane of elastic symmetry. Thus, for the orthotropic case, the z axis must be a principal material direction. If in addition the x and y axes are principal directions, then $E_{13} = E_{31} = E_{23} = E_{32} = 0$. Formulas for the special case of isotropy are given as Eqs. 4.4.7.

The analysis capabilities of the finite element method far exceed the knowledge of material behavior on which the analysis must be based. In the absence of test data, one must often estimate elastic constants for orthotropic materials and postulate the nature of any inelastic behavior.

1.4   Behavior of Plates and Shells

A flat plate is the two-dimensional analogue of a straight beam. Both are intended to support transverse loads. A piece cut from a flat plate is shown in Fig. 1.4.1, and Fig. 1.4.1a shows the stresses that act on one of the cut surfaces. The shear stress τ_{xy} is not present in beam theory, but τ_{zx} and σ_x would exist in a beam lying along the x axis. As in beam theory, σ_x and τ_{zx} are assumed to vary linearly and parabolically, respectively, through the thickness. Stress τ_{xy} is also assumed to vary linearly. A distributed surface load q acts in the z direction. As the plate is thin in comparison with its span, any body force may be converted to an equivalent load q. Loading in the xy plane is assumed to be absent.

The stresses shown in Fig. 1.4.1a give rise to the following forces:

$$M_x = \int_{-t/2}^{t/2} \sigma_x z\, dz, \qquad Q_x = \int_{-t/2}^{t/2} \tau_{zx}\, dz, \qquad M_{xy} = \int_{-t/2}^{t/2} \tau_{xy} z\, dz \qquad (1.4.1)$$

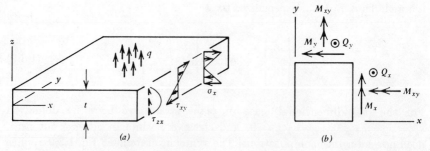

Figure 1.4.1.   (a) Stresses in a homogeneous, linearly elastic plate. (b) Forces and moments produced by the stresses.

With the exception of M_{xy}, which is absent from beam theory, these relations are identical to those that prevail in elementary beam theory—subject to the understanding that a unit width in the y direction has been assumed in Eqs. 1.4.1. Thus, M is a moment per unit length and Q a shear force per unit length. Forces M_y, Q_y, and M_{xy} appear on a surface perpendicular to the y axis and are shown in vector form in Fig. 1.4.1b.

Equilibrium equations may be derived as follows. Draw a free body diagram of a differential element, presuming that the Q's and M's are functions of x and y, so that, for example, Q_x acts on one side of the element and $Q_x + Q_{x,x}\,dx$ on the opposite side. Sum forces in the z direction and moments about x and y axes. The resulting equations reduce to (1 of Chapter Nine)

$$Q_{x,x} + Q_{y,y} + q = 0$$
$$M_{x,x} + M_{xy,y} = Q_x \qquad (1.4.2)$$
$$M_{xy,x} + M_{y,y} = Q_y$$

To explain plate behavior in the simplest way, let us assume for the remainder of this section that initial strains $\{\epsilon_o\}$ are absent and that transverse shear deformation may be neglected. An arbitrary point P in Fig. 1.4.2 experiences lateral displacement w and also in-plane displacements

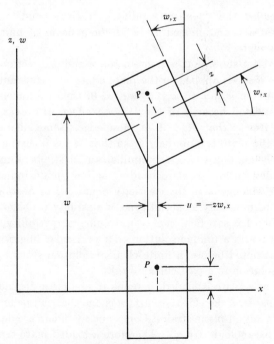

Figure 1.4.2. Deflection of a small plate element.

u and v because of the rotation of the plate element. Accordingly, we find the strain-displacement relations to be

$$
\begin{aligned}
\epsilon_x &= u_{,x} = -zw_{,xx} \\
\epsilon_y &= v_{,y} = -zw_{,yy} \\
\gamma_{xy} &= u_{,y} + v_{,x} = -2zw_{,xy}
\end{aligned}
\tag{1.4.3}
$$

Thus, with transverse shear deformation neglected, the lateral displacement field $w = w(x, y)$ completely describes the deformation state.

Stress σ_z is considered negligible in comparison with σ_x and σ_z; therefore Eqs. 1.2.8 are the appropriate Hooke's law for an isotropic plate. If Eqs. 1.4.3 are substituted into Eqs. 1.2.8, and the resulting stresses into Eqs. 1.4.1, there results (1 of Chapter Nine)

$$
\begin{aligned}
M_x &= -D(w_{,xx} + \nu w_{,yy}) \\
M_y &= -D(w_{,yy} + \nu w_{,xx}) \\
M_{xy} &= -\frac{Gt^3}{6} w_{,xy} = -(1 - \nu)Dw_{,xy} \\
D &= \frac{Et^3}{12(1 - \nu^2)}
\end{aligned}
\tag{1.4.4}
$$

Here D is called the "flexural rigidity." It corresponds to the bending stiffness EI of a beam; indeed $D = EI$ for a beam of unit width and if $\nu = 0$ for the plate.

Equations 1.4.4 show that actions in the x and y directions are coupled. For example, if M_x is applied to the plate edges $x =$ constant to create the cylindrical surface $w_{,xx} \neq 0$, $w_{,yy} = w_{,xy} = 0$, then a moment M_y is automatically generated because of the Poisson effect. However, if edges $y =$ constant are free so that $M_y = 0$ at these edges, some edge curvature $w_{,yy}$ will exist. If the plate is quite narrow so that $M_y \approx 0$ throughout, we have essentially a beam, and we see the familiar anticlastic bending.

The foregoing outline of plate bending action should indicate that plate equations are analogous to the elasticity equations of Section 1.2. Indeed, the plate problem is only a special type of elasticity problem. Therefore, as noted in Section 1.2, satisfaction of equilibrium, compatibility, and boundary conditions provides a unique solution to a particular linear plate problem. And, again as noted before, a finite element solution seeks to approximate this unique solution in a particular way.

Generally, as a plate is loaded, it tends to assume a displaced shape that is not developable. A developable surface is one that can be made from a flat sheet without any in-plane strains. Cones and cylinders are developable, but spheres and paraboloids are not. Therefore a loaded plate tends to develop in-plane strains and hence in-plane forces. These forces, called membrane

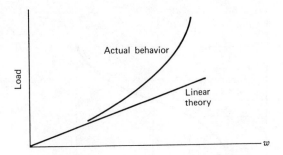

Figure 1.4.3. Load versus displacement for a typical plate.

forces, carry a part of the load, just as membrane forces support the pressure load in a balloon. As load increases, so do the membrane forces, and the plate appears to become stiffer (Fig. 1.4.3). The problem is a nonlinear one, and to solve it, higher-order terms must be added to Eqs. 1.4.3. Linear theory may err by more than 50 % for $w = t$. The size of error is, of course, problem-dependent.

Shell behavior resembles plate behavior in that the forces of Fig. 1.4.1*b* are again present. However, membrane forces are also present because of the curved geometry of the shell. More specifically, the stresses of Fig. 1.4.1*a* still exist, and while σ_x and τ_{xy} may still be taken as linear in z, they do not vanish at $z = 0$. We therefore have, in addition to the forces of Fig. 1.4.1*b*, the membrane forces per unit length

$$N_x = \int_{-t/2}^{t/2} \sigma_x \, dz, \qquad N_y = \int_{-t/2}^{t/2} \sigma_y \, dz, \qquad N_{xy} = \int_{-t/2}^{t/2} \tau_{xy} \, dz \qquad (1.4.5)$$

Bending and membrane actions interact in a shell, and behavior is often not easily visualized or analyzed. The geometric complexities make shell analysis difficult.

A given amount of material will support much more load in direct tension or compression than if loaded predominantly in bending. Similarly, the most effective shell would carry all loads by membrane action alone. This ideal situation is rarely possible because of practical constraints. Nevertheless, a shell structure represents a very efficient use of material.

In subsequent chapters we will present further details of plate and shell theory where these details are needed.

1.5 *Interpolation and Representation of Curves*

In an element based on assumed displacement fields, the displacement of any point within the element is a function of the coordinates of the point

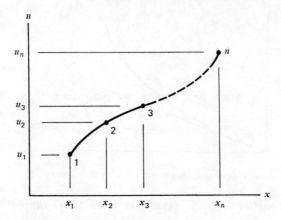

Figure 1.5.1.   Curve fitted through n points.

and the nodal displacements of the element. In other words, displacements u, v, w are governed by nodal displacements through an interpolation scheme. Differently based finite elements, such as those based on stress fields, also define field quantities in terms of nodal values. Interpolation, particularly by polynomials, is very much the "name of the game" in element formulation.

Suppose that some variable u has known values at n points (u_1, u_2, ..., u_n). The points need not be uniformly spaced. The points may be used to define a polynomial of degree $n-1$ by forcing the curve to pass through each point (Fig. 1.5.1). In the simplest case, $n = 2$, and we have the linear interpolation commonly used between two tabulated values. Lagrange's interpolation formula accomplishes the curve-fitting of Fig. 1.5.1. The formula is (4 of Chapter Five)

$$u = N_1 u_1 + N_2 u_2 + N_3 u_3 + \cdots + N_n u_n \qquad (1.5.1)$$

where

$$N_1 = \frac{(x - x_2)(x - x_3) \ldots (x - x_n)}{(x_1 - x_2)(x_1 - x_3) \ldots (x_1 - x_n)}$$

$$N_2 = \frac{(x - x_1)(x - x_3) \ldots (x - x_n)}{(x_2 - x_1)(x_2 - x_3) \ldots (x_2 - x_n)} \qquad (1.5.2)$$

$$N_n = \frac{(x - x_1)(x - x_2) \ldots (x - x_{n-1})}{(x_n - x_1)(x_n - x_2) \ldots (x_n - x_{n-1})}$$

Each of the N_i is a polynomial of degree $n - 1$. Note that each $N_i = 1$ for $x = x_i$, and that $N_i = 0$ for $x = x_j$, where $i \neq j$. This behavior is that demanded of an interpolation scheme. For any x between two points where u is defined, u is dependent on *all* u_i. The "true curve," if there is one, may

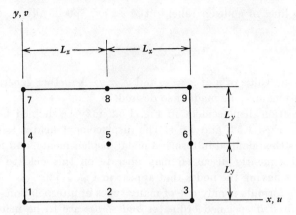

Figure 1.5.2. Plane rectangle.

not exactly match the curve given by Eq. 1.5.1 for all x. In a finite element context this statement means that a finite element solution for u may be slightly different than the displacements u of a theory of elasticity solution.

The two-dimensional analogue of Eq. 1.5.1 may be explained by the following example. Suppose that displacement u is to be interpolated from nodal values $u_1, u_2, \ldots, u_9$ in Fig. 1.5.2. Direct application of Eq. 1.5.1 yields

$$\text{on line 1–3,} \quad u = u_{13} = N_1(x)u_1 + N_2(x)u_2 + N_3(x)u_3$$
$$\text{on line 4–6,} \quad u = u_{46} = N_1(x)u_4 + N_2(x)u_5 + N_3(x)u_6 \quad (1.5.3)$$
$$\text{on line 7–9,} \quad u = u_{79} = N_1(x)u_7 + N_2(x)u_8 + N_3(x)u_9$$

Hence, along any y-parallel line,

$$u = N_1(y)u_{13} + N_2(y)u_{46} + N_3(y)u_{79} \quad (1.5.4)$$

where, with s representing x or y,

$$N_1(s) = \frac{(s - L_s)(s - 2L_s)}{(-L_s)(-2L_s)}$$

$$N_2(s) = \frac{(s)(s - 2L_s)}{(L_s)(-L_s)} \quad (1.5.5)$$

$$N_3(s) = \frac{(s)(s - L_s)}{(2L_s)(L_s)}$$

Substitution of Eqs. 1.5.3 into Eq. 1.5.4 yields the desired formula. This formula may be generalized to accommodate m lines of nodes parallel to the

y axis and n lines of nodes parallel to the x axis. Specifically,

$$u = \sum_{i=1}^{m} \sum_{j=1}^{n} N_i(x) N_j(y) u_{ij} \qquad (1.5.6)$$

where u_{ij} is the value of u at $x = x_i$ and $y = y_j$. A further generalization to three dimensions may be made if so desired.

The y-direction displacement in Fig. 1.5.2 may be defined by replacing u's by v's in Eqs. 1.5.3 and 1.5.4. The displacement field $u = u(x, y)$ and $v = v(x, y)$ is then defined in terms of nodal displacements, and formulation methods subsequently discussed may operate on this field to generate a finite element having the nodes that appear in Fig. 1.5.2.

Lagrange's formula is only one of many types of interpolation. A different type is required if specified ordinates *and slopes* are to be met at certain points. The name "Hermitian" is given to this type of interpolation (4 of Chapter Five). In Fig. 1.5.3a, two displacements and two rotations are

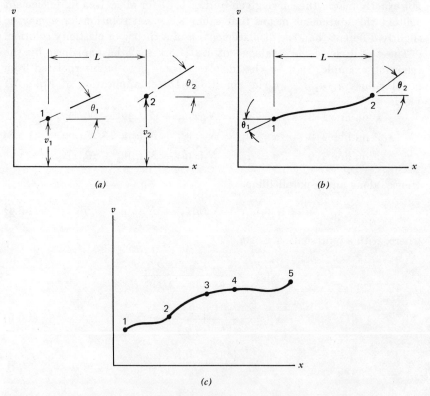

Figure 1.5.3. (a) Specified displacements and rotations at two points. (b) Cubic interpolation curve. (c) Cubic interpolation in each of four spans.

prescribed. Interpolation may be accomplished by means of a cubic polynomial, as a cubic is fully defined by four parameters. A straight beam has cubic displacements under end loading, so that the interpolated curve (Fig. 1.5.3b) may be viewed as a beam that carries end forces and end moments of such magnitude as to yield the necessary end displacements and rotations.

In Fig. 1.5.3c we presume that a displacement and a rotation are prescribed at each of the five points. Interpolation may be accomplished by fitting a different cubic to each of the four spans. Thus, within each span, displacement v is governed by nodal values at the ends of the span, but not by any nodal values outside the span. This is a characteristic of finite element behavior; displacements within an element are governed by only those nodes associated with that element. Stated another way, a finite element model of a structure involves a *piecewise* polynomial interpolation; nodal displacements define several different displacement fields that are laid side by side; there is no single polynomial that defines displacements of the entire structure. Note that in Fig. 1.5.3c displacements and slopes are continuous across nodes, but curvatures $v_{,xx}$ are not. Such behavior is typical of piecewise interpolation.

Fourier series may be used to represent loads and displacements that are periodic. A particular application is in the analysis of solids and shells of revolution under asymmetric loads. Here the period is 2π rad. A Fourier series for some parameter $u = u(x)$ may be written

$$u = \sum_{n=1}^{\infty} a_n \sin nx + \sum_{n=0}^{\infty} b_n \cos nx \qquad (1.5.7)$$

The sine functions are called *odd* or *antisymmetric*, as $u(x) = -u(-x)$. The cosine functions are called *even* or *symmetric*, as $u(x) = u(-x)$. Both a_n and b_n are functions of n. How a_n and b_n are to be chosen in order to represent a given periodic function is a straightforward process, but one that we will leave to other texts (3 of Chapter Five). However, as an example of the

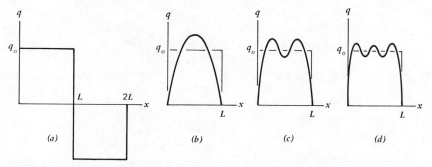

Figure 1.5.4. (a) Square wave and Fourier series representation using (b) one, (c) two, and (d) three terms.

manner in which a function is represented, consider the square wave of Fig. 1.5.4a. The wave repeats itself with a period of $x = 2L$, and could be used to represent a uniformly distributed load of intensity q_o on a span of length L. The series is found to be

$$q = \frac{4q_o}{\pi} \sum_{n \text{ odd}} \frac{1}{n} \sin \frac{n\pi x}{L} \tag{1.5.8}$$

As suggested by the latter parts of Fig. 1.5.4, the desired pattern is matched more and more closely as more terms of the series are used.

PROBLEMS

1.1.   For the beam element of Fig. 1.1.2, sketch the three deflected shapes $\{d\} = \{0\ 1\ 0\ 0\}$, $\{d\} = \{0\ 0\ 1\ 0\}$, and $\{d\} = \{0\ 0\ 0\ 1\}$. In each case show properly directed nodal forces (as in Fig. 1.1.2b).

1.2.   For a three-dimensional state of stress in cartesian coordinates, derive equations analogous to Eqs. 1.2.2.

1.3.   By equilibrium considerations instead of coordinate transformation, derive the form of Eqs. 1.2.2 appropriate to polar coordinates.

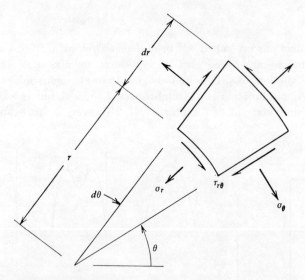

1.4.   (a)   Derive Eqs. 1.2.3.
        (b)   Derive Eq. 1.2.7.
        (c)   Derive Eqs. 1.2.9.

1.5. (a) Write Eq. 1.3.1 for an isotropic material, then derive Eq. 1.3.8 from it.

(b) Similarly, derive Eqs. 4.4.7.

(c) Derive the terms in $[E]$ of Eq. 1.3.9 from Eq. 1.3.6.

1.6. (a) Derive Eqs. 1.4.2.

(b) Derive Eqs. 1.4.4.

1.7. (a) Write out Eq. 1.5.6 for the problem of Fig. 1.5.2.

(b) Write out Eq. 1.5.6 for a plane rectangle having one node at each corner.

1.8. Establish the specific cubic polynomial needed in Fig. 1.5.3b.

1.9. Derive the equilibrium equations in cylindrical coordinates (Eqs. 7.3.5).

CHAPTER TWO

The Stiffness Method and the Plane Truss

2.1 Introduction

In Chapter One we merely introduced the idea of a stiffness matrix. We must now examine the idea more fully and explain how to assemble elements to form a structure, how to treat displacement boundary conditions, and how to solve the algebraic equations of the system. The plane truss is chosen as the vehicle for this development. It is a very simple structure, but one that is still practical. The concepts and procedures developed apply equally well to structures of other types.

A truss member is so simple that we have no trouble in developing its stiffness matrix by direct physical argument. This is not the case for most elements, and the necessary background for a general method of stiffness matrix derivation is discussed in Chapter Three.

In what follows, each member is assumed to be uniform, pin-connected at its ends, linearly elastic, and axially loaded. Displacements shown in sketches are *greatly exaggerated;* actual displacements are assumed to be small. Within the bounds of these assumptions, the following analysis is exact, not approximate.

We consider only the so-called *stiffness* method, in which stiffness coefficients are formed and displacements are the primary unknowns to be determined. The following discussion of structure and member stiffness resembles that of the 1956 paper by Turner, Clough, Martin, and Topp (1). References 1 and 4 might be regarded as the original works on finite elements in engineering.

2.2   Structure Stiffness Equations

The three-bar truss of Fig. 2.2.1 is used as an example. Nodes and elements (members) have been numbered arbitrarily. Let the stiffness of the elements be k_1, k_2, and k_3; that is, for any element i,

$$P_i = k_i e_i = \frac{A_i E_i}{L_i} e_i \qquad (2.2.1)$$

where

$$P_i = \text{force in element } i$$

$$e_i = \text{elongation of element } i$$

$$A_i, E_i, L_i = \text{cross-sectional area, elastic modulus}$$
$$\text{and length of element } i$$

Let us imagine that each of the three nodes is displaced a small amount, first in the x direction and then in the y direction, while all other nodal displacements are prohibited. In each of these six cases we may calculate the external forces that must be applied to maintain static equilibrium in the displaced configuration. The first two cases are shown in Fig. 2.2.2. The factors $1/2$ appear because member 2 is inclined at 45 degrees in this particular truss. As an example of the force computation, consider force $k_2 u_1/2$. The elongation of element 2 is $u_1/\sqrt{2}$, which produces an element force of

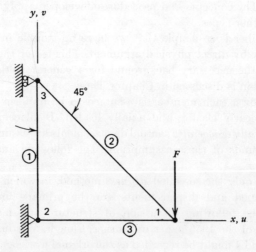

Figure 2.2.1.   Plane truss.

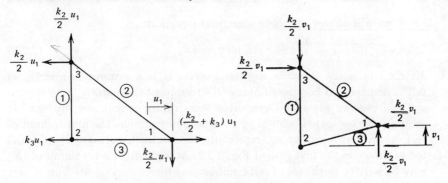

Figure 2.2.2. Forces induced by nodal displacements.

$k_2 u_1/\sqrt{2}$. The x and y direction components of this force are

$$(1/\sqrt{2})(k_2 u_1/\sqrt{2}) = k_2 u_1/2.$$

From the collection of six free-body diagrams, we may determine the equilibrating force required at any node in any displacement state. Designate the vector of nodal displacements in the positive coordinate directions as

$$\{D\} = \{u_1\, v_1\, u_2\, v_2\, u_3\, v_3\} \tag{2.2.2}$$

and the corresponding positively directed forces applied to the nodes as

$$\{R\} = \{p_1\, q_1\, p_2\, q_2\, p_3\, q_3\} \tag{2.2.3}$$

With this notation, the individual forces are

$$p_1 = \left(\frac{k_2}{2} + k_3\right)u_1 - \frac{k_2}{2}v_1 - k_3u_2 - \frac{k_2}{2}u_3 + \frac{k_2}{2}v_3$$

$$q_1 = -\frac{k_2}{2}u_1 + \frac{k_2}{2}v_1 + \frac{k_2}{2}u_3 - \frac{k_2}{2}v_3, \text{ etc.} \tag{2.2.4}$$

And finally, in matrix form these equilibrium equations are

$$
\begin{bmatrix}
\dfrac{k_2}{2} + k_3 & -\dfrac{k_2}{2} & -k_3 & 0 & -\dfrac{k_2}{2} & \dfrac{k_2}{2} \\[2mm]
-\dfrac{k_2}{2} & \dfrac{k_2}{2} & 0 & 0 & \dfrac{k_2}{2} & -\dfrac{k_2}{2} \\[2mm]
-k_3 & 0 & k_3 & 0 & 0 & 0 \\[2mm]
0 & 0 & 0 & k_1 & 0 & -k_1 \\[2mm]
-\dfrac{k_2}{2} & \dfrac{k_2}{2} & 0 & 0 & \dfrac{k_2}{2} & -\dfrac{k_2}{2} \\[2mm]
\dfrac{k_2}{2} & -\dfrac{k_2}{2} & 0 & -k_1 & -\dfrac{k_2}{2} & k_1 + \dfrac{k_2}{2}
\end{bmatrix}
\begin{Bmatrix} u_1 \\ v_1 \\ u_2 \\ v_2 \\ u_3 \\ v_3 \end{Bmatrix}
=
\begin{Bmatrix} p_1 \\ q_1 \\ p_2 \\ q_2 \\ p_3 \\ q_3 \end{Bmatrix}
\tag{2.2.5}
$$

or, as we will abbreviate these structural equations,

$$[K]\{D\} = \{R\} \qquad (2.2.6)$$

Matrix $[K]$ is the *structure stiffness matrix*. It is a symmetric matrix, as might be expected because of Maxwell's reciprocal theorem.[1]

The foregoing procedure generates the stiffness matrix row by row. It may also be generated column by column. For example, the first column of $[K]$, when multiplied by u_1, represents the particular array $\{R\}$ of nodal forces shown in the first case of Fig. 2.2.2. Accordingly, each column of $[K]$ may be written as the set of self-equilibrating forces that results from a unit displacement of a single node.

The above analysis may be applied to any structure, regardless of how many elements it has, and regardless of its degree of static indeterminacy. One always obtains as many equations as there are independent nodal displacements. If the structure is not a truss or frame, the stiffness properties of the elements must be somehow approximated, as is done in Chapters Four through Ten.

Each column of $[K]$ in Eq. 2.2.5 sums to zero, because each column represents an equilibrium set of nodal forces produced by unit displacement of one nodal degree of freedom. Each diagonal term K_{ii} is positive; if this were not so, a force and its corresponding displacement would be oppositely directed, which is physically unreasonable.

The stiffness matrix of Eq. 2.2.5 is singular; its order is six, but its rank only three. The physical reason is that no displacement boundary conditions have yet been imposed, and the structure is free to move as a rigid body. For any plane structure there are three independent rigid body motions; two may be translation and one rotation. Each is associated with zero forces $\{R\}$. Thus, for our three-bar truss, with displacement vectors $\{D\}$ of $\{1 \ 0 \ 1 \ 0 \ 1 \ 0\}$, $\{1 \ 1 \ 1 \ 1 \ 1 \ 1\}$, or $\{0 \ 0 \ 0 \ 1 \ 1 \ 1\}$, the product $[K]\{D\}$ is zero. The reader should sketch these displacements and verify that $[K]\{D\} = 0$. As suggested by these examples, an infinite number of rigid body motions are possible, but (for a plane structure) only three are independent. At least three nodal displacements must be prescribed in order to prevent rigid body motion of a plane structure.

In Fig. 2.2.1 the imposed conditions on force and displacement are

$$p_1 = 0, \qquad q_1 = -F, \qquad q_3 = 0$$
$$u_2 = v_2 = u_3 = 0 \qquad (2.2.7)$$

The remaining three forces and three displacements are as yet unknown.

---

[1] See a text on theory of elasticity or one of the many texts on analysis of framed structures by conventional and matrix methods.

Formal solution for the unknowns may proceed as follows. Let subscript o designate the known quantities of Eq. 2.2.7 and subscript s designate the remaining unknown quantities. By rearrangement of terms Eq. 2.2.5 may be partitioned:

$$\begin{bmatrix} K_{11} & K_{12} \\ \hline K_{21} & K_{22} \end{bmatrix} \begin{Bmatrix} D_s \\ D_o \end{Bmatrix} = \begin{Bmatrix} R_o \\ R_s \end{Bmatrix} \tag{2.2.8}$$

Consequently,

$$[K_{11}]\{D_s\} + [K_{12}]\{D_o\} = \{R_o\}$$
$$[K_{21}]\{D_s\} + [K_{22}]\{D_o\} = \{R_s\} \tag{2.2.9}$$

The first of Eqs. 2.2.9 yields the unknown displacements

$$\{D_s\} = [K_{11}]^{-1}(\{R_o\} - [K_{12}]\{D_o\}) \tag{2.2.10}$$

Hence, the second of Eqs. 2.2.9 yields the forces $\{R_s\}$. In the present example truss, $\{D_o\} = 0$; however, this need not be true in general.

2.3   Element Stiffness Equations

In Section 2.2 the structure matrix $[K]$ was formed by a direct attack on the entire structure. This procedure serves well in explaining the nature of $[K]$, but is too tedious for practical use. If instead we begin with a single representative element, we are led to a systematic and easily automated method for building $[K]$ by assembly of the individual elements.

Figure 2.3.1 shows a single element, arbitrarily placed in the xy plane. Let A, E, and L designate the (constant) cross-sectional area, the elastic modulus, and the element length. The sine and cosine of ϕ are denoted by s and c and may be evaluated from the nodal coordinates.

$$s = \sin \phi = \frac{(y_j - y_i)}{L}$$

$$c = \cos \phi = \frac{(x_j - x_i)}{L} \tag{2.3.1}$$

$$L = [(x_j - x_i)^2 + (y_j - y_i)^2]^{1/2}$$

Now, as in Section 2.2, each node is displaced a small amount in each coordinate direction, while all other nodal displacements are prohibited. The first of these four cases is shown in Fig. 2.3.2. The axial shortening cu_i produces an axial compressive force $cu_i AE/L$. The x and y components of this force must be equilibrated by the external forces p_i, q_i, p_j, and q_j.

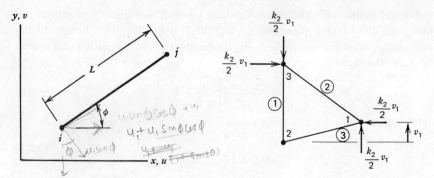

Figure 2.3.1.   Truss element.   *Figure 2.3.2.*   Deformed truss element.

Thus, with $AE/L = k$,

$$
k \begin{Bmatrix} c^2 \\ cs \\ -c^2 \\ -cs \end{Bmatrix} u_i = \begin{Bmatrix} p_i \\ q_i \\ p_j \\ q_j \end{Bmatrix}
\tag{2.3.2}
$$

After similar analysis using displacements v_i, u_j, and v_j, the results may be gathered in matrix form:

$$
k \begin{bmatrix} c^2 & cs & -c^2 & -cs \\ cs & s^2 & -cs & -s^2 \\ -c^2 & -cs & c^2 & cs \\ -cs & -s^2 & cs & s^2 \end{bmatrix} \begin{Bmatrix} u_i \\ v_i \\ u_j \\ v_j \end{Bmatrix} = \begin{Bmatrix} p_i \\ q_i \\ p_j \\ q_j \end{Bmatrix}
\tag{2.3.3}
$$

The square matrix, including the coefficient k, is the element stiffness matrix. We will abbreviate Eq. 2.3.3 as

$$
[k]\{d\} = \{\bar{r}\}
\tag{2.3.4}
$$

In Section 2.5 it will be necessary to distinguish between forces applied *to* nodes and forces applied *by* nodes. In Eq. 2.3.4, $\{\bar{r}\}$ indicates forces applied *by* the nodes *to* the element.

2.4   Assembly of Element Stiffness Matrices

The assembly process will be described with reference to the plane truss. However, the overall procedure applies to any element, regardless of its type, size, shape, and number of nodes. The important concepts are: (a) there are two numbering systems, one for the structure and another assigned to the element at hand; and (b) the relation between the two numbering systems determines the locations in $[K]$ to which coefficients in $[k]$ are assigned.

Let us consider how the structure stiffness matrix of Eq. 2.2.5 may be formed by assembly of the three element stiffness matrices. We first write the stiffness matrix for each of the three elements. Because element properties remain the same if the element is turned end for end, nodal labels i and j of Fig. 2.3.1 may be assigned arbitrarily. Thus, using Eqs. 2.3.1 and 2.3.3, we obtain:

$$
\begin{matrix} i=2 \quad j=3 \\ \phi = 90° \\ c = 0 \\ s = 1 \end{matrix}
\qquad
[k]_1 =
\begin{bmatrix}
0 & 0 & 0 & 0 \\
0 & k_1 & 0 & -k_1 \\
0 & 0 & 0 & 0 \\
0 & -k_1 & 0 & k_1
\end{bmatrix}
\begin{matrix} u_2 \\ v_2 \\ u_3 \\ v_3 \end{matrix}
\qquad (2.4.1)
$$

$$
\begin{matrix} i=3 \quad j=1 \\ \phi = 315° \\ c = \dfrac{1}{\sqrt{2}} \\ s = -\dfrac{1}{\sqrt{2}} \end{matrix}
\qquad
[k]_2 =
\begin{bmatrix}
\dfrac{k_2}{2} & -\dfrac{k_2}{2} & -\dfrac{k_2}{2} & \dfrac{k_2}{2} \\
-\dfrac{k_2}{2} & \dfrac{k_2}{2} & \dfrac{k_2}{2} & -\dfrac{k_2}{2} \\
-\dfrac{k_2}{2} & \dfrac{k_2}{2} & \dfrac{k_2}{2} & -\dfrac{k_2}{2} \\
\dfrac{k_2}{2} & -\dfrac{k_2}{2} & -\dfrac{k_2}{2} & \dfrac{k_2}{2}
\end{bmatrix}
\begin{matrix} u_3 \\ v_3 \\ u_1 \\ v_1 \end{matrix}
\qquad (2.4.2)
$$

$$
\begin{matrix} i=1 \quad j=2 \\ \phi = 180° \\ c = -1 \\ s = 0 \end{matrix}
\qquad
[k]_3 =
\begin{bmatrix}
k_3 & 0 & -k_3 & 0 \\
0 & 0 & 0 & 0 \\
-k_3 & 0 & k_3 & 0 \\
0 & 0 & 0 & 0
\end{bmatrix}
\begin{matrix} u_1 \\ v_1 \\ u_2 \\ v_2 \end{matrix}
\qquad (2.4.3)
$$

where the displacements u_1, v_1, etc., are appended merely to identify the degrees of freedom associated with each element.

We may imagine that the structure is built by adding elements one by one, with each element being placed in a preassigned location. As elements are added to the structure, contributions are made to the structure stiffness matrix. We may, in fact, add up the element stiffness matrices to obtain the structure stiffness matrix, provided that element matrices are all of "structure size" and operate on identical displacement vectors. Expansion to "structure size" is accomplished by simply adding rows and columns of zeros, so that each $[k]_i$ is made as big as $[K]$. For example, for element 1 of our example truss, expansion yields

$$[k]_1 = \begin{bmatrix} 0 & 0 & 0 & 0 & 0 & 0 \\ 0 & k_1 & 0 & -k_1 & 0 & 0 \\ 0 & 0 & 0 & 0 & 0 & 0 \\ 0 & -k_1 & 0 & k_1 & 0 & 0 \\ 0 & 0 & 0 & 0 & 0 & 0 \\ 0 & 0 & 0 & 0 & 0 & 0 \end{bmatrix} \begin{matrix} u_2 \\ v_2 \\ u_3 \\ v_3 \\ u_1 \\ v_1 \end{matrix} \qquad (2.4.4)$$

This expansion may be justified as follows. There is no way that arbitrary nodal displacements can cause element 1 to apply a force to node 1, thus accounting for the added rows of zeros. Also, displacement of node 1 cannot cause element 1 to apply a force to any node, thus accounting for the added columns of zeros.

As the last modification before addition, each element matrix $[k]$ must be rearranged so that each operates on the same array of nodal displacements. Rearrangement is accomplished by row and column interchanges; this is perhaps most easily seen by writing out the element relation $[k]\{d\} = \{\vec{r}\}$ as a set of equations, rearranging terms, and then writing the result in matrix form. Rearrangement of $[k]_1$ yields

$$[k]_1 = \begin{bmatrix} 0 & 0 & 0 & 0 & 0 & 0 \\ 0 & 0 & 0 & 0 & 0 & 0 \\ 0 & 0 & 0 & 0 & 0 & 0 \\ 0 & 0 & 0 & k_1 & 0 & -k_1 \\ 0 & 0 & 0 & 0 & 0 & 0 \\ 0 & 0 & 0 & -k_1 & 0 & k_1 \end{bmatrix} \begin{matrix} u_1 \\ v_1 \\ u_2 \\ v_2 \\ u_3 \\ v_3 \end{matrix} \qquad (2.4.5)$$

One may easily verify that simple addition of the expanded and rearranged element stiffness matrices produces the structure stiffness matrix of Eq. 2.2.5.

An alternative explanation of assembly of stiffness matrices may be helpful. Figure 2.4.1 shows a portion of an arbitrary plane truss. Structure node 5 is common to elements 1, 2, and 3. Node labels i and j for each element are assigned arbitrarily. For each element we have

$$[k]_n \{u_i \, v_i \, u_j \, v_j\}_n = \{p_i \, q_i \, p_j \, q_j\}_n \qquad (2.4.6)$$

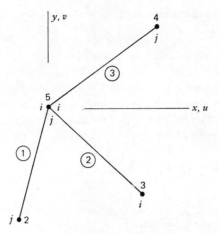

Figure 2.4.1. Node of arbitrary truss.

where $n = 1, 2, 3$. Now consider, for example, the net horizontal force p_5 that must be applied to node 5 to maintain equilibrium under an arbitrary set of nodal displacements. A contribution to p_5 is made by each of the three elements.

$$p_5 = (p_i)_1 + (p_j)_2 + (p_i)_3 \tag{2.4.7}$$

Substitution for $(p_i)_1$, $(p_j)_2$, and $(p_i)_3$ from Eq. 2.4.6 yields

$$
\begin{aligned}
p_5 = {} & (k_{11})_1 u_5 + (k_{12})_1 v_5 + (k_{13})_1 u_2 + (k_{14})_1 v_2 \\
& + (k_{31})_2 u_3 + (k_{32})_2 v_3 + (k_{33})_2 u_5 + (k_{34})_2 v_5 \\
& + (k_{11})_3 u_5 + (k_{12})_3 v_5 + (k_{13})_3 u_4 + (k_{14})_3 v_4
\end{aligned}
\tag{2.4.8}
$$

Upon gathering terms we obtain

$$p_5 = [(k_{11})_1 + (k_{33})_2 + (k_{11})_3]u_5 + [(k_{12})_1 + (k_{34})_2 + (k_{12})_3]v_5$$
$$+ \text{(terms involving } u_2, v_2, u_3, v_3, u_4, \text{ and } v_4) \tag{2.4.9}$$

$$p_5 = (K_{9,9})u_5 + (K_{9,10})v_5 + (\cdots)$$

The structure stiffness terms have subscripts 9,9 and 9,10 because u_5 and v_5 are the 9th and 10th terms in the structural displacement vector $\{D\}$.

Equation 2.4.9 shows again that the structure stiffness matrix is built by addition of terms from element stiffness matrices. It also shows that the locations in $[K]$ to which terms of an element $[k]$ are assigned depend only on the *structure* node point numbering scheme. Thus we see that element node labels i and j are used only for convenience in formulation of element properties; any other letters or numbers would serve just as well.

2.5   Assembly of Equilibrium Equations

In practice it is convenient to assemble the structure stiffness matrix and the equations of nodal equilibrium as parts of a single operation. Before formulating this operation let us introduce the following additional loads. We elect to consider these loads as applied *to* the nodes *by* the element. Equal and opposite forces are applied by the nodes to the element.

With gravity acting in the $-y$ direction, a uniform truss member of weight density γ applies forces $\{r_\gamma\}$ to the nodes.

$$\{r_\gamma\} = \frac{\gamma AL}{2} \{0 \quad -1 \quad 0 \quad -1\} \tag{2.5.1}$$

We will ignore the slight nonuniformity of stress in each member, which arises from self-weight loading.

If nodal displacements are prohibited, the heating of a member T degrees produces a compressive axial force of αEAT, where α is the coefficient of thermal expansion. Accordingly, the following forces are applied to the nodes:

$$\{r_T\} = \alpha EAT\{-c \quad -s \quad c \quad s\} \tag{2.5.2}$$

where again $c = \cos \phi$ and $s = \sin \phi$ (Fig. 2.3.2). The same forces $\{r_T\}$ would arise from the force-fitting of a member αLT units too long. (In some computer programs it is convenient to simulate lack of fit or prestress by means of a temperature change.) We will again consider forces $\{r_T\}$ in the stress computations of Section 2.7.

The forces $[k]\{d\}$ arising from nodal displacements $\{d\}$ have been defined as forces applied *to* the element. Accordingly, the net force $\{r\}$ applied to the nodes *by* an element is

$$\{r\} = -[k]\{d\} + \{r_\gamma\} + \{r_T\} \tag{2.5.3}$$

The *structural* equations are a set of nodal equilibrium equations. They may be formed by isolating the nodes as free bodies and writing the equations of static equilibrium for each. (Physically, nodes of a truss may be viewed as pins that join the members.) Nodal loads consist of externally applied forces $\{P\}$ and forces $\{r\}$ applied by each of the several elements. Accordingly, the entire set of equilibrium equations, two for each node, is

$$\{P\} + \sum_{n=1}^{m} \{r\}_n = 0 \tag{2.5.4}$$

where $\{r\}$ for each element is given by Eq. 2.5.3, and the summation extends over all m elements of the structure. Substitution of Eq. 2.5.3 into Eq.

```
                    DO 500 N=1,M
                    CALL ELEMNT
                    KK(2) = 2*NODI(N)
                    KK(4) = 2*NODJ(N)
                    KK(1) = KK(2) - 1
                    KK(3) = KK(4) - 1
                    DO 400 I=1,4
                    II = KK(I)
                    R(II) = R(II) + RE(I)
                    DO 400 J=1,4
                    JJ = KK(J)
                    S(II,JJ) = S(II,JJ) + SE(I,J)
                400 CONTINUE
                500 CONTINUE
```

Figure 2.5.1.   Assembly of element matrices.

2.5.4 yields

$$\left(\sum_{n=1}^{m} [k]_n \right) \{D\} = \{P\} + \sum_{n=1}^{m} (\{r_\gamma\}_n + \{r_T\}_n) \tag{2.5.5}$$

or

$$[K]\{D\} = \{R\}$$

In making this substitution we have supposed that the element matrices $[k]_n$, $\{r_\gamma\}_n$, and $\{r_T\}_n$ have been expanded to "structure size," and their terms rearranged as necessary, so that the displacement vector $\{d\}$ of each element has been replaced by the structure displacement vector $\{D\}$. The summations in Eq. 2.5.5 symbolize the specific arguments of the preceding sections regarding assembly of matrices.

It is now apparent that construction of structural equations and assembly of element matrices are essentially the same process. The Fortran statements of Fig. 2.5.1 perform the assembly of M elements. Arrays S, R, SE, and RE contain the matrices $[K]$, $\{R\}$, $[k]$, and $\{r_\gamma\} + \{r_T\}$, respectively. Structure node numbers associated with element node labels i and j are stored in arrays NODI and NODJ. Subroutine ELEMNT is assumed to return the element matrix $[k]$ in SE and nodal loads $\{r\}$ due to weight, temperature change, etc. in RE. Data in SE and RE are repeatedly created and destroyed, as these two arrays are used for each element in turn. External loads $\{P\}$, if any, must subsequently be added to array R.

We see that the expansion of element arrays to "structure size" is purely symbolic; the actual operation is that of properly locating terms by means of the subscripting array KK. Subscripting is made slightly obscure by the fact that two displacements, not one, are associated with each node of a truss. To clarify the matter, consider element 2 of the three-bar truss, as shown in Fig. 2.5.2. The subscripts shown are computed according to the foregoing Fortran statements. We see that these subscripts identify the row and column numbers in the structural equations to which element quantities must be assigned. Thus, the treatment described in Section 2.4 is

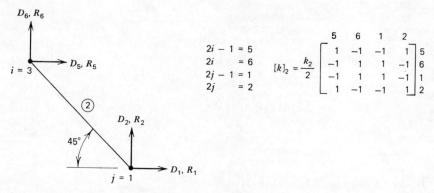

$$
\begin{array}{cccc}
 & 5 & 6 & 1 & 2 \\
\end{array}
$$

$$
\begin{array}{c}
2i - 1 = 5 \\
2i \quad = 6 \\
2j - 1 = 1 \\
2j \quad = 2
\end{array}
\qquad
[k]_2 = \frac{k_2}{2}
\begin{bmatrix}
1 & -1 & -1 & 1 \\
-1 & 1 & 1 & -1 \\
-1 & 1 & 1 & -1 \\
1 & -1 & -1 & 1
\end{bmatrix}
\begin{array}{c}
5 \\
6 \\
1 \\
2
\end{array}
$$

Figure 2.5.2. Subscripting scheme for nodal degrees of freedom.

accomplished without actually writing the expanded and rearranged element matrices.

2.6 Node Point Numbering To Exploit Sparsity

Nodes of a structure may be numbered arbitrarily. For a given structure, all numbering schemes lead to the same size of stiffness matrix $[K]$ and the same number of nonzero terms; however, different numbering schemes lead to different *arrangements* of nonzero terms. We wish to obtain the arrangement known as a banded stiffness matrix, since we may then use simple and efficient techniques for storage of $[K]$ and solution of the structural equations.

Figure 2.6.1 shows a small truss with "proper" node numbering. The structure stiffness matrix is shown in Fig. 2.6.2 with nonzero terms identified by the symbol X. The matrix is sparse and has all its nonzero terms clustered in a band along the diagonal. This band includes, in any row, terms on both sides of the diagonal. As we deal here with symmetric matrices, it is convenient to designate the "semibandwidth" as B (see Fig. 2.6.2). The total

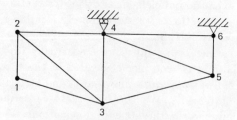

Figure 2.6.1. Example of node point numbering.

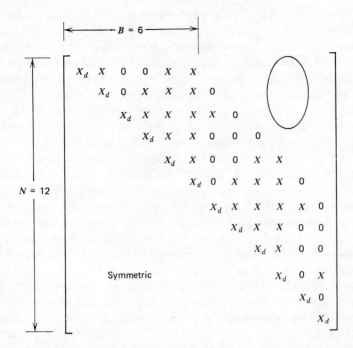

Figure 2.6.2. Full matrix, X = nonzero term, X_d = diagonal term.

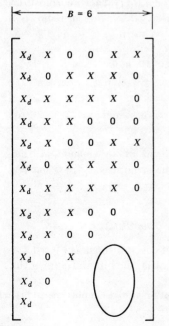

Figure 2.6.3. Band form storage.

matrix bandwidth is $2B - 1$. In a truss with more degrees of freedom, the matrix sparsity and the narrowness of the band would be more striking.

Because of symmetry and the triangular block of zeros in Fig. 2.6.2, in any row i we need retain only columns i through $i + 5$ in order to convey the full information content of the stiffness matrix. A convenient storage format for this information is shown in Fig. 2.6.3, where each column j in row i of Fig. 2.6.2 has been shifted $i - 1$ spaces to the left. Thus, in Fig. 2.6.3 all diagonal terms of the actual matrix are stored in the first column of the array.

Band form storage of $[K]$ requires only NB locations as compared with about $N^2/2$ locations for storage of the upper or lower triangle of a full symmetric matrix of order N. Also of importance is the saving in execution time that band operations permit. A band-form Gauss elimination solver is discussed subsequently. By simply counting operations,[2] one may conclude that solution time is approximately proportional to $NB^2/2$, as compared with approximately $N^3/6$ for a full but symmetric matrix (2). The advantage of band form is therefore obvious in practical problems, where N may be 20 times larger than B.

A small bandwidth is usually obtained simply by numbering nodes along the shorter dimension of a structure, as in Fig. 2.6.1. For plane trusses B is equal to 2 plus twice the maximum node number difference in an element. In Fig. 2.6.1, for example, we have $B = 2 + 2[abs(1 - 3)] = 6$. This computation is easily programmed. Analogous formulas apply to other types of finite elements.

To program the assembly of $[K]$ in band form, only the innermost DO loop in Fig. 2.5.1 must be changed. The revised statements are shown in

```
        DO 400 J=1,4
        IF (KK(J) .LT. II)  GO TO 400
        JJ = KK(J) - II + 1
        S(II,JJ) = S(II,JJ) + SE(I,J)
400 CONTINUE
```

Figure 2.6.4. Construction of banded structure stiffness matrix.

Fig. 2.6.4. The IF statement avoids using terms of an element stiffness matrix that would fall below the diagonal of the structure stiffness matrix.

2.7   Boundary Conditions and Stress Computation

Prior to application of displacement boundary conditions, the structure stiffness matrix $[K]$ is positive semidefinite. That is, the quadratic form that

---

[2] An "operation" in this context means one multiplication or division plus one addition or subtraction.

represents strain energy (see Eq. 4.2.13)

$$U = \tfrac{1}{2}\{D\}^T[K]\{D\} \qquad (2.7.1)$$

is either positive or zero. If $\{D\}$ represents a rigid body motion, $U = 0$. For any other $\{D\}$, $U > 0$. In order to solve for $\{D\}$, U must be made positive definite; that is, $U > 0$ for all $\{D\}$. This is done by imposition of displacement boundary conditions.

The boundary condition treatment of Section 2.2 is not well adapted to computer programming, especially if the band storage format is adopted, because rearrangement is time consuming and the necessary bookkeeping is awkward. Accordingly, other methods have been devised. We will discuss two of these methods.

The first method is described with reference to the equations of Fig. 2.7.1a. The matrices are *arbitrary* and not intended to represent any particular physical structure. Suppose that displacement D_1 must have the known value D_1^*. Rearrangement of the equations is avoided by substitution of D_1^* into equations 2 and 3 and by writing the first equation as the trivial equation $D_1 = D_1^*$ (Fig. 2.7.1b). No unknowns appear on the right side. The equations are then solved for D_1, D_2, and D_3. The solution, of course, yields $D_1 = D_1^*$. In structural work the special case $D_1^* = 0$ is most common.

The foregoing treatment may be repeated for other d.o.f., so that any or all displacements may be assigned numerical values. The programming of this technique is slightly difficult when matrices are stored in band format, but only about 30 Fortran statements are needed. Two disadvantages are associated with this method. First, a large number of trivial equations $D_i = D_i^*$ is wasteful of storage space. Second, information in rows in $[K]$ is destroyed, so that after solving for $\{D\}$, we cannot find the force R_i associated with D_i^* by the row-column multiplication $R_i = \sum_j K_{ij}D_j$. The necessary information could, of course, be placed in auxiliary storage and recalled when needed.

$$\begin{bmatrix} K_{11} & K_{12} & K_{13} \\ K_{21} & K_{22} & K_{23} \\ K_{31} & K_{32} & K_{33} \end{bmatrix} \begin{Bmatrix} D_1 \\ D_2 \\ D_3 \end{Bmatrix} = \begin{Bmatrix} R_1 \\ R_2 \\ R_3 \end{Bmatrix},$$

(a)

$$\begin{bmatrix} 1 & 0 & 0 \\ 0 & K_{22} & K_{23} \\ 0 & K_{32} & K_{33} \end{bmatrix} \begin{Bmatrix} D_1 \\ D_2 \\ D_3 \end{Bmatrix} = \begin{Bmatrix} D_1^* \\ R_2 - K_{21}D_1^* \\ R_3 - K_{31}D_1^* \end{Bmatrix}$$

(b)

Figure 2.7.1. A method of treating specified displacements.

The second method for treatment of boundary conditions appears to be more in favor. The method will be described in two parts: (a) treatment of specified zero displacements where the associated support reaction is not required, and (b) where the support reaction *is* required or where there are specified nonzero displacements (5).

Where support reactions associated with zero displacements are not required, the equations corresponding to these displacements are simply discarded. In other words, for each $D_i^* = 0$, the terms D_i, R_i and row i and column i of $[K]$ are omitted from the structural equations. In this way we arrive at Eq. 2.2.10 with $\{D_o\} = 0$.

We are then immediately confronted with the programming problem of how to arrange bookkeeping so that element stiffnesses are properly located when forming the structural stiffness, and computed displacements in $\{D\}$ may be associated with the proper node and direction. The scheme adopted in Ref. 5 will now be outlined. Let a two-dimensional array ID be filled as node point data is read. If d.o.f. I at node N carries load and hence has an unknown displacement, we input ID(N, I) = 0. If this particular d.o.f. is fixed at zero displacement, we input ID(N, I) = 1. We now enter the routine of Fig. 2.7.2. NUMNP = total number of nodes in the structure. NDOF = number of d.o.f. per node (NDOF = 2 for the plane truss). Array ID now identifies by number the structural equation corresponding to each nodal d.o.f. After solution of equations, ID may be used to extract from $\{D\}$ the particular d.o.f. associated with a given element. To describe assembly of structural equations, we consider again the plane truss. Figure 2.5.1 is modified to yield Fig. 2.7.3. The first IF statement causes the "zero-displacement equations," corresponding now to ID(N,I) = 0, to be discarded. The second IF statement functions as in Fig. 2.6.4, and also serves to discard the column of $[K]$ corresponding to a zero displacement (as here KK(J) = 0, so that KK(J) < II).

It might be noted that Ref. 5 allocates storage space more cleverly than suggested by Fig. 2.7.3. For example, node point information resides in core, and element matrices and location information (array KK of Fig. 2.7.3) are formed and written on auxiliary storage as soon as data for an

```
            NEQ = 0
            DO 60 N=1,NUMNP
            DO 60 I=1,NDOF
            IF (ID(N,I) .GT. 0)  GO TO 58
            NEQ = NEQ + 1
            ID(N,I) = NEQ
            GO TO 60
         58 ID(N,I) = 0
         60 CONTINUE
```

Figure 2.7.2. Establishment of relation between nodal freedoms and equation numbers.

```
        DO 500 N=1,M
        CALL ELEMNT
        I = NODI(N)
        J = NODJ(N)
        KK(1) = ID(I,1)
        KK(2) = ID(I,2)
        KK(3) = ID(J,1)
        KK(4) = ID(J,2)
        DO 400 I=1,4
        IF (KK(I) .LE. 0)  GO TO 400
        II = KK(I)
        R(II) = R(II) + RE(I)
        DO 400 J=1,4
        IF (KK(J) .LT. II)  GO TO 400
        JJ = KK(J) - II + 1
        S(II,JJ) = S(II,JJ) + SE(I,J)
400     CONTINUE
500     CONTINUE
```

Figure 2.7.3. Alternate assembly of plane truss equations in banded format.

element is read. When all elements have been treated, node point information is no longer needed, so this space is used to store the structure matrices,
which are built by recalling element information from auxiliary storage.

If support reactions are required, extra members are added to the structure (5). For example, let the truss of Fig. 2.2.1 be augmented by members
4, 5, and 6, as shown in Fig. 2.7.4. If the new members are orders of magnitude
stiffer than the others, the rigid support conditions of Fig. 2.2.1 are closely
approximated. Zero-displacement conditions at the supported nodes 4, 5,
and 6 of the new members may be imposed as in Figs. 2.7.2 and 2.7.3. Thus,

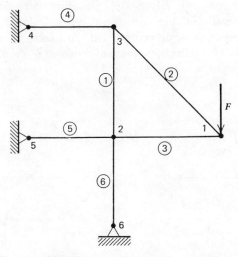

Figure 2.7.4. Support elements.

the structure stiffness matrix remains 6 by 6. Computed displacements u_2, v_2, and u_3 are quite small, but when multiplied by the large stiffnesses of members 4, 5, and 6 yield forces of appropriate magnitude. These forces are the desired support reactions.

Clearly the comparatively rigid "support elements" may be used in three dimensions as well as two, and could have torsional stiffness if a rotational d.o.f. is involved. They could also be used to approximate a support where displacement is permitted in a single direction not parallel to the coordinate directions (Fig. 2.7.5). However, one must guard against making the structure stiffness matrix ill-conditioned by improper use of support elements. A very stiff element should be connected to a rigid support; it should not be connected only to comparatively flexible members (see Section 16.5). Where a very stiff element exerts a force parallel to a global displacement d.o.f., only the corresponding diagonal coefficient of $[K]$ is made large, and no trouble arises. But if used as shown in Fig. 2.7.5, large off-diagonal coefficients also appear, and numerical error is likely (5). The latter situation may be avoided by use of new d.o.f. at the node in question (Section 11.4), or, where circumstances permit, by use of support members of only moderate stiffness.

If a specified nonzero displacement is to be imposed, the support elements may again be used (5). Suppose that some d.o.f. D_i is to be assigned the value D_i^*. To do this the diagonal term K_{ii} of $[K]$ is augmented by a number C, which is several orders of magnitude larger than K_{ii}, and on the right side R_i is augmented by C times D_i^*. Physically, this device amounts to adding to the structure a spring of high stiffness C that resists displacement D_i, then loading the spring by a large force CD_i^*. The spring deflects practically D_i^* units, dragging the comparatively flexible structure along with it. Because the structure has small but finite resistance, the calculated D_i will not be exactly D_i^*. However, the difference will probably be so small that $C(D_i^* - D_i)$ will not be a reliable estimate of the force R_i required to produce displacement D_i^* in the actual structure. In such a case we could resort to auxiliary storage and the operation $R_i = \sum_j K_{ij} D_j$.

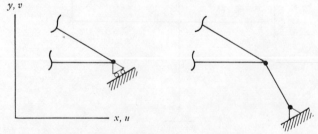

Figure 2.7.5.   Simulation of skew support in a plane truss.

Finally, to compute stresses in elements of the truss we proceed as follows. In the notation of Figs. 2.3.1 and 2.3.2, the elongation of an element is

$$e = (u_j - u_i) \cos \phi + (v_j - v_i) \sin \phi \tag{2.7.2}$$

which may be computed when all nodal displacements have been determined. Stress caused by elongation e must be superposed on any initial stress that may be present. If, for example, the member was heated T degrees and has expansion coefficient α, the net stress is

$$\sigma = E\left(\frac{e}{L} - \alpha T\right) \tag{2.7.3}$$

where E = elastic modulus and L = member length. Thus, if the member were free to expand, we would find $e = \alpha L T$ and $\sigma = 0$.

2.8 Direct Solution of Equations

We will briefly describe the well-known Gauss elimination method (2), as applied to equations $[K]\{D\} = \{R\}$, stored in band form, and with $[K]$ symmetric. In this method the first equation is solved for D_1, which is then substituted into the remaining equations. Then the second equation is similarly treated, then the third, etc. In this forward reduction process $[K]$ is reduced to upper triangular form, with 1.0 as each diagonal term, and the right side $\{R\}$ is also modified. Unknowns are then found by back-substitution, so that D_1 is the last unknown to be determined.

As implied above, the ith equation is used to eliminate the ith unknown; i.e., all pivots are diagonal terms of $[K]$, thus avoiding a search for the largest pivot in each elimination. This simple approach is appropriate when the coefficient matrix $[K]$ is a stiffness matrix, since very small pivots K_{ii} do not arise unless the structure is nearly unstable or has been improperly modeled (see also Chapter 16).

For a numerical example of Gaussian elimination, consider the following set of three equations.

$$\begin{bmatrix} 200 & -100 & 0 \\ -100 & 200 & -100 \\ 0 & -100 & 100 \end{bmatrix} \begin{Bmatrix} u_1 \\ u_2 \\ u_3 \end{Bmatrix} = \begin{Bmatrix} -8 \\ -8 \\ -8 \end{Bmatrix} \tag{2.8.1}$$

The first equation is divided by 200, then multiplied by -100 and subtracted from the second equation. Thus,

$$\begin{bmatrix} 1 & -0.5 & 0 \\ 0 & 150 & -100 \\ 0 & -100 & 100 \end{bmatrix} \begin{Bmatrix} u_1 \\ u_2 \\ u_3 \end{Bmatrix} = \begin{Bmatrix} -0.04 \\ -12 \\ -8 \end{Bmatrix} \tag{2.8.2}$$

Similarly, the new second equation is divided by 150, then multiplied by -100 and subtracted from the third. A final division of the new third equation by its diagonal coefficient of 33.3 yields

$$\begin{bmatrix} 1 & -0.5 & 0 \\ 0 & 1 & -0.667 \\ 0 & 0 & 1 \end{bmatrix} \begin{Bmatrix} u_1 \\ u_2 \\ u_3 \end{Bmatrix} = \begin{Bmatrix} -0.04 \\ -0.08 \\ -0.48 \end{Bmatrix} \tag{2.8.3}$$

By back-substitution we find first $u_3 = -0.48$, then $u_2 = -0.08 - (-0.667)u_3 = -0.40$, and finally $u_1 = -0.04 - (-0.5)u_2 = -0.24$.

The following observations should be made.

- After each reduction, the remaining equations remain symmetric and banded.

- With B the semibandwidth, each reduction involves only B rows and no more than B coefficients in each row ($B = 2$ in the above example).

- The same vector of unknowns satisfies the equations at each stage of reduction.

A physical interpretation of Gauss elimination is revealing. Elimination of an equation frees the corresponding unknown d.o.f. to move. Similarly, elimination of all d.o.f. at a node frees the node to move as dictated by the remainder of the structure. This behavior may be seen in the foregoing example. Equation 2.8.1 represents an assembly of three springs, each of stiffness $k = 100$, carrying a load of 8 units at each node (see Fig. 16.7.1a, but consider only three nodes with a force of 8 at each). When u_1 is eliminated, node 1 is released; the adjacent springs now act in series with a combined stiffness of 50. Thus, the diagonal stiffness coefficient associated with u_2 drops from $100 + 100 = 200$ in Eq. 2.8.1 to $100 + 50 = 150$ in Eq. 2.8.2. Similarly, elimination of u_2 yields the diagonal coefficient 33.3 in the third equation, which is the stiffness of three springs in series; d.o.f. u_1 and u_2 have both been released and "ride along" as demanded by the elastic properties of the system.

Study of the method reveals that reduction of $\{R\}$ may be performed subsequent to reduction of $[K]$ to upper triangular form. Information needed consists of the value of each diagonal term K_{ii} just prior to elimination of the ith equation, plus the remaining reduced terms K_{ij}, $j > i$, in the triangularized matrix $[K]$. Because we know that 1.0 is the "proper" final value of each diagonal term, we may easily retain the needed K_{ii}'s simply by not replacing each diagonal term K_{ii} by unity when elimination of the ith equation is begun by dividing this equation by the reduced diagonal K_{ii}.

The motivation for separating the two reduction operations is to permit efficient analysis of two or more different loadings on a given structure.

```
C       FORWARD REDUCTION OF MATRIX (GAUSS ELIMINATION)
700 DO 790 N=1,NSIZE
    DO 780 L=2,MBAND
    IF (S(N,L) .EQ. 0.)  GO TO 780
    I = N + L - 1
    C = S(N,L)/S(N,1)
    J = 0
    DO 750 K=L,MBAND
    J = J + 1
750 S(I,J) = S(I,J) - C*S(N,K)
    S(N,L) = C
780 CONTINUE
790 CONTINUE
C       FORWARD REDUCTION OF CONSTANTS (GAUSS ELIMINATION)
800 DO 830 N=1,NSIZE
    DO 820 L=2,MBAND
    IF (S(N,L) .EQ. 0.)  GO TO 820
    I = N + L - 1
    R(I) = R(I) - S(N,L)*R(N)
820 CONTINUE
830 R(N) = R(N)/S(N,1)
C       SOLVE FOR UNKNOWNS BY BACK-SUBSTITUTION
    DO 860 M=2,NSIZE
    N = NSIZE + 1 - M
    DO 850 L=2,MBAND
    IF (S(N,L) .EQ. 0.)  GO TO 850
    K = N + L - 1
    R(N) = R(N) - S(N,L)*R(K)
850 CONTINUE
860 CONTINUE
```

Figure 2.8.1.   Gauss elimination equation solver, banded format.

Forward reduction of $[K]$ need be done only once, while forward reduction of the load vector $\{R\}$ and back-substitution are done as many times as there are loading conditions. In this way roughly $B/2$ load vectors may be reduced for the cost of a single reduction of $[K]$, where B is the semibandwidth of $[K]$ (shown in Fig. 2.6.2). Instead of repeatedly filling the single vector $\{R\}$ with new information, one might prefer to have an array $[R]$ of several columns and treat all loadings simultaneously. The latter option requires more storage space.

The necessary Fortran statements (3) are given in Fig. 2.8.1. Matrix $[K]$ is stored in band form in the first NSIZE rows and MBAND columns of array S. Thus diagonal terms K_{ii} occupy locations S(I,1). Load vector $\{R\}$ is stored in the first NSIZE locations of array R. (Modification to permit simultaneous treatment of several loads in an array $[R]$ is left to the reader as an exercise.) As unknowns are found by back-substitution they replace the existing entries in array R. The first loading is treated by starting at statement 700. Each subsequent loading on the same structure is treated by entering the routine at statement 800.

A brief explanation of this routine is as follows. The DO 790 and DO 830 loops cause each equation to be treated. The DO 780 and DO 820 loops

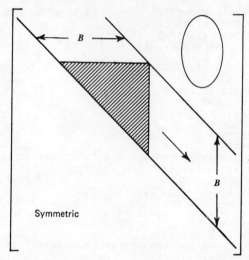

Figure 2.8.2.   Active portion of stiffness array.

substitute an equation into the succeeding equations. The DO 750 loop treats each stiffness term in a succeeding equation. The "proper" value of unity for each fully reduced diagonal term in the stiffness matrix is implicitly assumed in the back-substitution.

During reduction, index L runs to MBAND rather than NSIZE, because only MBAND subsequent equations can be affected by elimination of any unknown. A similar comment applies to back substitution. During elimination of the Nth unknown, the number K of terms that are modified in the $(N + L - 1)$th equation is $K = B - L + 1$. Accordingly, index K runs L to MBAND. Effectively, we have a "triangle of operations" that moves a step along the stored band as each unknown is eliminated (Fig. 2.8.2).

The IF statements serve two purposes and may be better understood by considering the algorithm as applied to the matrix of Fig. 2.6.3. First, useless computation (in which no terms are altered) is avoided if S(N,L) happens to be zero. Second, and most important, use of subscripts larger than NSIZE is prevented, owing to the presence of the triangular block of zeros in the lower right-hand corner of S. Obviously array S must be cleared to zero before terms of $[K]$ are added into it.

It is worth noting that matrix inversion requires roughly three times as many operations as a single Gauss elimination solution (2), and that the inverse of a band matrix is full, not banded.

The obvious advantage of small bandwidth has led to the development of algorithms that rearrange node numbers in an attempt to minimize or significantly reduce bandwidth (6, 7, 8, Fortran statements in 32). The efficiency of these algorithms varies widely, but all offer a major benefit to the user; he may number nodes in any convenient way, and leave effective numbering to the program. Internal manipulations are based on the new numbering, but output may be in terms of the user's original numbering. It should also be mentioned that there are efficient routines that permit a few nonzero coefficients to be scattered outside of the band (23 of Chapter Twelve).

Despite the relative compactness of band form storage, computer core space may be inadequate for storage of the entire stiffness matrix of a large structure. However, as suggested by Fig. 2.8.2, only an "active triangle" of data need be in core storage at any given time. After reduction of a row, the reduced row may be transferred to auxiliary storage, then the "active triangle" shifted in core storage to make way for additional data to be read from auxiliary storage. Schemes similar to this have been widely used. For example, Ref. 3 of Chapter Seven retains two blocks of $[K]$, each of size MBAND by MBAND, in core storage. When the upper block has been reduced, it is transferred to auxiliary storage, the lower block shifted up in core, and a new lower block read in. Reference 9 makes use of several blocks and liberal auxiliary storage to solve very large systems. References 5 and 31 use blocks whose lengths are automatically determined to suit the problem at hand. Reference 30 discusses still more sophisticated schemes, and comments on the desirable properties of large-scale equation solvers. Reference 33 gives a listing of a Gauss-Jordan equation solver, which uses a sparse-matrix storage scheme and permits the coefficient matrix to be unsymmetric.

The Choleski method for solution of simultaneous equations (2, 10) differs little from the Gauss elimination method in that it requires the same number of operations and adapts readily to band form and "blocking" for solution of very large systems. Neither method seems to have a marked advantage over the other (11), so that the choice may depend largely on personal preference.

2.9 Further Comments on Direct Solution

Equation solution and other operations on sparse matrices are widely used, structural mechanics being only one field of application. Accordingly, the problem of efficient operations on sparse matrices has attracted considerable attention. In this section some of the work is summarized.

Use of bandedness is one of the simplest sparse matrix operations. A disadvantage is that the user is burdened with numbering a structure so that bandwidth is small (although nodes could be numbered arbitrarily if an automatic bandwidth reduction routine is available). Another disadvantage is that sparsity is not fully exploited, because zeros within the band do not all remain zero as elimination proceeds. For example, elimination of the third row of Fig. 2.6.2 fills in the seventh column of rows 4, 5, and 6. Physically, each fill-in represents a new linking of d.o.f., created as the unknowns are successively "released" by elimination. These fill-in terms must be processed in subsequent steps of elimination. If zeros remained zero their processing could be skipped (see the first IF statement in Fig. 2.8.1). Hence advanced forms of equation solvers, while often using the Gauss elimination method, may ignore banding in favor of arranging terms in the structure stiffness matrix in such a way that fill-in is minimized or greatly reduced. Node point numbers may be assigned entirely arbitrarily, since the program will always consider rearrangement in order to reduce fill-in. References 12 to 18 and 34 deal with this topic. References 29 and 35 present some technical details and extensive bibliographies.

As the nonzero terms in $[K]$ are now scattered throughout the matrix, some compact storage scheme is needed. A simple scheme involves a two-dimensional "pointer array," say NDX. For example, suppose that the 9th nonzero entry in the 200th row of the upper triangle of $[K]$ is in column 289. We could store the numerical value of $K_{200,289}$ in array S in position S(200,9) and the "true" subscript 289 in array NDX in position NDX(200,9). Thus, operations involving array S must also involve "true" column numbers stored in array NDX.

However, the bookkeeping scheme must be more sophisticated than this, since *some* fill-in will occur and provision must be made for storage of these terms. It is possible for the rearrangement routine to count the number of fill-ins that will occur, so that adequate space may be reserved. Bookkeeping becomes still more complicated with large systems where data transfers must be made between core and auxiliary storage. Indeed, bookkeeping may comprise an appreciable fraction of the solution time. Reference 19 advises that a sophisticated strategy, intended to be quite efficient, may actually be quite inefficient because of peculiarities of the computer operating system. Hence, as computer hardware and operating systems become more complicated, it becomes more advisable for the engineer to cooperate with the systems programmer.[3]

Published information suggests that the bookkeeping expense of near-optimal ordering is more than repaid by shorter solution time. The effectiveness of near-optimal ordering is more pronounced if there are many d.o.f. per node, and also in branched structures such as electrical and piping networks where a compact band is hard to achieve. The effort becomes more worthwhile in nonlinear problems, where reordering need be done only once for a set of equations that will be solved repeatedly.

---

[3] On some computers, equations are solved much faster if innermost loops are coded in assembly language instead of Fortran.

2.10 Indirect Solution of Equations

In contrast to direct methods such as Gauss elimination, indirect methods require an indefinite number of operations, and are terminated when some convergence test is satisfied. They are less popular than direct methods but have some attractive aspects. Here we will summarize their principal advantages and disadvantages.

Gauss-Seidel iteration for solution of equations $[K]\{D\} = \{R\}$ is a well-known method that is guaranteed to converge if matrix $[K]$ is symmetric and positive definite (2) (see below Eq. 2.7.1). In any iterative cycle, a new value D_i' is found for unknown D_i by the operation

$$D_i' = D_i + \frac{\beta}{K_{ii}}\left(R_i - \sum_{j=1}^{N} K_{ij} D_j\right) \qquad (2.10.1)$$

where N = number of equations and β = an overrelaxation factor. The quantity in parentheses may be regarded as the unbalance between nodal load R_i and the resistance to R_i generated by the deformed structure. The object of iteration is to reduce the unbalance to zero. Each new estimate of an unknown, D_i', replaces the old value D_i as soon as D_i' is calculated. Thus the most recent estimate of each D_i is used in each operation. If $\beta = 1.0$ we have "pure" Gauss-Seidel iteration. If the banding of Figs. 2.6.2 and 2.6.3 is present, index j may span not more than B different values.

The speed of convergence and the optimum value of β in Eq. 2.10.1 are problem-dependent. Use of $\beta = 1.8$ speeds convergence by roughly a factor of 10 as compared with $\beta = 1.0$ (20). Use of a "group relaxation" (20) may further speed convergence by a factor of roughly 3. One assumes that $\{D'\} = c\{D\}$ is a good estimate of the actual displacements, where c is a scalar to be determined. Each unknown in $\{D\}$ is revised by being multiplied by c. To find c, we equate work done by gradually applied loads to work stored as strain energy. Hence,

$$\tfrac{1}{2}c\{D\}^T\{R\} = \tfrac{1}{2}c^2\{D\}^T[K]\{D\} \qquad (2.10.2)$$

Scalar c is computed according to Eq. 2.10.2 and is applied to $\{D\}$ roughly every tenth iterative cycle.

Additional ways of speeding convergence include solving for two or more unknowns at each step (21, 22). Thus the scalar parameters in Eq. 2.10.1 are replaced by submatrices of the entire system. Also, the complete system can be reduced to an intermediate number of d.o.f. (reduction is similarly done in "eigenvalue economizers," discussed in Section 12.4). Results of an iterative solution of the smallest system are used as initial estimates for an iterative solution of the next larger system, and so on until the full system is treated

(21). Indeed, the intermediate solutions need not be carried to full convergence; they may be stopped when "close enough." Unfortunately, some methods, namely conjugate-gradient solutions, may be worthless unless fully converged (28).

Nodes may be numbered at random, and $[K]$ stored in some compact format, using perhaps the "pointer array," mentioned in Section 2.9. No space for "fill-in" need be reserved, as terms in $[K]$ are used repeatedly and without alteration. These conveniences are common to most indirect methods.

Energy minimization is another indirect solution method. Displacements that prevail in stable equilibrium are those that render the total potential energy of the system a minimum. Thus one may regard equation solving as a problem of minimization or optimization of a function, and use one of the many methods of mathematical programming that have been devised for this purpose. Sample references in structural mechanics are Refs. 23 to 28. A feature of energy minimization methods is that computer storage demands are quite small, because structural equations need not be assembled in order to evaluate the total energy; one may instead sum the strain energies of the several elements and subtract the work done by external loads (26). One may pay for this advantage by increased use of auxiliary storage and computer time.

Indirect methods seem best suited to nonlinear problems where a solution is obtained by a step-by-step process, because the solution $\{D\}$ from one load level serves as a good estimate for beginning an indirect solution at the next load level. Indirect methods seem ill-suited to linear problems where several load vectors $\{R\}$ are to be studied; unless the loadings are similar, solution $\{D\}_n$ is not a good starting estimate for solution $\{D\}_{n+1}$, and in contrast to direct methods, the *complete* solution process must be repeated for each loading. Convergence rates may be acceptable in problems of two- and three-dimensional solids, but too slow in problems of beams, plates, and shells. Iterative methods converge more rapidly when the coefficient matrix has large diagonal terms.

PROBLEMS

2.1. With reference to the truss of Fig. 2.2.1, complete the set of six free body diagrams begun in Fig. 2.2.2, and write the remaining four equations of Eqs. 2.2.4.

2.2. (a) Consider the unrestrained stiffness matrix of Eq. 2.2.5, and let the corresponding truss have the rigid body motion u_2, v_2, θ (a motion of node 2 plus a rigid body rotation). Show that the nodal forces $\{\bar{r}\} = [k]\{d\}$ are indeed zero.

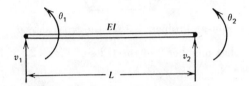

(b) A beam element has the nodal displacement vector $\{d\} = \{v_1 \ \theta_1 \ v_2 \ \theta_2\}$, where θ_1 and θ_2 are end rotations (see sketch). Will columns of its 4 by 4 element stiffness matrix sum to zero? Explain.

2.3. Let $k_1 = k_2 = k_3 = 2$ in the truss of Fig. 2.2.1. Carry out the solution for $\{D_s\}$ and $\{R_s\}$ of Eqs. 2.2.9. Check the computed $\{R_s\}$ by use of static equilibrium conditions.

2.4. Write Fortran statements that generate the element stiffness matrix of Section 2.3.

2.5. (a) Number the nodes of the truss in such a way that the structure stiffness matrix $[K]$ will have as small a bandwidth as possible.

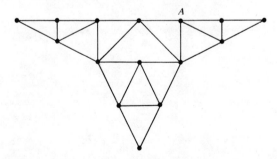

(b) Consider the diagonal terms of $[K]$ that correspond to node A of the truss. In your numbering scheme, what are the row and column numbers of these terms? Which elements make nonzero contributions to the first of these terms? And which to the second?

(c)   Consider terms in the ith row of $[K]$ to the right of the diagonal term K_{ii}. In your numbering scheme, which of these K_{ij} are nonzero in each of the two rows containing diagonal terms corresponding to node A?

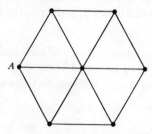

2.6.   (a, b, c)   Repeat Problem 2.5 for the truss shown.

2.7.   Suppose that member 3 of the truss of Fig. 2.2.1 is heated by some amount T. Also let $k_1 = k_2 = k_3 = 10^6$, $L_1 = L_3 = L_2/\sqrt{2} = 10$, and $A_1 = A_3 = A_2/\sqrt{2} = 1$. Elementary analysis shows that all members remain stress free and that node 1 has displacements $u_1 = v_1 = 10\alpha T$. Show that the same result is produced by the matrix methods of this chapter.

2.8.   Suppose that several load vectors are stored in the first NCOL columns of an array $[R]$. Modify the program of Fig. 2.8.1 so that all load vectors may be treated in a single pass through the routine.

2.9.   Assemble the Fortran statements of this chapter and write additional statements, where necessary, to produce a working program for analysis of plane trusses by the stiffness method. Make the program capable of treating mechanical and thermal loads, nonzero displacement boundary conditions, and multiple mechanical loads on a given truss.

CHAPTER THREE

Potential Energy and the Rayleigh-Ritz Method

3.1 Introduction

Direct physical argument was used in Chapter Two to generate equilibrium equations of the form $[K]\{D\} = \{R\}$. In this chapter a "total potential" function is introduced; it is a function of displacements and when minimized with respect to the displacements, it produces a set of equilibrium equations. Thus, as an alternate to the direct argument of Chapter Two, a "minimum potential" approach may be used to define a state of static equilibrium. Indeed, for the truss problem, the two methods produce identical equations $[K]\{D\} = \{R\}$. Why, then, is another method necessary? Elements more complicated than two-force members and beams are difficult if not impossible to formulate properly by a direct approach. An approach based on minimum potential is more general and more reliable. In this chapter the minimum potential argument is explained, and in subsequent chapters it is used to formulate finite elements.

The finite element method may be viewed as a form of the Rayleigh-Ritz method, which in turn is a technique that approximates the equilibrium configuration of the structure under study. Usually, but not necessarily, a polynomial is used to describe the approximate displacement field. In a given problem many polynomials are possible, but a particular one is selected by the solution method. Simply put, the solution method selects the polynomial that simultaneously makes the potential energy a minimum and satisfies equilibrium conditions approximately. The nature of the approximation bears careful study.

A mathematical explanation of the finite element method need not be

based on the minimum potential concept; for example, the finite element
method may be regarded as a form of the Galerkin method. But to date the
minimum potential concept has been most useful for problems of stress
analysis, so we limit the discussion to this approach.

Physical insight is valuable in using the finite element method. A theoretical
understanding is also valuable, because it leads to some effective techniques
that are not suggested by purely physical reasoning. It also suggests, and
quite correctly, that the method may be applied to problems other than
stress analysis, and that functions other than potential energy may be used
as a basis.

3.2   Total Potential Energy

Let a *system* be an elastic structure plus the loads that act on it. The system
is *conservative* if, in order to deform the system from any configuration A
and then lead it back to A again, zero net work is done regardless of the
path taken. The potential energy of a system is contained in its elastic distortions and the capacity of its loads to do work when they move. If we grasp
the system and move it, the work we do, be it positive or negative, is (alge-
braically) added to the existing potential energy of a conservative system.

The *principle of minimum potential energy* states that:

---

*Among all displacement configurations that satisfy internal
compatibility and kinematic boundary conditions, those that
also satisfy the equations of equilibrium make the potential
energy a stationary value. If the stationary value is a minimum,
the equilibrium is stable.*

Elastic systems that come most readily to mind have only one equilibrium
position, and this position is stable. But the principle as stated is valid for
nonlinear (but conservative) systems with one or more equilibrium configurations. Note that we may not consider any and all configurations, but
only those that satisfy compatibility conditions. These are termed "admissible" configurations. As examples, a continuum is not permitted to
split internally, and a cantilever beam is not permitted to "kink" or its
fixed end to rotate or displace.

As a simple example, consider the spring-force system of Fig. 3.2.1.
Admissible configurations are defined by the single displacement D. Let
$D = 0$ when the spring is unstretched. The potential energy Π_p is

$$\Pi_p = \tfrac{1}{2}kD^2 - PD \qquad (3.2.1)$$

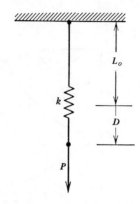

Figure 3.2.1. Spring-force system.

The negative sign appears in Eq. 3.2.1 because P loses some of its capacity for doing work when it displaces in the direction it acts. Note also that if we define the zero-energy state as $D = 0$, the external work we must do to create displacement D is $+kD^2/2$ against the spring and $-PD$ against load P. This work is stored as potential energy.

The displacement producing static equilibrium makes Π_p stationary with respect to virtual displacement δD. Thus

$$\delta \Pi_p = (kD - P)\, \delta D = 0 \tag{3.2.2}$$

and the equilibrium position is

$$D_{eq} = \frac{P}{k} \tag{3.2.3}$$

Equation 3.2.2 is an expression of the *virtual work principle*. It states that work done by internal force kD plus work done by external force P equals zero when the system is given an infinitesimal displacement δD from its equilibrium position. One could also say that an observer must do zero net work in changing the configuration from D_{eq} to $D_{eq} \pm \delta D$.

Plots of Π_p and its internal and external energy components are given in Fig. 3.2.2. We see that the stationary value of Π_p is a minimum, hence the equilibrium is stable. Note also that the zero datum for potential of load P is arbitrary. (For example, we could write $-P(L_o + D)$ for this potential instead of $-PD$. Thus, the Π_p curve would be shifted downward, but D_{eq} is still the equilibrium displacement.) The load has potential energy $-PD$ instead of $-PD/2$, because this potential arises from the magnitude of force and its capacity to displace; it is not in any way dependent on the linear properties of the spring. True, we could bypass the principle of minimum potential energy by noting that if the force is *gradually increased* to level P, the work it does is stored as strain energy. Thus

$$\frac{1}{2}PD = \frac{1}{2}kD^2, \qquad D_{eq} = \frac{P}{k} \tag{3.2.4}$$

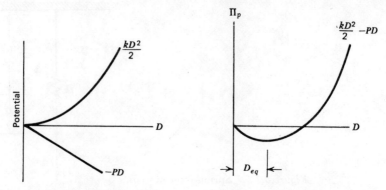

Figure 3.2.2. Components of Π_p.

However, this approach yields only one equation, even when there are several degrees of freedom, and it is therefore not very helpful.

3.3 Several Degrees of Freedom

If n independent quantities are required to define the configuration of a system, the system is said to have n degrees of freedom (d.o.f.). The n quantities are also called *generalized coordinates*. The d.o.f. in a given problem need not be all of the same type; for example, some may be linear displacements, and others may be angular displacements (rotations) or even strains.

Let potential energy Π_p of a system be written as a function of generalized coordinates D_i. Thus, $\Pi_p = \Pi_p(D_1, D_2, \ldots, D_n)$. Then, by differentiation,

$$\delta\Pi_p = \frac{\partial\Pi_p}{\partial D_1}\delta D_1 + \frac{\partial\Pi_p}{\partial D_2}\delta D_2 + \cdots + \frac{\partial\Pi_p}{\partial D_n}\delta D_n = \left\{\frac{\partial\Pi_p}{\partial D}\right\}^T\{\delta D\} \quad (3.3.1)$$

For a stationary condition, $\delta\Pi_p = 0$ for *any* admissible set of infinitesimal displacements $\{\delta D\}$. This can happen only if the coefficient of each δD_i vanishes. Thus,

$$\frac{\partial\Pi_p}{\partial D_i} = 0, \quad i = 1, 2, 3, \ldots, n; \quad \text{or} \quad \left\{\frac{\partial\Pi_p}{\partial D}\right\} = 0 \quad (3.3.2)$$

According to the principle of minimum potential energy, Eqs. 3.3.2 define the equilibrium configuration of the system. The unknowns are the n generalized coordinates. There are as many equations as there are unknowns, hence we can expect to obtain a solution.

As a simple example, consider the two d.o.f. system of Fig. 3.3.1. The bar is considered rigid and weightless. Generalized coordinates are translation D and rotation θ. Let $D = \theta = 0$ when the springs are unstretched.

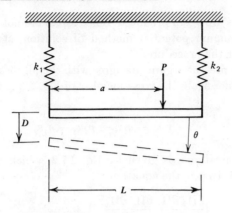

Figure 3.3.1. Two d.o.f. system.

For small rotation θ we may write

$$\Pi_p = \frac{1}{2} k_1 D^2 + \frac{1}{2} k_2 (D + \theta L)^2 - P(D + a\theta) \qquad (3.3.3)$$

Hence, from Eq. 3.3.2 we have

$$\begin{bmatrix} (k_1 + k_2) & k_2 L \\ k_2 L & k_2 L^2 \end{bmatrix} \begin{Bmatrix} D \\ \theta \end{Bmatrix} - \begin{Bmatrix} P \\ Pa \end{Bmatrix} = 0 \qquad (3.3.4)$$

The square matrix is a stiffness matrix. The terms P and Pa may be called *generalized forces*. The solution of Eq. 3.3.4 for D and θ agrees with the solution obtained by elementary methods.

From this example we draw the following conclusions which are true in general.

1. The stiffness matrix of a system with linear load versus displacement properties is symmetric; that is, $k_{ij} = k_{ji}$. This is true because each symmetrically located pair of off-diagonal terms arises from a single term in Π_p whose form is a constant times $D_i D_j$.

2. Each equation $\partial \Pi_p / \partial D_i = 0$ is an equilibrium equation. In the present example Eq. 3.3.4 may be obtained by considering equilibrium of vertical forces and moments about the left end.

3. The product of each generalized displacement and its corresponding generalized force has units of work or energy. Indeed, premultiplication of Eq. 3.3.4 by the solution row vector $[D \quad \theta]$ yields a scalar work-energy balance equation corresponding to that in Eq. 3.2.4.

If a third spring is introduced, the foregoing example becomes statically indeterminate. However, the d.o.f. D and θ still suffice, and two equations in these two unknowns are again produced. Since forces are not the primary unknowns in the minimum potential method of solution, static indeterminacy does not affect the procedure.

To confirm that the present arguments agree with the physical arguments of Chapter Two, consider again the truss member of Fig. 2.3.2. Its potential is

$$\Pi_p = \frac{1}{2} k e^2 - p_i u_i - q_i v_i - p_j u_j - q_j v_j \qquad (3.3.5)$$

where $k = AE/L$. Elongation e is given by Eq. 2.7.2, which may be substituted into Eq. 3.3.5. Hence, the equations

$$\left\{ \frac{\partial \Pi_p}{\partial u_i} \frac{\partial \Pi_p}{\partial v_i} \frac{\partial \Pi_p}{\partial u_j} \frac{\partial \Pi_p}{\partial v_j} \right\} = 0 \qquad (3.3.6)$$

are found to be the same as Eqs. 2.3.3.

3.4   Initial Strain and Stress

Consider a unit volume of linearly elastic material (Fig. 3.4.1) that is distortion-free at temperature $T = 0$. We wish to write the expression for strain energy when both thermal expansion and uniaxial stress are applied.

Let α be the coefficient of thermal expansion. When heated an amount T, free thermal expansion αT occurs. Applied stress increases the deformation by an amount σ/E. Only the latter distortion requires the input of mechanical work, which is stored as strain energy. Thus the strain energy per unit volume is

$$U_o = \frac{1}{2} \sigma \frac{\sigma}{E} = \frac{1}{2} E \left(\frac{\sigma}{E} \right)^2 = \frac{1}{2} E (\epsilon - \alpha T)^2$$

$$U_o = \frac{1}{2} E \epsilon^2 - E \epsilon \alpha T + \frac{1}{2} E \alpha^2 T^2 \qquad (3.4.1)$$

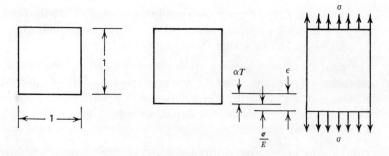

Figure 3.4.1.   Mechanical and thermal strain.

The last term in Eq. 3.4.1 represents strain energy present if expansion ϵ is prohibited. This term may be discarded, since it would disappear anyway when the energy expression is differentiated in the process of forming the equations $\partial \Pi_p / \partial D_i = 0$. Note that with $\Pi_p = U_o - \sigma \epsilon$, we obtain the familiar equation $\sigma = E(\epsilon - \alpha T)$ from $\partial \Pi_p / \partial \epsilon = 0$.

Let us define ϵ_o and σ_o as

$$\epsilon_o = \alpha T, \qquad \sigma_o = -E\epsilon_o = -E\alpha T \qquad (3.4.2)$$

Thus, after discarding the superfluous term, we may write Eq. 3.4.1 in the forms

$$U_o = \frac{1}{2} E\epsilon^2 - E\epsilon_o \epsilon = \frac{1}{2} E\epsilon^2 + \sigma_o \epsilon \qquad (3.4.3)$$

The *initial strain* ϵ_o may arise from thermal action, initial lack of fit in a structural member, or other causes. The *initial stress* σ_o arises when expansion is prohibited or when the expanded or overlength member is forced back to some initial reference size. Thus, as implied by Eq. 3.4.3, we may deal with initial stresses and strains arising from various causes.

3.5 General Expression for Potential Energy

We now consider the case where all six stress components act on in a linearly elastic material. The notation for stresses and strains is

$$\{\sigma\} = \{\sigma_x \quad \sigma_y \quad \sigma_z \quad \tau_{xy} \quad \tau_{yz} \quad \tau_{zx}\}$$
$$\{\epsilon\} = \{\epsilon_x \quad \epsilon_y \quad \epsilon_z \quad \gamma_{xy} \quad \gamma_{yz} \quad \gamma_{zx}\} \qquad (3.5.1)$$

where the engineering definition of shear strain is used; for example, $\gamma_{xy} = \partial u / \partial x + \partial v / \partial y$. Displacements in the x, y, z coordinate directions are

$$\{f\} = \{u \quad v \quad w\} \qquad (3.5.2)$$

The stress-strain relation may be written in either of two ways:

$$\{\sigma\} = [E](\{\epsilon\} - \{\epsilon_o\}) = [E]\{\epsilon\} + \{\sigma_o\} \qquad (3.5.3)$$

where $[E]$ is a *symmetric* matrix of elastic constants that may describe an isotropic or anisotropic material. Initial stresses $\{\sigma_o\}$ arise if, for example, mechanical strain $\{\epsilon\}$ is suppressed as a body is heated.

The increment in strain energy per unit volume owing to infinitesimal distortion of a unit volume is

$$\delta U_o = \{\sigma\}^T \{\delta \epsilon\} = \sigma_x \, \delta \epsilon_x + \sigma_y \, \delta \epsilon_y + \cdots + \tau_{zx} \, \delta \gamma_{zx} \qquad (3.5.4)$$

Changes in $\{\sigma\}$ arising from strain changes $\{\delta \epsilon\}$ are discarded because they produce higher order terms. For example, $(\sigma_x + \delta \sigma_x) \, \delta \epsilon_x \approx \sigma_x \, \delta \epsilon_x$. From

Eq. 3.5.4 we conclude that

$$\frac{\partial U_o}{\partial \epsilon_x} = \sigma_x, \quad \frac{\partial U_o}{\partial \epsilon_y} = \sigma_y, \ldots, \frac{\partial U_o}{\partial \gamma_{zx}} = \tau_{zx} \qquad (3.5.5)$$

Expressing Eq. 3.5.5 in matrix form and using Eq. 3.5.3, we obtain the six equations

$$\left\{\frac{\partial U_o}{\partial \epsilon}\right\} = \{\sigma\} = [E]\{\epsilon\} + \{\sigma_o\} \qquad (3.5.6)$$

Integration with respect to the strains yields

$$U_o = \frac{1}{2}\{\epsilon\}^T[E]\{\epsilon\} + \{\epsilon\}^T\{\sigma_o\} \qquad (3.5.7)$$

That the integration is correct may be seen by differentiating Eqs. 3.5.7 with respect to the strains.[1] As in Eq. 3.4.3, a superfluous integration constant has been discarded.

Equation 3.4.3 is a special case of Eq. 3.5.7. Other special forms (for plane stress, plane strain, axisymmetric solids, etc.) are obtained by setting known zero values of stress or strain into the stress-strain law, Eq. 3.5.6, and collecting terms to yield a new and smaller matrix $[E]$.

Before proceeding we must define body forces per unit volume

$$\{F\} = \{F_x \quad F_y \quad F_z\} \qquad (3.5.8)$$

and surface tractions (force per unit area) on the boundary of the body

$$\{\Phi\} = \{\Phi_x \quad \Phi_y \quad \Phi_z\} \qquad (3.5.9)$$

Components of both $\{F\}$ and $\{\Phi\}$ are considered positive when acting in the positive coordinate directions.

---

[1] Useful rules of matrix manipulation are as follows. Let $[A]$ be a symmetric matrix and $[B]$ an arbitrary rectangular matrix. Define column vectors $\{X\} = \{X_1 \, X_2 \ldots X_n\}$ and $\{Y\} = \{Y_1 \, Y_2 \ldots Y_m\}$. Define scalar $\phi = \frac{1}{2}\{X\}^T[A]\{X\}$ and scalar $\psi = \{X\}^T[B]\{Y\}$.

a. $\{X\}^T\{X\}$, $[B]^T[B]$, and $[B]^T[A][B]$ are symmetric.

b. $\left\{\dfrac{\partial \phi}{\partial X_1} \dfrac{\partial \phi}{\partial X_2} \cdots \dfrac{\partial \phi}{\partial X_n}\right\} = [A]\{X\}$.

c. $\left\{\dfrac{\partial \psi}{\partial X_1} \dfrac{\partial \psi}{\partial X_2} \cdots \dfrac{\partial \psi}{\partial X_n}\right\} = [B]\{Y\}$.

d. Since ψ is a scalar, $\psi = \psi^T = \{Y\}^T[B]^T\{X\}$.

e. $\left\{\dfrac{\partial \psi}{\partial Y_1} \dfrac{\partial \psi}{\partial Y_2} \cdots \dfrac{\partial \psi}{\partial Y_m}\right\} = [B]^T\{X\}$.

Using Eq. 3.5.7, we write the potential energy expression as follows:

$$\Pi_p = \int_{vol} \left(\frac{1}{2} \{\epsilon\}^T [E] \{\epsilon\} + \{\epsilon\}^T \{\sigma_o\} \right) dV$$

$$- \int_{vol} \{f\}^T \{F\} \, dV - \int_{surface} \{f\}^T \{\Phi\} \, dS \quad (3.5.10)$$

The first integral represents the work of internal stresses, this work being stored as strain energy. The last two integrals represent work done (hence potential lost) when body forces and surface forces are displaced in the direction they tend to move. Displacements $\{f\}$ in the surface integral are evaluated at the boundary.

3.6 Properties of Stationary Potential Energy

This section has two purposes. One is to introduce the calculus of variations, which is essential to a fundamental study of the finite element method and to formulation of element properties using certain bases other than assumed displacement fields. The second purpose is to demonstrate that a solution based on stationary potential energy tends to satisfy equilibrium conditions as an automatic consequence of the procedure, provided that compatibility conditions are enforced. Our treatment is brief and relies on an example problem. A general treatment is beyond the scope of this book.

Consider a bar under distributed axial load Q (Fig. 3.6.1a), where Q is a

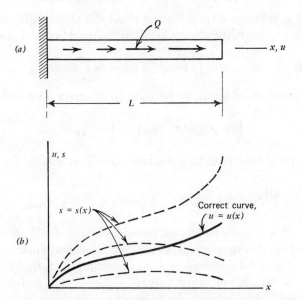

Figure 3.6.1.   (a) Bar under axial loading. (b) Correct and "varied" shapes.

function of x and has dimensions of force per unit length. The state of stress is uniaxial. The expression for potential energy is therefore

$$\Pi_p = \int_0^L \frac{E}{2}\, u_{,x}^2 A\, dx - \int_0^L Qu\, dx \tag{3.6.1}$$

where $A = $ cross-sectional area of the bar. It is assumed that u is continuous and single-valued, so that the integration is meaningful. Specifically, $u_{,x}$ must not become infinite at any point. By calculus of variations, we now seek the "correct curve" $u = u(x)$, which makes Π_p stationary. What is, in fact, obtained is a differential equation for u and certain boundary conditions to be imposed on its solution.

A valid solution must satify $u = 0$ at $x = 0$. This is a kinematic or "forced" boundary condition. Let $s = s(x)$ be a function that satisfies the same forced boundary condition, $s = 0$ at $x = 0$, but is otherwise independent of u. Some possible curves $s = s(x)$ are shown as dashed lines in Fig. 3.6.1b. If e is an arbitrary number, then $u + se$ is an arbitrary shape, and $e = 0$ defines the "correct" shape. Also, with δe a small number, $u + s\,\delta e$ defines a small alteration of the correct curve $u = u(x)$. We ask, what must $u = u(x)$ be in order that the change in Π_p is vanishingly small for small departures from this configuration? To answer the question we first write the potential of a "varied" shape; it is

$$\Pi_p + \delta\Pi_p = \int_0^L \frac{E}{2}\,(u_{,x} + s_{,x}\,\delta e)^2 A\, dx - \int_0^L Q(u + s\,\delta e)\, dx \tag{3.6.2}$$

If Eq. 3.6.1 is subtracted from Eq. 3.6.2 and the higher order term $s_{,x}^2(\delta e)^2$ is neglected, we obtain the change in Π_p in the neighborhood of $e = 0$ as

$$\delta\Pi_p = \delta e \int_0^L E u_{,x} s_{,x} A\, dx - \delta e \int_0^L Qs\, dx \tag{3.6.3}$$

To permit gathering of terms, we perform an integration by parts:

$$\int_0^L u_{,x} s_{,x}\, dx = s u_{,x}\Big|_0^L - \int_0^L s u_{,xx}\, dx \tag{3.6.4}$$

Because of the forced boundary condition, $s = 0$ at $x = 0$, Eqs. 3.6.3 and 3.6.4 yield

$$\frac{\delta\Pi_p}{\delta e} = -\int_0^L (AE u_{,xx} + Q)s\, dx + AE(u_{,x})_L (s)_L \tag{3.6.5}$$

As Π_p is to be stationary for the configuration $u = u(x)$, $\delta\Pi_p/\delta e$ must vanish regardless of whether the stationary value is a minimum, a neutral point, or a maximum (Fig. 3.6.2). Because $s = s(x)$ is *arbitrary*, both for $0 < x < L$ and at $x = L$, $\delta\Pi_p/\delta e$ can vanish only if

$$AE u_{,xx} + Q = 0 \quad\text{and if}\quad AE(u_{,x})_L = 0 \tag{3.6.6}$$

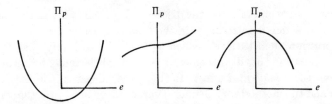

Figure 3.6.2. Minimum, neutral, and maximum values of Π_p at $e = 0$.

The former equation is the governing differential equation of the problem. Note that it is an equilibrium equation (see Fig. 16.4.2). The latter equation is called a "natural" boundary condition because it develops out of the variational procedure; it says that no axial stress exists in the bar at $x = L$. We now have a second-order differential equation and two boundary conditions, one forced and one natural; hence the solution $u = u(x)$ may be established.

How does this discussion relate to the finite element method? The bar of Fig. 3.6.1a might be modeled by a series of n truss elements, each of length $a = L/n$, with nodal loads $Q_i a$, where Q_i is the value of Q at node i. With the left end fixed, arbitrary nodal displacements u_i prescribe a "varied" shape, analogous to a dashed curve in Fig. 3.6.1b. For any n, the smooth curve $u = u(x)$ is approximated by a series of straight line segments, that is, by piecewise polynomial interpolation. Strain $u_{,x}$ changes abruptly from one segment to another but is never infinite. Thus far, then, in both the continuum problem and its discretized analogue, we have satisfied compatibility conditions and displacement boundary conditions by considering only "admissible" configurations. If in addition the differential equations of equilibrium are satisfied for all x and if the stress boundary condition (the "natural" boundary condition) is met, the correct solution has been found.

The solution of the continuum problem satisfies the equilibrium equations 3.6.6, but the solution of the discretized problem does not. The discretized solution is the set of nodal displacements u_i satisfying the *nodal* equilibrium equations $\partial \Pi_p'/\partial u_i = 0$. The symbol Π_p' is used because the continuum and discretized problems are different and have different potentials. Indeed, the u_i of the discretized solution should not be expected to lie on the continuum solution $u = u(x)$. As the number of elements increases without limit, the smooth varied shape $s = s(x)$ is permitted, Π_p' approaches Π_p, and displacements u_i coincide with $u = u(x)$. Also, the nodal equilibrium equations are converted into Eqs. 3.6.6, and the discretized solution becomes exact instead of approximate.

To summarize, we emphasize that if only admissible displacements are permitted, then the method of stationary Π_p satisfies equilibrium conditions implicitly, as a natural outcome of the solution. With only a change in terminology, our remarks about the discretized system apply to the Rayleigh-Ritz method, which is introduced in the following section.

An expression such as Π_p is called a *functional* to imply that Π_p is not simply a function of displacements and displacement derivatives but depends on their *integrated* effect. If Π_p is specialized to the case of plane elasticity and the methods of calculus of variations applied to it, the stationary functional condition $\delta\Pi_p = 0$ is found to demand that:

● The differential equations of equilibrium (Eqs. 1.2.2), must be satisfied within the body.

● On portions of boundary where displacements are not prescribed, the boundary equilibrium equations (Eqs. 1.2.3) must be satisfied.

Provided that the displacement field is compatible, these equilibrium conditions are implicitly contained in the statement $\delta\Pi_p = 0$.

3.7 The Rayleigh-Ritz Method

Continuum problems have an infinite number of d.o.f., because independent displacements are associated with every point. One way to approach such a problem is to attempt to generate and solve the appropriate set of partial differential equations. The difficulties of this approach may be avoided, and the problem reduced to a finite number of degrees of freedom, if a governing functional such as Π_p is available and if we are content with an approximate solution. Instead of generating partial differential equations we generate algebraic equations, equal in number to the generalized coordinates.

In what follows we will use only Π_p as the governing functional. The Rayleigh-Ritz method may of course be used with other functionals as well.

Let us consider a problem in two-dimensional elasticity. A body is subjected to certain boundary tractions and displacements. We are to find an approximate solution for displacements at every point. To do this by the Rayleigh-Ritz method, we begin by constructing approximations for the displacements.

$$u = a_1 g_1(x, y) + a_2 g_2(x, y) + \cdots + a_m g_m(x, y)$$
$$v = b_1 h_1(x, y) + b_2 h_2(x, y) + \cdots + b_n h_n(x, y)$$

$$(3.7.1)$$

Each function g_i and h_i is assumed to have continuous derivatives and must have such a form that all forced boundary conditions of the problem at hand are satisfied, regardless of what numerical values may be assigned to the parameters $a_1, a_2, \ldots, a_m,\ b_1, b_2, \ldots, b_n$. Internal compatibility is automatically satisfied because strains are formed from displacements by the usual relations $\epsilon_x = u_{,x}$, etc. Hence, potential energy Π_p may be expressed as a function of the a's and b's. The a_i and b_i govern Π_p, just as do the D_i in Eq. 3.3.1; indeed, the a's, b's, and D's are all *generalized coordinates*. Accordingly, following arguments made with reference to Eqs. 3.3.1 and 3.3.2, the a's and b's that make Π_p stationary are determined from the set of $m + n$ algebraic equations

$$\frac{\partial \Pi_p}{\partial a_i} = 0, \qquad \frac{\partial \Pi_p}{\partial b_i} = 0 \qquad (3.7.2)$$

The a_i and b_i determined from Eq. 3.7.2 may be substituted into Eqs. 3.7.1. The displacement field is then fully defined, and strains and stresses may be computed from it.

The differential equations of equilibrium should be satisfied at every point of a continuous body. Equations 3.7.2 "do their best" to accomplish this but usually cannot, because they involve only a finite number of d.o.f. Thus our solution is only an approximation of the exact result but should improve if more parameters a_i and b_i are used.

As usual, Eqs. 3.7.2 yield a set of stiffness equations, which may be written $[K]\{D\} = \{R\}$, where $\{D\}$ contains the a's and b's. Not all entries in $\{D\}$ have units of displacement, as is apparent in Eq. 3.7.4. The specific form of $\{R\}$ is governed by the displacement field assumed as well as by the loading on the structure. Vector $\{R\}$ contains *generalized forces*, and while not all R_i have units of force, the R_i and D_i have units such that each product $R_1 D_1$, $R_2 D_2$, etc., has units of work. This is consistent with the virtual work concept, which states that if a system is in equilibrium, then during an arbitrary small displacement from the equilibrium position, the virtual work of applied loads equals the virtual work of internal forces. Symbolically, with $\{\delta D\}$ the arbitrary virtual displacement, we write $\{\delta D\}^T \{R\} = \{\delta D\}^T [K]\{D\}$ as the virtual work equation.

For a specific numerical example of the Rayleigh-Ritz method, consider the cantilever beam of Fig. 3.7.1. With thermal and distributed loads omitted and shear deformation neglected, the potential energy of a beam carrying

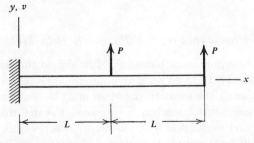

Figure 3.7.1. Cantilever beam.

concentrated forces P_1, P_2, etc. is

$$\Pi_p = \int \frac{EI}{2} (v,_{xx})^2 \, dx - \sum P_i v_i \qquad (3.7.3)$$

Here, an admissible displacement field must have $v = v,_x = 0$ at $x = 0$. Hence let us assume

$$v = a_1 x^2 + a_2 x^3 + \cdots + a_n x^{n+1} \qquad (3.7.4)$$

If only the first term is retained, $d\Pi_p/da_1 = 0$ yields $a_1 = 5PL/8EI$. Using the first two terms, we write $\partial\Pi_p/\partial a_1 = \partial\Pi_p/\partial a_2 = 0$ and obtain $a_1 = 11PL/8EI$ and $a_2 = -P/4EI$. Table 3.7.1 compares exact and approximate

Table 3.7.1. Exact and Rayleigh-Ritz Results

	$EIv_{x=L}$	$EIv_{x=2L}$	$M_{x=0}$	$M_{x=L}$	$M_{x=2L}$
Exact	$1.167PL^3$	$3.50PL^3$	$3PL$	PL	0
One term	$0.625PL^3$	$2.50PL^3$	$1.25PL$	$1.25PL$	$1.25PL$
Two terms	$1.125PL^3$	$3.50PL^3$	$2.75PL$	$1.25PL$	$-0.25PL$

results. Moments are computed as $M = EIv,_{xx}$. As might be expected, displacements are predicted more accurately than moments; clearly two curves $v = v(x)$ may be similar, while their second derivatives may be rather different. Accuracy improves if more terms in Eq. 3.7.4 are retained, but the labor of doing so quickly becomes excessive.

If a distributed load q is applied, Eq. 3.7.3 must be modified. Since each force $q \, dx$ is associated with v units of displacement, the integral of $qv \, dx$ must be subtracted from the right side of Eq. 3.7.3.

Finally, we note again that assumed displacement fields need not be polynomials. Sine series are conveniently applied to problems of simply supported beams and plates.

3.8 Further Comments on the Rayleigh-Ritz Method

Two significant questions may be raised. How should the trial functions of Eq. 3.7.1 be selected? And how good an approximate solution do we obtain?

As noted previously the assumed functions must be continuous and satisfy all forced (displacement) boundary conditions. Beyond this, there seems to be no simple and useful rule that tells us what sort of functions to select and at what point to terminate the series.

Let us suppose that we use the Rayleigh-Ritz process several times in a given problem, each time using one more term in our series of trial functions. In this way we generate a sequence of approximations to the exact potential energy Π_p. If the limit of this sequence is the exact Π_p, it is reasonable to expect that the limit of the corresponding sequence of trial displacement functions will be the exact displacement function. In order to assure convergence to the exact Π_p it is necessary that the trial functions be "relatively complete." Relative completeness is achieved if the trial functions will approximate the true functions, and their derivatives that appear in Π_p, arbitrarily closely as more terms are added to the trial functions. It follows that an infinite series capable of representing *any* continuous displacement field within the domain will be satisfactory, because it contains the true functions as a special case.

We may state that trial functions should be truncated by omitting higher order terms instead of low order terms. A heuristic argument is as follows. Suppose that the cantilever beam of Fig. 3.7.1 is loaded by only a moment at the right end. The exact displacement is of the form $v = a_1x^2$. Indeed, if we use Eq. 3.7.4 as the trial function, we will obtain the exact value of a_1 and also $a_2 = a_3 = \cdots = a_n = 0$. If instead we use the trial function

$$v = a_2x^3 + a_3x^4 + a_5x^5 + \cdots + a_nx^{n+1} \qquad (3.8.1)$$

we may be able to approximate the actual shape $v = a_1x^2$ fairly well, but only by using many terms. But we cannot guarantee convergence to the exact Π_p as n approaches infinity, because omission of the term a_1x^2 prevents the series from being relatively complete. Convergence to an *incorrect* Π_p is to be expected.

A more general statement, related to the foregoing example, pertains to higher derivatives of the function. Higher derivatives may not converge to correct results, even if the trial function is relatively complete and correct convergence is obtained for Π_p, the function itself and its lower derivatives (7, 9). In finite element work this defect appears to be troublesome only when computing transverse shear forces from third derivatives, as, for example, $Q = EI\,d^3v/dx^3$ in a straight beam. In pathological cases— but not in finite elements used in practice—the sequence of approximate Π_p values may converge to the correct value, while the corresponding sequence of trial functions converges to an incorrect result or even diverges.

Regarding the type of approximation provided by the Rayleigh-Ritz method, we can say that the approximate solution is too stiff. The structure is permitted to displace only into shapes that may be formed by superposition of the various terms in the assumed displacement functions. The true displacement shape is therefore excluded, except in quite simple or exceptional cases. Effectively, additional restraints that stiffen the structure have been imposed. This argument suggests the viewpoint that we have obtained

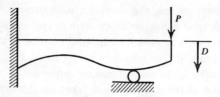

Figure 3.8.1. Elastic beam.

the exact solution of a different problem; a structure different in elastic properties or loading from the actual structure and in effect created by the solution method.

To make some specific statements, it is helpful to refer to an example structure (Fig. 3.8.1). In terms of load P and its corresponding static displacement D, the potential energy of the equilibrium configuration is

$$\Pi_p = \frac{1}{2}PD - PD = -\frac{1}{2}PD \tag{3.8.2}$$

This relation is applicable to either the actual structure or to the "substitute" structure created by an approximate solution. If we define the stiffness k under load P as $k = P/D$, we may write

$$\text{stiffness} = k = -\frac{P^2}{2\Pi_p}$$

$$\text{strain energy} = \frac{1}{2}PD = \frac{P^2}{2k} = -\Pi_p \tag{3.8.3}$$

If P is prescribed and the resulting D has been approximated by a Rayleigh-Ritz solution, Eqs. 3.8.3 and Fig. 3.2.2 indicate that:

1. Because $D_{\text{approx}} \neq D_{\text{exact}}$ and $(\Pi_p)_{\text{exact}}$ is a minimum, $(\Pi_p)_{\text{approx}} > (\Pi_p)_{\text{exact}}$. In other words, we have obtained an upper bound on Π_p. In magnitude, the approximate Π_p is too small.

2. The approximate stiffness k is too large.

3. The approximate strain energy is too small; that is, we have obtained a lower bound on strain energy.

Where there is a single load, as in Fig. 3.8.1, we can state that because strain energy is underestimated, $D_{\text{approx}} < D_{\text{exact}}$. But if several loads act, we can state only that the product $\{P_1 \, P_2 \ldots P_n\}^T \{D_1 \, D_2 \ldots D_n\}$ is underestimated.

If displacements are prescribed instead of forces, the approximate strain energy is overestimated (why?). In problems where both forces and displacements are prescribed, a bound cannot be established.

The foregoing conclusions are not valid in the presence of initial stresses or strains. In such problems the approximate displacement field is used to

generate the forces resulting from the initial conditions as well as to approximate the structure stiffness. These forces may be underestimated or overestimated by the approximation.

Note that no conclusion has been made regarding stresses. We may argue
that because stresses are calculated from displacements, stresses will be
underestimated if displacements are underestimated. But as seen in Table
3.7.1, stresses may be too high in some places and too low in others, so that
this argument is of slight practical help.

3.9   Finite Element Form of the Rayleigh-Ritz Method

Let us examine the problem of Fig. 3.7.1 again. Consider the displacement
fields

$$v = a_1 + a_2 x + a_3 x^2 + a_4 x^3 \qquad \text{for} \qquad 0 \leq x \leq L$$
$$v = a_5 + a_6 x + a_7 x^2 + a_8 x^3 \qquad \text{for} \qquad L \leq x \leq 2L$$

$$(3.9.1)$$

A Rayleigh-Ritz solution requires that we meet forced boundary conditions
and satisfy compatibility with the trial functions. Therefore the first of Eqs.
3.9.1 must be modified so that $v = v_{,x} = 0$ at $x = 0$. Also, at $x = L$ both
polynomials must exhibit the same displacement v and the same slope $v_{,x}$.
Thus a total of four constraints are imposed, which reduce the number of
independent a's in Eq. 3.9.1 from eight to four. Hence Eq. 3.7.3 yields Π_p as a
function of four a_i, with one integral spanning 0 to L and a second spanning
L to $2L$. Next the four equations $\partial \Pi_p / \partial a_i = 0$ yield the four independent
a's; the displacement field is completely defined, and stresses may then be
computed.

The solution outlined above is a Rayleigh-Ritz solution. *It is also a finite
element solution*, because we have used a piecewise fit to the displaced shape.
In other words, the classical Rayleigh-Ritz method uses a single trial function,
often a lengthy expression, which spans the entire structure; the finite
element form of the Rayleigh-Ritz method uses several trial functions, each
usually simple, and each defined over only a portion of the structure. This
is piecewise polynomial interpolation, and it is the essence of the finite
element method.

A solution based on Eqs. 3.9.1 is not entirely satisfactory. The a's do not
all have an obvious physical meaning. Also, some algebraic work is needed
to make the two polynomials match at $x = L$. We therefore do what is
always done in finite element work; we exchange the a's for nodal displacements. It is this exchange that makes it comparatively easy to automate the
method by writing a generally applicable computer program. Let us now
consider some details of this approach.

Figure 3.9.1 shows a straight, uniform beam element of flexural stiffness

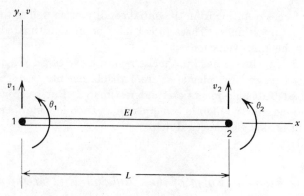

Figure 3.9.1.   Beam element.

EI. The four nodal d.o.f. are displacements v_1, v_2 and rotations θ_1, θ_2, where $\theta = v_{,x}$. It is reasonable to assume a cubic displacement field, because a cubic is uniquely defined by four parameters. We therefore start with Eq. 3.9.1 and write it in the form

$$v = [1 \quad x \quad x^2 \quad x^3] \begin{Bmatrix} a_1 \\ a_2 \\ a_3 \\ a_4 \end{Bmatrix}, \quad \text{or} \quad v = [X]\{a\} \qquad (3.9.2)$$

Differentiation gives $\theta = [0 \quad 1 \quad 2x \quad 3x^2]\{a\}$. Hence, by use of the four boundary conditions $v = v_1$ and $\theta = \theta_1$ at $x = 0$, $v = v_2$ and $\theta = \theta_2$ at $x = L$, we obtain

$$\begin{Bmatrix} v_1 \\ \theta_1 \\ v_2 \\ \theta_2 \end{Bmatrix} = \begin{bmatrix} 1 & 0 & 0 & 0 \\ 0 & 1 & 0 & 0 \\ 1 & L & L^2 & L^3 \\ 0 & 1 & 2L & 3L^2 \end{bmatrix} \begin{Bmatrix} a_1 \\ a_2 \\ a_3 \\ a_4 \end{Bmatrix} \quad \text{or} \quad \{d\} = [DA]\{a\} \qquad (3.9.3)$$

Hence v may be defined in terms of nodal d.o.f. $\{d\}$:

$$\{a\} = [DA]^{-1}\{d\} \quad \text{and} \quad v = [X][DA]^{-1}\{d\} \qquad (3.9.4)$$

where

$$[DA]^{-1} = \begin{bmatrix} 1 & 0 & 0 & 0 \\ 0 & 1 & 0 & 0 \\ -3/L^2 & -2/L & 3/L^2 & -1/L \\ 2/L^3 & 1/L^2 & -2/L^3 & 1/L^2 \end{bmatrix} \qquad (3.9.5)$$

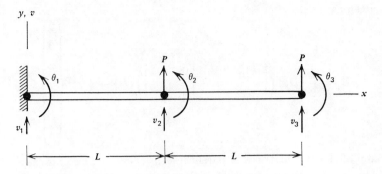

Figure 3.9.2.   Nodal d.o.f. for beam problem.

We have only begun the solution, as Eq. 3.9.4 merely defines an assumed displacement field $v = v(x)$ in terms of parameters $\{d\}$. But subsequent steps are straightforward and may be summarized as follows.

The problem of Fig. 3.7.1 is shown again in Fig. 3.9.2, with the d.o.f. appropriate to the finite element solution. The forced boundary condition at $x = 0$ is met by setting $v_1 = \theta_1 = 0$, and this may be done after the governing algebraic equations are assembled. Compatibility of elements at node 2 is enforced by use of v_2 and θ_2 as d.o.f. in both the left element and the right element. There remain four d.o.f. to be determined by the condition $\delta\Pi_p = 0$, just as is the case with a solution based on Eqs. 3.9.1; indeed, the curves $v = v(x)$ defined by the two solutions are exactly the same.

The remainder of the finite element solution is outlined as follows. Element curvature is $v,_{xx} = [B]\{d\}$, where $[B] = [0 \quad 0 \quad 2 \quad 6x][DA]^{-1}$. Now $v,_{xx}^2 = [v,_{xx}]^T[v,_{xx}]$, so Eq. 3.7.3 yields

$$\Pi_p = \frac{1}{2}\{d\}_1^T \int_0^L [B]^T EI[B]\,dx\{d\}_1$$

$$+ \frac{1}{2}\{d\}_2^T \int_0^L [B]^T EI[B]\,dx\{d\}_2 - Pv_2 - Pv_3 \quad (3.9.6)$$

where $\{d\}_1 = \{v_1 \quad \theta_1 \quad v_2 \quad \theta_2\}$ and $\{d\}_2 = \{v_2 \quad \theta_2 \quad v_3 \quad \theta_3\}$. In Chapter Four we will identify each integral of $[B]^T EI[B]$ as an element stiffness matrix $[k]$. Here each element $[k]$ is 4 by 4, and because the two elements are identical, $[k]_1 = [k]_2$. Expansion to "structure size" is accomplished by adding two rows and two columns of zeros to the end of $[k]_1$ and to the beginning of $[k]_2$. Thus, with $\{D\} = \{v_1 \quad \theta_1 \quad v_2 \quad \theta_2 \quad v_3 \quad \theta_3\}$, Eq. 3.9.6 may be written as

$$\Pi_p = \frac{1}{2} \{D\}^T \left(\begin{bmatrix} [k]_1 & 0 \\ 0 & 0 \end{bmatrix}_{6\times 6} + \begin{bmatrix} 0 & 0 \\ 0 & [k]_2 \end{bmatrix}_{6\times 6} \right) \{D\} - \{D\}^T \begin{Bmatrix} 0 \\ 0 \\ P \\ 0 \\ P \\ 0 \end{Bmatrix} \qquad (3.9.7)$$

The sum of the two matrices in parentheses is the structure stiffness matrix $[K]$. The set of six equations $\{\partial\Pi_p/\partial D\} = 0$ defines the $\{D\}$ that makes Π_p stationary; it is

$$[K]\{D\} - \{0 \quad 0 \quad P \quad 0 \quad P \quad 0\} = 0 \qquad (3.9.8)$$

By setting $v_1 = \theta_1 = 0$ we impose the forced boundary condition and prevent rigid body motion. Methods for doing this are described in Sections 2.2 and 2.7. The resulting equations may be solved for v_2, θ_2, v_3, and θ_3.

The solution outlined above is *exact*, owing to the fact that the beam of Fig. 3.9.2 actually does deflect into a shape defined by two cubic polynomials. If additional loading were *distributed* along the beam, quartic curves would define the exact deflected shape, and the finite element solution would be approximate. The treatment of distributed loads in the finite element method is straightforward and is described in Chapter Four. The derivation of $[k]$ for a beam element, begun in Eq. 3.9.6, is completed in Problem 4.6.

3.10   Concluding Remarks

Mathematical rigor and further details regarding energy methods, variational principles, and the Rayleigh-Ritz method may be found in several references. Some are given in the reference list for this chapter.

Finally, recall that compatibility of displacements has been assumed throughout this chapter. In the finite element form of the Rayleigh-Ritz method, we encounter many elements that violate the compatibility condition but nevertheless work very well. When using these elements we must modify our statements of how the Rayleigh-Ritz method behaves.

PROBLEMS

3.1.   Temporarily remove the supports from the plane truss of Fig. 2.2.1. Write the total potential as a function of $u_1, v_1, \ldots, v_3$. Hence, show that the six equations $\partial\Pi_p/\partial u_i = \partial\Pi_p/\partial v_i = 0$ yield Eqs. 2.2.5.

3.2. Work out the two-term solution of Table 3.7.1 and verify that the results given are correct.

3.3. Replace loads P in Fig. 3.7.1 by a uniformly distributed load. Use the first two terms of Eq. 3.7.4 in a Rayleigh-Ritz solution. Compare computed displacements and moments with the exact values.

3.4. A cantilever beam is fixed at the end $x = 0$ and subjected to a specified displacement $v = c$ at the end $x = L$. Is the following assumption suitable for a Rayleigh-Ritz solution? Explain.

$$v = \frac{cx^2}{L^2} + \sum a_i \sin \frac{i\pi x}{L}$$

3.5. The two-dimensional body is fixed along $x = 0$. The right edge is given linearly varying displacement u, but no value of v is prescribed. Write admissible displacement field assumptions for u and v in the form of infinite polynomial series.

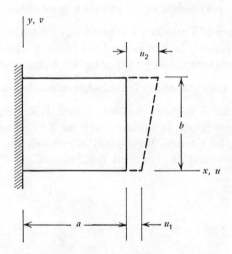

3.6. Insofar as possible, justify the statements that follow Eq. 3.8.3 for the general case where there are several loads and several d.o.f.

3.7. A stepped bar (modulus E, thermal coefficient α) is of cross-sectional area A in one half and $2A$ in the other. Let the end displacement be D and make the (crude) approximation that $u = xD/L$. Thus, as will be justified in Chapter four, the Rayleigh-Ritz approximation for D caused by heating the bar an amount T is

$$\left(\int_0^L \frac{E}{L^2} dV \right) D = \frac{3AE}{2L} D = \int_0^L \frac{E\alpha T}{L} dV$$

where $3AE/2L$ is the approximate stiffness. Hence, assume that the left half only, then the right half only, is heated an amount T. In each case compare the exact and approximate displacements D. (The results should show that no bound can be predicted, as noted in the latter part of Section 3.8.)

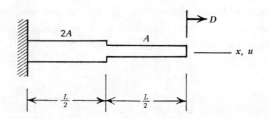

3.8.   A beam is simply supported at its ends $x = 0$ and $x = L$. Write series for lateral displacement v that are acceptable for a Rayleigh-Ritz solution, using (a) polynomial series and (b) trigonometric series.

3.9.   If the beam of Problem 3.8 is uniformly loaded, solve for the maximum displacement using (a) one term of the polynomial series, (b) one term of the trigonometric series, and (c) the complete trigonometric series.

3.10.   (a–e)   Verify the matrix formulas given as a footnote in Section 3.5.

3.11.   In the finite element method, forced boundary conditions are imposed after equilibrium equations (such as 3.9.8) have been written. Explain why, in the classical Rayleigh-Ritz method, these conditions are imposed on the assumed displacement field instead.

CHAPTER FOUR

Use of Assumed Displacement Fields

4.1 *Introduction*

We have thus far viewed the finite element method as a form of the Rayleigh-Ritz method. An alternative and historically earlier viewpoint is the physical one. Just as the motivation for linear interpolation is that a small enough piece of any smooth curve is practically a straight line, the physical motivation for finite elements is that a small enough piece of any continuum may have a simple state, such as constant strain. Or, if we wish to use larger pieces, a linear strain state might be assumed. If the "interpolation pieces" (elements) can be properly joined, it is reasonable to expect a satisfactory approximation of physical reality.

From either the physical or mathematical point of view, a necessary and important step is that of choosing the trial function in each element. There are various finite element procedures, some that use an assumed displacement field as the trial function, some that use an assumed stress field, and some that use a combination of the two. Elements based on assumed displacement fields have been most successful and versatile, so we will consider this method only. Accordingly, this chapter presents the general formulation for finite elements based on assumed displacement fields and offers general comments on the behavior to be expected from such elements.

Any finite element procedure is capable of producing many elements applicable to a given class of problem. It is not easy to decide which element is to be preferred, and the question will be avoided for the time being.

4.2 Formulation Based on Assumed Displacements

The general procedure is simple. Because displacements are the primary unknowns, the appropriate functional is that of total potential energy Π_p (Eq. 3.5.10). First, Π_p is expressed in terms of nodal d.o.f. Then, as required by the principle of stationary potential energy, nodal d.o.f. must assume such values that Π_p is stationary, thus yielding a set of simultaneous algebraic equations in the nodal d.o.f. The reader may note that the question of compatibility has not been mentioned here. The question is an important one, inasmuch as Chapter Three has implied that the solution will not satisfy equilibrium if the basic assumptions do not satisfy compatibility. The matter requires some discussion and is taken up in Section 4.6.

Let displacements of an arbitrary point within an element be given by

$$\{f\} = \{u \quad v \quad w\} = [N]\{d\} \tag{4.2.1}$$

where $\{d\}$ is the array of nodal generalized coordinates of the element. Some of these nodal d.o.f. may be displacements, and others may be displacement derivatives, such as strain or rotation. Matrix $[N]$ defines the nature of the assumed displacement field, but for the time being, we need not say how it is to be constructed.

Strains are expressed in terms of $\{d\}$ by applying to $\{f\}$ the usual strain-displacement relations:

$$\epsilon_x = u_{,x}, \qquad \epsilon_y = v_{,y}, \qquad \gamma_{xy} = u_{,y} + v_{,x}, \text{ etc.} \tag{4.2.2}$$

The resulting strains are gathered into matrix form and symbolized as

$$\{\epsilon\} = [B]\{d\} \tag{4.2.3}$$

In general $\{\epsilon\}$ contains all six strains. The number may be less in special cases, for example, three suffice for plane stress problems.

Substitution of Eqs. 4.2.1 and 4.2.3 into Eq. 3.5.10 yields the total potential of one element:

$$\Pi_{pe} = \frac{1}{2}\{d\}^T\left(\int_{\text{vol}} [B]^T[E][B]\,dV\right)\{d\} + \{d\}^T\int_{\text{vol}} [B]^T\{\sigma_o\}\,dV$$

$$- \{d\}^T\int_{\text{vol}} [N]^T\{F\}\,dV - \{d\}^T\int_{\text{surface}} [N]^T\{\Phi\}\,dS \tag{4.2.4}$$

If desired, we may replace $\{\sigma_o\}$ by $-[E]\{\epsilon_o\}$ (see Eq. 3.5.3). Indeed, the initial effects integral is a sort of "catch-all," as an element might be simultaneously subjected to initial stress from one cause and initial strain from another.[1]

---

[1] It may be noted that the interference present in a shrink-fit problem may be simulated by applying a temperature change to one of the mating parts.

The total potential of the entire structure is the sum of the potentials of the individual elements. At this stage let us explicitly include the potential of concentrated loads $\{P\}$, which are applied to nodes of the structure by some external agency. A component of $\{P\}$ is taken as positive when acting in the same direction as the corresponding nodal d.o.f. Thus, for a structure of m elements,

$$\Pi_p = \left(\sum_1^m \Pi_{pe}\right) - \{D\}^T\{P\} \tag{4.2.5}$$

As described in Chapter Two, each component of each element vector $\{d\}$ appears in the structure vector $\{D\}$. Accordingly, $\{d\}$ for each element may be replaced by $\{D\}$ if the remaining element matrices in the Π_{pe} expression are enlarged and where necessary their terms rearranged. In other words, the summation of Eq. 4.2.5 implies the expansion of element matrices to "structure size," followed by summation of overlapping terms. Thus, Eqs. 4.2.4 and 4.2.5 yield

$$\Pi_p = \frac{1}{2}\{D\}^T\left(\sum_1^m \int_{\text{vol}}[B]^T[E][B]\,dV\right)\{D\} + \{D\}^T\sum_1^m\left(\int_{\text{vol}}[B]^T\{\sigma_o\}\,dV\right.$$

$$\left. - \int_{\text{vol}}[N]^T\{F\}\,dV - \int_{\text{surface}}[N]^T\{\Phi\}\,dS\right) - \{D\}^T\{P\} \tag{4.2.6}$$

The actual structure has now been replaced by a finite element model whose total potential is governed by the nodal d.o.f. $\{D\}$. The static equilibrium configuration is therefore that set of values $\{D\}$ that satisfies the equations

$$0 = \frac{\partial\Pi_p}{\partial D_1} = \frac{\partial\Pi_p}{\partial D_2} = \cdots = \frac{\partial\Pi_p}{\partial D_n}, \quad \text{or} \quad \left\{\frac{\partial\Pi_p}{\partial D}\right\} = 0 \tag{4.2.7}$$

Thus Eq. 4.2.6 yields the equilibrium equations

$$\left(\sum_1^m\int_{\text{vol}}[B]^T[E][B]\,dV\right)\{D\}$$

$$= \sum_1^m\left(-\int_{\text{vol}}[B]^T\{\sigma_o\}\,dV\right.$$

$$\left. + \int_{\text{vol}}[N]^T\{F\}\,dV + \int_{\text{surface}}[N]^T\{\Phi\}\,dS\right) + \{P\} \tag{4.2.8}$$

We identify each of the m integrals on the left side of Eq. 4.2.8 as an element stiffness matrix $[k]$:

$$[k] = \int_{\text{vol}}[B]^T[E][B]\,dV \tag{4.2.9}$$

Similarly, the group of three integrals on the right side is identified as the array of element nodal forces $\{r\}$ produced by initial stress, body force, and surface force in the element.

$$\{r\} = -\int_{\text{vol}} [B]^T\{\sigma_o\}\,dV + \int_{\text{vol}} [N]^T\{F\}\,dV + \int_{\text{surface}} [N]^T\{\Phi\}\,dS \quad (4.2.10)$$

Any or all contributions to $\{r\}$ may vanish. In particular, the surface integral is nonzero only for those element boundaries that also are part of a structure boundary subjected to externally applied distributed loading.

Let $\{p\} = \{p_x\ p_y\ p_z\}$ be a concentrated load that acts on an element but not at a nodal point. If several such loads act on an element, the nodal load $\{r_p\}$ they produce may be found by viewing the last integral of Eq. 4.2.10 as a summation of forces $\{\Phi\}\ \Delta S$, where large tractions $\{\Phi\}$ act on small separated areas ΔS. Thus,

$$\{r_p\} = [N]_1^T\{p\}_1 + [N]_2^T\{p\}_2 + \cdots + [N]_n^T\{p\}_n \quad (4.2.11)$$

where matrix $[N]$ is evaluated at the locations of $\{p\}_1$, $\{p\}_2$, etc.

Forces $\{r\}$ and $\{r_p\}$ of Eqs. 4.2.10 and 4.2.11 may be called "kinematically consistent generalized nodal forces." The term "generalized" indicates that some terms in $\{r\}$ may be moments or even higher order quantities if nodal d.o.f. include strains or curvatures. The forces are "kinematically consistent," because they satisfy the virtual work equation. That is, virtual work $r_i\ \delta d_i$ is done by a particular generalized force r_i when the corresponding displacement δd_i permitted while other nodal displacements are prohibited. The work $r_i\ \delta d_i$ is equal to the work done by the distributed forces or non-nodal loads in moving through the displacements dictated by δd_i and the assumed displacement field (see Problem 4.8). The symbolic statement of this relation, involving all forces $\{r\}$ instead of just one, is obtained by pre-multiplying Eqs. 4.2.10 and 4.2.11 by $\{\delta d\}^T$. Thus, for example, in Eq. 4.2.10 the last integral involves $\{\delta d\}^T[N]^T\{\Phi\}\,dS = \{\delta f\}^T\{\Phi\}\,dS$, which is work done by forces $\{\Phi\}\,dS$ during displacements $\{\delta f\}$.

Equations 4.2.9 through 4.2.11 constitute the "recipe" for generation of element matrices based on assumed displacement fields. The above derivation parallels that found in various papers, for example Ref. 1.

Assembly of element matrices as indicated by Eq. 4.2.8 yields

$$\left(\sum_1^m [k]\right)\{D\} = \left(\sum_1^m \{r\}\right) + \{P\}, \quad \text{or} \quad [K]\{D\} = \{R\} \quad (4.2.12)$$

This is the same result as Eq. 2.5.5, which was obtained by physical arguments. Note from Eqs. 4.2.6 through 4.2.12 that a concise expression for the total potential energy of the structure is

$$\Pi_p = \frac{1}{2}\{D\}^T[K]\{D\} - \{D\}^T\{R\} \quad (4.2.13)$$

Most elements considered in this book are conveniently formulated in terms of nodal displacements d_i, using matrices $[N]$ and $[B]$ of Eqs. 4.2.1 and 4.2.3. But it is also possible, and occasionally convenient, to work in terms of the coefficients a_i of assumed polynomials such as Eqs. 4.4.1 and 4.4.2. Corresponding to Eqs. 4.2.1 and 4.2.3, we then have $\{f\} = [N_a]\{a\}$ and $\{\epsilon\} = [B_a]\{a\}$. The "$\{d\}$ basis" and "$\{a\}$ basis" formulations are related as follows. The equation $\{d\} = [DA]\{a\}$ is established by substitution of nodal coordinates into the assumed polynomial, just as was done in Eq. 3.9.3. Thus,

$$\{f\} = [N_a]\{a\} = [N_a][DA]^{-1}\{d\} = [N]\{d\}$$
$$\{\epsilon\} = [B_a]\{a\} = [B_a][DA]^{-1}\{d\} = [B]\{d\}$$

(4.2.14)

We may therefore substitute $[N] = [N_a][DA]^{-1}$ and $[B] = [B_a][DA]^{-1}$ in all equations. Corresponding to Eqs. 4.2.9 and 4.2.10, we obtain

$$[k_a] = \int_{\text{vol}} [B_a]^T [E][B_a] \, dV, \qquad \{r_a\} = -\int_{\text{vol}} [B_a]^T \{\sigma_o\} \, dV + \cdots \quad (4.2.15)$$

Now as a *final* step in element formulation we exchange $\{a\}$ for $\{d\}$:

$$[k] = [DA]^{-T}[k_a][DA]^{-1}, \qquad \{r\} = [DA]^{-T}\{r_a\} \qquad (4.2.16)$$

The essential difference in approach is merely that replacement of $\{a\}$ by $\{d\}$ is delayed until the end of element formulation. One should be warned, however, that matrix $[DA]$ may be singular for some elements (8). When this happens it indicates that the assumed displacement field does not permit the nodal d.o.f. to be independent of one another. Further comments regarding the "$\{a\}$ basis" formulation appear in Section 17.7.

4.3   An Example of Element Formulation

To give the formulation of Section 4.2 some physical appeal and substantiate its validity, let us consider a uniform two-force (truss) member of cross-sectional area A and elastic modulus E. For simplicity let the bar lie along the x axis, as shown in Fig. 4.3.1.[2]

The x direction displacement u is the only one that need be considered here. We will assume that u varies linearly with x.

$$\{f\} = \{u\} = u = a_1 + a_2 x = [1 \quad x]\{a_1 \quad a_2\} \qquad (4.3.1)$$

[2] The reader may note that element nodes are now numbered, while in Chapter Two they are lettered. The letters or numbers are "dummy indices," independent of structural node numbering, and serving only as convenient labels for a typical element. Thus, no confusion need arise.

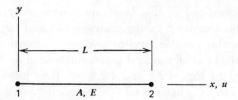

Figure 4.3.1.   Truss member.

This assumption is exact for this particular problem. Next, as in Section 3.9, coefficients a_1 and a_2 are replaced by nodal d.o.f. (u_1 and u_2). Evaluation of Eq. 4.3.1 at the two ends yields

$$\{d\} = \begin{Bmatrix} u_1 \\ u_2 \end{Bmatrix} = \begin{bmatrix} 1 & 0 \\ 1 & L \end{bmatrix} \begin{Bmatrix} a_1 \\ a_2 \end{Bmatrix} = [DA]\{a\} \qquad (4.3.2)$$

Hence

$$\{f\} = \begin{bmatrix} 1 & x \end{bmatrix}[DA]^{-1}\{d\} = \left[\left(1 - \frac{x}{L}\right) \quad \frac{x}{L} \right]\{d\} = [N]\{d\} \qquad (4.3.3)$$

which is the form of Eq. 4.2.1 appropriate to this problem. Equation 4.2.3 becomes

$$\{\epsilon\} = \epsilon_x = u,_x = \left[-\frac{1}{L} \quad \frac{1}{L} \right]\{u_1 \quad u_2\} = [B]\{d\} \qquad (4.3.4)$$

Equation 4.2.9 may be specialized for the present problem by modifying matrix $[E]$ so that it represents a condition of uniaxial stress. But for this simple problem it happens to be easier to note that (a) the strain energy per unit volume in uniaxial stress is $U_o = E\epsilon_x^2/2$, and (b) in general, element strain energy is $U = \{d\}^T[k]\{d\}/2$ (see Eq. 4.2.13 and consider a one-element structure). Accordingly, for U and the element stiffness matrix $[k]$ we write,

$$U = \int_0^L \frac{1}{2} E\epsilon_x^2 A \, dx = \frac{1}{2} \int_0^L \epsilon_x^T E \epsilon_x A \, dx = \frac{1}{2} \{d\}^T \int_0^L [B]^T E[B] A \, dx \{d\}$$

$$(4.3.5)$$

$$[k] = \int_0^L [B]^T E[B] A \, dx = \int_0^L \begin{Bmatrix} -\dfrac{1}{L} \\ \dfrac{1}{L} \end{Bmatrix} E \left[-\frac{1}{L} \quad \frac{1}{L} \right] A \, dx = \frac{AE}{L} \begin{bmatrix} 1 & -1 \\ -1 & 1 \end{bmatrix}$$

To construct a vector $\{r\}$, let us suppose that the bar is initially too long in the amount ΔL so that the initial strain is $\epsilon_o = \Delta L/L$, and is heated T degrees so that with thermal expansion αT prohibited the initial stress is $\sigma_o = -E\alpha T$. Also suppose that there is a body force in the negative x direction of γ units of force per unit volume, and a concentrated force p in the positive x direction at $x = 2L/3$. As noted below Eq. 4.2.4, the "initial effects" integral may be used to account for both σ_o and ϵ_o. Thus Eqs. 4.2.10 and 4.2.11 yield

$$\{r\} = \int_0^L \left\{ \begin{array}{c} -\dfrac{1}{L} \\[2mm] \dfrac{1}{L} \end{array} \right\} E \frac{\Delta L}{L}\,(A\,dx) - \int_0^L \left\{ \begin{array}{c} -\dfrac{1}{L} \\[2mm] \dfrac{1}{L} \end{array} \right\} (-E\alpha T)(A\,dx)$$

$$+ \int_0^L \left\{ \begin{array}{c} 1 - \dfrac{x}{L} \\[2mm] \dfrac{x}{L} \end{array} \right\} (-\gamma)(A\,dx) + \left\{ \begin{array}{c} 1 - \dfrac{2}{3} \\[2mm] \dfrac{2}{3} \end{array} \right\} p \qquad (4.3.6)$$

$$\{r\} = EA\frac{\Delta L}{L}\left\{ \begin{array}{c} -1 \\ 1 \end{array} \right\} + EA\alpha T\left\{ \begin{array}{c} -1 \\ 1 \end{array} \right\} + \frac{\gamma AL}{2}\left\{ \begin{array}{c} -1 \\ -1 \end{array} \right\} + p\left\{ \begin{array}{c} \dfrac{1}{3} \\[2mm] \dfrac{2}{3} \end{array} \right\} \qquad (4.3.7)$$

Apparently, $\{r\}$ is a vector of loads applied *by* the element *to* the nodes.

When nodal displacements have been determined, element stresses follow from Hooke's law (Eq. 3.5.3):

$$\sigma = E(\epsilon - \epsilon_o) + \sigma_o = E[B]\{d\} - E\frac{\Delta L}{L} - E\alpha T \qquad (4.3.8)$$

Again we have made no fuss regarding the distinction between σ_o and ϵ_o; they may be used interchangeably or simultaneously, as happens to be convenient and appropriate.

The reader may check that these results are reasonable and in accord with the physical arguments of Chapter Two. This observation may suggest that "common sense" is sufficient for formulation of finite element matrices, but this is not the case in more complicated situations.

4.4 The Plane Constant-Strain Triangle

A plane triangle having six d.o.f. is shown in Fig. 4.4.1. This element is widely used, owing perhaps to its simplicity and early widespread adoption. For this reason it deserves comment. Also, the element and equations associated with it are useful in subsequent discussions. However, better elements are now available. A quadrilateral built of four such triangles (Fig. 6.1.1) behaves rather like element $Q4$ of Table 6.4.1.

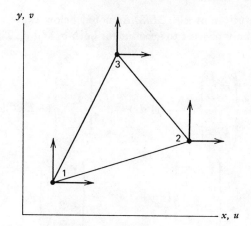

Figure 4.4.1. Plane triangle with six d.o.f.

The assumed displacement field is linear:

$$\begin{Bmatrix} u \\ v \end{Bmatrix} = \begin{bmatrix} 1 & x & y & 0 & 0 & 0 \\ 0 & 0 & 0 & 1 & x & y \end{bmatrix} \begin{Bmatrix} a_1 \\ a_2 \\ \cdot \\ \cdot \\ \cdot \\ a_6 \end{Bmatrix} \tag{4.4.1}$$

Thus $\epsilon_x = a_2$, $\epsilon_y = a_6$, and $\gamma_{xy} = a_3 + a_5$; hence the name, "constant strain triangle."

To exchange the a's for nodal displacements we need only examine displacement u, since the transformation of displacement v is identical. As in Eq. 3.9.3 we evaluate u at each node. Thus,

$$\begin{Bmatrix} u_1 \\ u_2 \\ u_3 \end{Bmatrix} = \begin{bmatrix} 1 & x_1 & y_1 \\ 1 & x_2 & y_2 \\ 1 & x_3 & y_3 \end{bmatrix} \begin{Bmatrix} a_1 \\ a_2 \\ a_3 \end{Bmatrix}, \quad \text{or} \quad \{u\} = [DA]\{a\} \tag{4.4.2}$$

By inversion we obtain $\{a\} = [DA]^{-1}\{u\}$, where

$$[DA]^{-1} = \frac{1}{2A} \begin{bmatrix} x_2 y_3 - x_3 y_2 & x_3 y_1 - x_1 y_3 & x_1 y_2 - x_2 y_1 \\ y_2 - y_3 & y_3 - y_1 & y_1 - y_2 \\ x_3 - x_2 & x_1 - x_3 & x_2 - x_1 \end{bmatrix} \tag{4.4.3}$$

and A is the area of the triangle. Area A is equal to one half the magnitude of the determinant of $[DA]$. The determinant becomes negative if nodes 1, 2, 3 are labeled clockwise around the element.

The element displacement field in terms of nodal displacements is therefore

$$u = [1 \quad x \quad y][DA]^{-1}\{u_1 \quad u_2 \quad u_3\} \tag{4.4.4}$$

$$v = [1 \quad x \quad y][DA]^{-1}\{v_1 \quad v_2 \quad v_3\} \tag{4.4.5}$$

Rearrangement of terms in Eqs. 4.4.4 and 4.4.5 yields the displacement field in the form $\{u \quad v\} = [N]\{d\}$, where $\{d\} = \{u_1 \quad v_1 \quad u_2 \quad v_2 \quad u_3 \quad v_3\}$.

The vector of strains is $\{\epsilon\} = \{\epsilon_x \quad \epsilon_y \quad \gamma_{xy}\} = [B]\{d\}$. Therefore strain-displacement matrix $[B]$ is given by differentiation of $[N]$:

$$[B] = \begin{bmatrix} \dfrac{\partial}{\partial x} & 0 \\ 0 & \dfrac{\partial}{\partial y} \\ \dfrac{\partial}{\partial y} & \dfrac{\partial}{\partial x} \end{bmatrix} [N] \tag{4.4.6}$$

Matrix $[B]$ is not a function of x or y, so that the element stiffness matrix is simply $[k] = [B]^T[E][B]tA$, where t = element thickness and A = element area. Matrix $[E]$ for an isotropic material is, for plane stress and plane strain conditions, respectively,

$$\frac{E}{1-\nu^2}\begin{bmatrix} 1 & \nu & 0 \\ \nu & 1 & 0 \\ 0 & 0 & \dfrac{1-\nu}{2} \end{bmatrix}, \quad \frac{E}{(1+\nu)(1-2\nu)}\begin{bmatrix} 1-\nu & \nu & 0 \\ \nu & 1-\nu & 0 \\ 0 & 0 & \dfrac{1-2\nu}{2} \end{bmatrix} \tag{4.4.7}$$

where E = elastic modulus and ν = Poisson's ratio. A temperature change T produces initial strains $\{\epsilon_o\} = \{\alpha T \quad \alpha T \quad 0\}$ in the plane stress case and $\{\epsilon_o\} = (1+\nu)\{\alpha T \quad \alpha T \quad 0\}$ in the plane strain case, where α = coefficient of thermal expansion.

A variable element thickness may be accommodated by interpolation from nodal values of thickness. Specifically, we may use Eq. 4.4.4 with u replaced by t and u_1, u_2, u_3 replaced by nodal thicknesses t_1, t_2, t_3. Integration of linear functions of x and y is now required in forming $[k]$. And, if body forces should happen to be linear in x and y, quadratic terms must be integrated to form nodal loads $\{r\}$. The following formulas are helpful (6 of Chapter Nine).

These formulas are valid if the origin of coordinates xy is placed at the centroid of the triangle. Again, A = element area.

$$\iint x \, dx \, dy = \iint y \, dx \, dy = 0$$

$$\iint x^2 \, dx \, dy = \frac{A}{12} \left(x_1{}^2 + x_2{}^2 + x_3{}^2 \right)$$

$$\iint y^2 \, dx \, dy = \frac{A}{12} \left(y_1{}^2 + y_2{}^2 + y_3{}^2 \right)$$
(4.4.8)

$$\iint xy \, dx \, dy = \frac{A}{12} \left(x_1 y_1 + x_2 y_2 + x_3 y_3 \right)$$

If elements are small and their x and y nodal coordinates are large, numerical error can creep in during formation of some element matrices (consider for example the first row of matrix $[DA]^{-1}$). A cheap and simple precaution is therefore to localize element coordinates before generating element matrices. In the case of the triangle, if X and Y represent global coordinates, a local centroidal system is introduced as follows. Locate the centroid

$$X_c = \frac{(X_1 + X_2 + X_3)}{3}, \qquad Y_c = \frac{(Y_1 + Y_2 + Y_3)}{3}$$
(4.4.9)

Nodal coordinates in a centroidal system are

$$x_1 = X_1 - X_c, \qquad x_2 = X_2 - X_c, \text{ etc.}$$
(4.4.10)

and similarly for the y coordinates. In the local centroidal system for the triangle,

$$x_1 + x_2 + x_3 = y_1 + y_2 + y_3 = 0$$
(4.4.11)

4.5 General Comments Regarding Element Displacement Fields

In our discussion it is convenient to refer to the constant strain triangle of Section 4.4 and to a rectangular element. The latter element has eight d.o.f. and is shown in Fig. 4.5.1. A suitable displacement field for this element is

$$\begin{Bmatrix} u \\ v \end{Bmatrix} = \begin{bmatrix} 1 & x & y & xy & 0 & 0 & 0 & 0 \\ 0 & 0 & 0 & 0 & 1 & x & y & xy \end{bmatrix} \begin{Bmatrix} a_1 \\ a_2 \\ \cdot \\ \cdot \\ \cdot \\ a_8 \end{Bmatrix}$$
(4.5.1)

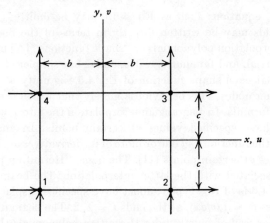

Figure 4.5.1. A plane rectangular element.

The usual procedure for exchanging generalized coordinates a_i for nodal displacements u_i and v_i requires the inversion of a 4 by 4 matrix $[DA]$. Clearly this procedure becomes time consuming and cumbersome as the number of element d.o.f. increases. An alternate and usually preferable method is suggested by the result obtained for the rectangle:

$$u = \frac{1}{4bc}[(b-x)(c-y)u_1 + (b+x)(c-y)u_2$$

$$+ (b+x)(c+y)u_3 + (b-x)(c+y)u_4] \quad (4.5.2)$$

A similar result is obtained for displacement v. Thus the displacement field $\{f\} = [N]\{d\}$ may be written

$$\begin{Bmatrix} u \\ v \end{Bmatrix} = \begin{bmatrix} N_1 & 0 & N_2 & 0 & N_3 & 0 & N_4 & 0 \\ 0 & N_1 & 0 & N_2 & 0 & N_3 & 0 & N_4 \end{bmatrix} \begin{Bmatrix} u_1 \\ v_1 \\ u_2 \\ \cdot \\ \cdot \\ \cdot \\ v_4 \end{Bmatrix} \quad (4.5.3)$$

where the "shape functions" are

$$N_1 = \frac{(b-x)(c-y)}{4bc} \qquad N_2 = \frac{(b+x)(c-y)}{4bc}$$

$$N_3 = \frac{(b+x)(c+y)}{4bc} \qquad N_4 = \frac{(b-x)(c+y)}{4bc} \quad (4.5.4)$$

The point to be made is that the latter equations represent an application of Lagrange's interpolation formula to a two-dimensional problem (1). In general, then, equations such as Eq. 4.5.1 may be omitted; assumed displacement fields may be written directly in terms of the nodal d.o.f. The necessary interpolation polynomials or "shape functions" $[N]$ may be devised by intuition, trial, and familiarity with similar but simpler elements.

We note that each shape function of Eq. 4.5.4 is unity at one node and zero at all other nodes. This behavior is merely that which is expected of an interpolation formula. In Lagrangian interpolation the interpolation function is forced to have specified values at certain points. In another kind of interpolation the function *and* one or more of its derivatives are forced to have specified values at certain points (11). The name "Hermitian polynomial" is sometimes associated with the latter interpolation. The beam displacement v of Problem 4.6 is of this type; v must have specified displacements v_i *and* specified slopes $\theta_i = (dv/dx)_i$ at the ends $i = 1, 2$. The beam displacement is said to be composed of fourth-order Hermitian polynomials (11).

What equilibrium and compatibility conditions might we expect in a finite element solution based on assumed displacement fields? With regard to equilibrium:

1. Equilibrium is usually not satisfied within elements. For example, Eqs. 4.5.1 do not satisfy the differential equations of equilibrium in terms of displacements (Eqs. 1.2.9). Equilibrium is satisfied within the triangle because of its extreme simplicity.

   The rectangle would satisfy equilibrium if $a_4 = a_8 = 0$, but this is the case only in a field of constant strain. One cannot conclude from this that in a general strain field the rectangle is inferior to the triangle; despite its nonsatisfaction of equilibrium, the rectangle may yield more accurate results.

2. Equilibrium is usually not satisfied between elements. This is clearly the case with the triangle, where stresses are constant within the element but differ from one element to another.

3. Equilibrium of nodal forces and moments is satisfied. In other words, the structural equations $\{R\} - [K]\{D\} = 0$ are a set of nodal equilibrium equations; therefore the solution vector $\{D\}$ is such that resultant forces and moments acting on each node are zero.

   This does not imply that the resulting stresses $\{\sigma\} = [E]([B]\{d\} - \{\epsilon_o\})$ satisfy interelement equilibrium or continuity (see, for example, the behavior of element $Q6$ in Section 6.4). However, in certain cases interelement stress continuity may exist. A common example is that of a uniform beam modeled by ordinary beam elements (Problem 4.6) and loaded at its nodes. Here computed stresses satisfy moment equilibrium at the nodes, because the assumed cubic displacement field is exact for this particular problem.

With regard to compatibility:

1. Compatibility is satisfied within elements if the assumed element displacement field is continuous.

2. Compatibility may or may not be satisfied between elements. In the elements of Figs. 4.4.1 and 4.5.1, both u and v are linear in x (or y) along element edges. Hence, for any nodal displacements the edges remain straight, and adjacent elements do not overlap or separate. Interelement compatibility is violated in other elements to be discussed later.

3. Compatibility is enforced at nodes by joining elements at these locations.

In a "proper" finite element solution any violations of equilibrium and compatibility tend to vanish as more and more elements are used to model a given structure.

4.6 Convergence and Compatibility Requirements

A sequence of solutions to a problem may be generated by using successively finer meshes of elements. The sequence may be expected to converge to the correct result if assumed element displacement fields satisfy certain criteria. The mandatory criteria are as follows.

1. The displacement field within an element must be continuous. This requirement is so easily satisfied that we need not mention it again.

2. When nodal d.o.f. are given values corresponding to a state of constant strain, the displacement field must produce the constant strain state throughout the element (2, 3). In the case of a thin plate element we might speak of "constant curvature" instead of "constant strain."

The motivation for Requirement 2 is stated in the first paragraph of this chapter and may be visualized by considering the distribution of axial strain ϵ_x in a bar carrying a distributed axial load (Fig. 4.6.1). Let the bar be divided into elements of length Δx. As Δx approaches zero, any variation of ϵ_x within an element becomes insignificant in comparison with the magnitude of ϵ_x. Accordingly, if for all x we are to approach exact values of displacements and strains arbitrarily closely as the mesh is refined, each element must be capable of representing constant strain when conditions so require (recall the need for "relative completeness" in assumed functions for the Rayleigh-Ritz method, Section 3.8). We could not expect convergence if each element could represent only $\epsilon_x = (\text{constant})(s/\Delta x)$, where $0 \leq s \leq \Delta x$. The requirement for constant strain capability is stated without reference

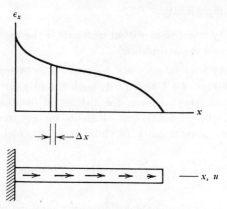

Figure 4.6.1.   Bar under arbitrary axial loading.

to element size, because it is clearly quite valuable for large elements to have the same capability; thus in areas of low strain gradient a single large element serves the function of many small ones.

A modified form of Requirement 2, known as the "patch test" (14), is easy to apply. Take an assembly of perhaps 10 elements, arranged so that at least one node is completely surrounded by elements. Such internal nodes may be positioned so that elements are of arbitrary shape. Apply to the boundary nodes either displacements or forces consistent with a constant strain state. Internal nodes are to be neither loaded not restrained. Hence computed displacements, strains, and stresses within elements should be consistent with the constant strain state. If not, the element type is invalid or at least suspect (it may happen that an element is valid in certain shapes only). An application of the patch test appears in Section 6.4.

Examples of difficulty arising from nonsatisfaction of Requirement 2 may be found in the literature. For example, in the bending of thin plates the lateral displacement w defines the displaced shape. An assumed polynomial for w that contains x^2 and y^2 terms but not the xy term permits constant curvatures $w_{,xx}$ and $w_{,yy}$ but excludes the constant twist $w_{,xy}$ (4). A state of constant twist will not exist in such an element even if nodal d.o.f. corresponding to constant twist are imposed. Mesh refinement usually produces convergence to an incorrect result (5). A further example appears in Problem 4.12.

Two more requirements that elements should satisfy are as follows. As the mesh of elements is refined, but not necessarily in larger elements:

3. Rigid body modes must be represented. That is, when nodal d.o.f. are given values corresponding to a state of rigid body motion, the element must exhibit zero strain and therefore zero nodal forces (1).

4. Compatibility must exist between elements; elements must not overlap or separate. In the case of beam, plate, and shell elements it is also required that there be no sudden changes in slope across interelement boundaries (1, 2, 12).

If Requirement 3 is violated, extraneous nodal forces appear, and thus the equations of nodal equilibrium are altered. Some successful shell elements satisfy this requirement only if elements are made vanishingly small. The elements have been widely used because the error remains small for the size of element used in practice. Nevertheless, other things being equal, better results are obtained when rigid body modes are included (7).

Requirement 4 is violated by many successful elements (2, 5); however, such elements do satisfy interelement compatibility in the limit of mesh refinement as each element approaches a state of constant strain.

Incompatible elements often work better than closely related compatible elements. The reason lies in the nature of our approximate solution. As mentioned in Section 3.8, use of assumed displacement fields yields an approximate structure that is stiffer than the actual structure. But the approximate structure is softened by allowing separation, overlapping, or "kinks" between elements. The two effects tend to cancel one another, often to good advantage.

Incompatible elements exist that are capable of representing constant strain states in some mesh arrangements but not in others (2, 3). In the "poor" arrangements, however, errors are quite small and may be neglected in practice.

Incompatible elements are apparently in harmony with Eq. 4.2.6, in which integration is performed only over each element. What happens between elements is ignored. However, because compatibility is a requirement for use of the Π_p expression, a solution based on incompatible elements is not a solution based on the principle of stationary potential energy. Therefore we cannot state that the approximate structure is too stiff; convergence of results may not be monotonic and may be from above or below (2, 5) (see Fig. 4.6.2).

In one of the first papers to recognize the variational nature of the finite element method, Melosh (1) argues that each refined mesh should be capable of representing all deformation states that were possible in the preceding coarser mesh. For example, each element in a mesh of plane rectangles could be divided into four smaller elements of the same type. If mesh subdivision is made in this way and Requirements 1 through 4 above are satisfied, *monotonic* convergence of results can be expected.[3] We note the similarity of

---

[3] It is perhaps obvious that the mesh layout should not misrepresent the volume of the structure. Also, as noted in Section 5.7, the structure should not be "softened" by low order numerical integration rules.

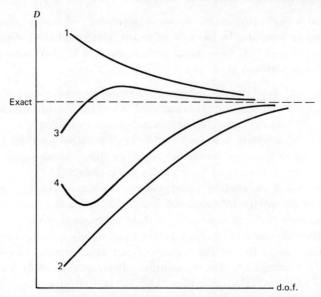

Figure 4.6.2. Convergence of displacement D with mesh refinement. 1, monotonic from above; 2, monotonic from below; 3 and 4, not monotonic.

this subdivision scheme to the process of adding terms to trial functions in the classical Rayleigh-Ritz method; lower order terms are retained, so that the old displacement field is contained in the new one.

Another requirement that an element should satisfy is:

5. The element should have no preferred directions. That is, under any set of loads having a fixed orientation with respect to the element, the element response should be independent of how it and its loads are oriented in global xy coordinates. "Response" means element strain energy or element strains in a coordinate system that moves with the element. Elements that satisfy this requirement will be called *invariant*. There is some opinion (3, 7) that invariance is mandatory if convergence to correct results is to be obtained. In any case it is a quite desirable attribute, and there is no necessity to violate it.

Invariance exists if complete polynomials are used for element displacement fields. "Complete" in the present context means containing all linear terms, or all linear and quadratic terms, etc. For example, the polynomial

$$w = a_1 + a_2 x + a_3 y + a_4 x^2 + a_5 xy + a_6 y^2 + a_7 x^3 + a_8 x^2 y + a_9 y^3$$

$$(4.6.1)$$

is an incomplete cubic, because the xy^2 term is missing. [Use of Eq. 4.6.1 in plate problems yields convergence to incorrect results (7).] Use of complete

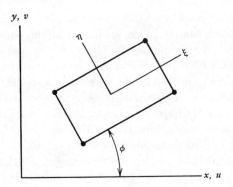

Figure 4.6.3.   Local coordinates.

polynomials guarantees that the displacement of any line in the element will be a complete polynomial in the linear coordinate regardless of how the line is oriented with respect to global coordinate axes (8).

However, elements may be invariant even when based on incomplete polynomials; indeed, most successful elements are of this type. Invariance is achieved if the element is formulated in a local coordinate system that moves with the element so that the system has the same orientation with respect to the element, regardless of how the element may be oriented in global coordinates. The element displacement field must be expressed in terms of the local coordinates and should have a "balanced" representation of terms if the polynomial is incomplete. Also, when numerical integration is used, it must be "uniform," that is, the same set of sampling points must be used for integration of all strain components (9).

Invariance will be discussed further in subsequent chapters. For the present, the element of Fig. 4.5.1 serves as an example. Let local coordinates $\xi\eta$ move with the element (Fig. 4.6.3). If element displacements u and v have the form

$$a_1 + a_2\xi + a_3\eta + a_4\xi\eta \tag{4.6.2}$$

then invariance will be achieved. The quadratic term $a_4\xi\eta$ favors neither ξ nor η and is therefore "balanced." Similarly, a polynomial that contains all linear and quadratic terms but only the cubic terms $a_7\xi^2\eta + a_8\xi\eta^2$ would also be considered "balanced." Such a polynomial could be used for an eight d.o.f. plane element having corner and midside nodes (Section 5.5).

4.7 Stress Computation

When element d.o.f. $\{d\}$ are known, the stress-strain relation and Eq. 4.2.3 yield element stresses:

$$\{\sigma\} = [E](\{\epsilon\} - \{\epsilon_o\}) = [E]([B]\{d\} - \{\epsilon_o\}) \tag{4.7.1}$$

Initial stresses $\{\sigma_o\}$, if used in addition to or in place of initial strains $\{\epsilon_o\}$, must also be included. Matrix $[B]$ is evaluated at the point where stresses are to be computed. For computational efficiency the operation $[E]([B]\{d\})$ is much better than $([E][B])\{d\}$. Stress calculation according to Eq. 4.7.1 is satisfactory unless the element is crude. Stresses in a mesh of crude elements, such as constant-strain triangles, might be more accurately computed by other methods (13; 4.4 of Chapter Eighteen; Problem 6.3). One such method is computation of nodal forces, followed by division by a reference area, with perhaps some curve fitting from adjacent stress values.

Stresses arising from internal forces deserve comment. If, for example, the beam of Fig. 4.7.1 is modeled by a single element having nodes 1 and 2, nodal displacements are zero, and Eq. 4.7.1 predicts zero stress at all points. This incorrect result suggests that computed stresses should include those stresses arising from internal forces acting on a fully restrained element. In practice such stresses are usually ignored. They are often awkward to compute or approximate, and they tend to vanish as elements are reduced in size. Elements having internal d.o.f. automatically account for internal load effects in an approximate way.

Computation of strain, and hence of stress, is based on nodal displacement *differences* in an element; rigid body motion of the element does not contribute. Accordingly, for accurate stress computation, nodal d.o.f. $\{d\}$ must be computed with sufficient accuracy that rigid body motion does not overwhelm the essential differences. Otherwise, the differences will have few if any correct digits. If displacements have been computed using double precision arithmetic, stress computation in an element may be preceded by removal of most rigid body motion, say by subtracting displacements of node 1 from the other nodal displacements. The net element displacements may then be truncated to single precision. Hence, computation of stress and strain may proceed in single precision with confidence that essential information has not been lost. Also, as $\{d\}$ now contains zeros in the node 1 positions, some computation may be skipped. Caution: In solids and shells of revolution, hoop strain $\epsilon_\theta = u/r$ must be based on the *total* radial displacement u.

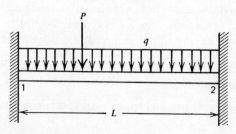

Figure 4.7.1.   Fixed-fixed beam.

4.8   Corner Nodes and Side Nodes

Elements may have corner nodes only or both corner nodes and side nodes. Since we will subsequently encounter both element types in various contexts, let us make a brief comparison at this time, based on the cost of solving a banded set of structural equations.

Figure 4.8.1 shows two meshes. One has m elements on each side and the other has n elements on each side. Node and element numbers are shown. For purposes of argument let there be k d.o.f. per node. Now if m and n are "large" but otherwise unspecified, there are roughly km^2 structural equations for the first mesh and $3kn^2$ for the second. Also, the semibandwidths B of these equations are roughly km and $3kn$ respectively. Therefore, as noted in Section 2.6, solution times for the two systems of equations will be proportional to k^3m^4 and $27k^3n^4$, respectively. If solution times are to be equal we must have $m \approx \sqrt[4]{27}\, n = 2.3n$. Thus, if two large meshes have the same number of d.o.f. per node, for the same solution effort we can use a mesh of corner-node elements more than twice as fine as the mesh of elements having both corner nodes and side nodes (10).

These arguments extend to elements having, say, two nodes per side, and solid elements having perhaps corner, edge, and face nodes. The general conclusion remains that corner nodes are to be preferred for the sake of solution efficiency. It may happen, of course, that in a given problem class, there are elements having side nodes that behave much better than competing corner-node elements and hence may be used in a coarse mesh with no overall penalty in computer time.

4.9   Finite Elements and Finite Differences

By this time the reader may wonder how the finite element method compares with the finite difference method. Both are discretization techniques. In both methods a continuum is represented by a set of nodal generalized coordinates, and both require solution of a set of simultaneous algebraic equations. Aside from these similarities the two methods may appear quite different, since what happens within a finite element is determined entirely by nodal displacements within the element and on its boundaries, while in a finite difference mesh there are nodes outside each "element." Also, the finite element approach is usually viewed as minimization of a functional without reference to differential equations, while the finite difference approach has usually been presented as a method for approximating the governing differential equations without reference to functionals. However, recently

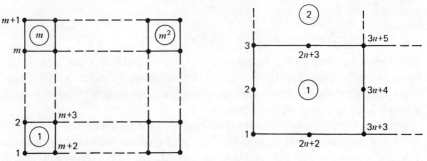

Figure 4.8.1. Meshes containing side and corner nodes.

it has been found profitable to derive finite difference models from functionals, using, for example, the same potential energy expression as used to generate finite element models (15). Thus, the two methods may be said to differ only in the choice of generalized coordinates and location of nodes. Indeed, it has been remarked that the finite element method can be viewed as a method for generating a set of finite difference equations (16).

Available information (15, 17, 18) suggests that there are types of problems to which finite differences are better suited than finite elements, and vice versa. There are accurate and user-oriented programs based on each method. It appears that neither method will wholly supplant the other. For a given number of d.o.f. both appear capable of about the same accuracy. Less computer time may be needed to generate structural equations by the finite difference method (15); however, comparisons inevitably depend on the type of problem, mesh regularity, and program organization as well as the basic analysis method. It might be noted that the finite difference method is not necessarily restricted to the regular meshes adopted in most explanations of the subject. However, the method tends to become awkward if a mathematical description of the structure geometry is not easy to write, an example case being a vehicle that is built of odd-shaped stiffened panels that join at various inclinations.

Generally, the advantages of the finite element method arise from its physical appeal (19) and the relative ease with which a complicated structure can be modeled and its boundary conditions treated. Finite elements are in a sense pieces of the actual structure; therefore the engineer may be aided by his structural intuition and a sense of physical reality when selecting what sort of elements to use, joining elements of different types or of different orientations in space, etc. Similarly, improvements in the properties of finite elements have often come from physical insight into element behavior. Physical insight is perhaps less readily applied to finite difference methods, because there is usually no definite "element" that can be visualized.

PROBLEMS

4.1.   If element nodal d.o.f. are given arbitrary infinitesimal displacements $\{\delta d\}$, element strains are changed according to $\{\delta\epsilon\} = [B]\{\delta d\}$. Element strain energy is incremented an amount $\{\delta\epsilon\}^T\{\sigma\}$, this energy being equal to the work increment performed by applied loads. By completing this virtual work argument, derive Eqs. 4.2.9 and 4.2.10.

4.2.   Derive the formula for stiffness given in Problem 3.7.

4.3.   Let the truss member of Section 4.3 have radius r and mass density ρ. Determine nodal loads $\{r\}$ resulting from:
(a)   A surface traction in the positive x direction of ρ_o units of force per unit of surface area.
(b)   Rotation of the bar about a point b units to the right of node 2 with constant angular velocity ω. Is the assumed displacement Eq. 4.3.1 valid for this problem? Explain.

4.4.   (a)   Consider a straight bar of length $2L$ that has nodes at $x = 0$, $x = L$, and $x = 2L$. Assume that the displacement along the bar may be taken as $u = a_1 + a_2 x + a_3 x^2$. Use Lagrange's interpolation formula to express u in terms of the nodal displacements u_1, u_2, and u_3.
(b)   Let the cross-sectional area of this bar vary linearly from A_o at $x = 0$ to $2A_o$ at $x = 2L$, and let the bar be heated uniformly T degrees. Hence, using the displacement assumption of Part a of this problem, determine the load vector $\{r\}$.
(c)   Determine the exact expression for $\{r\}$ and compare it with the result in Part b of this problem. Why is the result of Part b not exact?

4.5.   (a)   Using the result of Problem 4.4a and assuming that the bar has constant cross-sectional area A_o, determine the 3 by 3 stiffness matrix for the bar.
(b)   Assume that u_2 is an internal d.o.f., that is, node 2 is not to be connected to any other element. Hence, condense the stiffness matrix of Part a of this problem to size 2 by 2. Compare the result with Eq. 4.3.5. (Condensation is described in Section 6.1.)

4.6.   The formulation of a beam element is outlined in Section 3.9.
(a)   Verify the correctness of matrix $[DA]^{-1}$ in Eq. 3.9.5.

(b) Derive the element stiffness matrix $[k]$. The result is given below, with nodal d.o.f. arranged as $\{d\} = \{v_1 \quad \theta_1 \quad v_2 \quad \theta_2\}$.

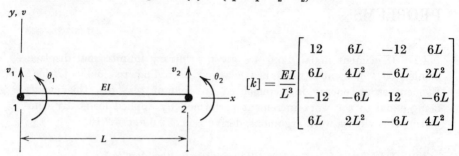

$$[k] = \frac{EI}{L^3} \begin{bmatrix} 12 & 6L & -12 & 6L \\ 6L & 4L^2 & -6L & 2L^2 \\ -12 & -6L & 12 & -6L \\ 6L & 2L^2 & -6L & 4L^2 \end{bmatrix}$$

4.7. (a) Let the beam element of Problem 4.6 carry a uniform downward load of intensity q units of force per unit of length. Determining the element load vector $\{r\}$ for this loading.

(b) Similarly, determine $\{r\}$ for a concentrated upward load P at midspan.

(c) Similarly, determine $\{r\}$ for a counterclockwise couple M_o applied at midspan.

4.8. Results in Problem 4.7 agree with exact values. This may be of some surprise, since the assumed cubic shape is not correct for these loadings. How may the correct results be explained? *Suggestion.* Consider the virtual work done by the applied loads in acting through displacements produced by virtual changes in nodal d.o.f. Solve Problem 4.7a again, finding each support reaction by letting each d.o.f. in turn have a unit displacement while the remaining three are held fixed.

4.9. (a) Evaluate the strain-displacement matrix $[B]$ of a constant-strain triangle of uniform thickness and elastic properties (Fig. 4.4.1). The displacement field is given by Eqs. 4.4.1.

(b) Hence, find the stiffness matrix of this element.

4.10. (a) Evaluate the strain-displacement matrix $[B]$ for the rectangular element of Fig. 4.5.1. Assume uniform thickness and constant elastic properties over the element.

(b) Hence, find the stiffness matrix of this element. Note that the integrand of Eq. 4.2.9 is now a function of the coordinates; however, integration is easily performed.

4.11. Let the element of Fig. 4.5.1 be 2 units on a side and located in a structure so that global d.o.f. are numbered as shown. Also let matrix $[E]$ be the diagonal matrix $[E \quad E \quad E/2]$. Determine the contribution of this element to:

(a) The 41st row and 43rd column of the structure stiffness matrix.
(b) The 36th row and 42nd column of the structure stiffness matrix.

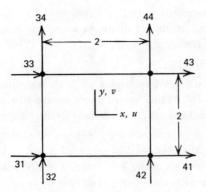

4.12. The elements shown are uniform two-force members having end nodes 1 and 2. In each case construct the element stiffness matrix using the given displacement field. Comment on the inadequacies displayed by the stiffness matrices and by the given displacement fields.

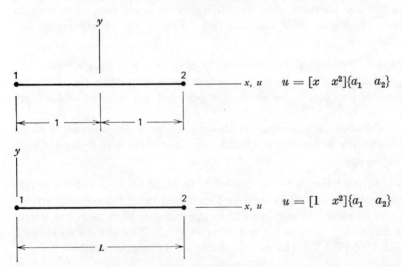

4.13. For the elements listed below, check that the constant-strain and rigid-body requirements of Section 4.6 are satisfied. Where possible examine the original displacement assumption, the shape function matrix $[N]$, the strain-displacement matrix $[B]$, and the element stiffness matrix $[k]$.

    (a)   Truss element of Section 4.3.

    (b)   Triangular element of Fig. 4.4.1.

    (c)   Rectangular element of Fig. 4.5.1.

    (d)   Bar element of Problem 4.5.

    (e)   Beam element of Problem 4.6.

4.14.   For the elements listed in Problem 4.13, show that the product $[k]\{d\}$ yields an equilibrium set of forces when any single d.o.f. in $\{d\}$ is set to unity with the remaining d.o.f. set to zero. What is the rank of each element stiffness matrix?

4.15.   In Fig. 4.6.3 let the origins of both coordinate systems coincide and let angle $\phi = 45°$. (Thus, coordinates xy are ill-suited to the element.) Express displacements u and v according to Eq. 4.6.2, then change these expressions by the substitutions $\xi = (x + y)/\sqrt{2}$ and $\eta = (-x + y)/\sqrt{2}$ so that displacements are expressed in terms of x and y. Hence, by comparing element strain expressions in the $\phi = 0$ and $\phi = 45°$ positions, show that the element is not invariant when derived in this way.

4.16.   Consider a rectangular thin-plate element having 12 d.o.f., 3 at each corner. The nodal d.o.f. are displacement w and the slopes $w_{,x}$ and $w_{,y}$. The lateral displacement w of this element in bending may be taken as a complete cubic plus the two quartic terms $a_{11}x^3y + a_{12}xy^3$. Thus, the 12 a's are uniquely determined by the 12 nodal d.o.f. By considering the slope $w_{,x}$ along a line $x = $ constant, show that this element is not compatible with its neighbor. *Suggestion.* For compatibility, what d.o.f. *should* completely define the slope?

4.17.   (a)   Following the argument of Section 4.8, compare mesh sizes for equal solution effort if one has corner nodes only while the other has corner nodes plus two nodes along each side. Let $k = $ number of d.o.f. per node.

(b)   Similarly, compare meshes of solid (three-dimensional) elements if one has corner nodes only while the other has corner nodes plus one node along each edge.

4.18.   Suppose that a beam element is to be based on a cubic $v = v(x)$ for lateral displacement and is to be formulated using four equally spaced nodes, one on each end and two within the element. Each node has lateral displacement only, so that $\{d\} = \{v_1 \quad v_2 \quad v_3 \quad v_4\}$. Why is such an element unacceptable? Which of the criteria of Section 4.6 does it violate?

CHAPTER FIVE

The Isoparametric Formulation

5.1 Introduction

This chapter introduces the "isoparametric" family of elements. The name derives from use of the same interpolation functions to define the element shape as are used to define displacements within the element. Isoparametric elements have been used with great success in various problem areas, including two- and three-dimensional elasticity, plates, and general shells. Their popularity derives in part from the fact that when one element has been thoroughly understood, it is not difficult to extend one's understanding to other isoparametric elements. Linear elements have straight sides, but quadratic and higher order isoparametric elements may have either straight or curved sides, which makes them very useful for modeling of curved structures.

"Isoparametric" coordinates are a type of "intrinsic" or "natural" coordinate system. For a simple example of a natural coordinate system, consider a bar of length L that lies along an x axis with its midpoint at the origin $x = 0$. The two ends are described by $x = -L/2$ and $x = +L/2$. If another lengthwise coordinate ξ is defined as $\xi = 2x/L$, then the ends are defined by $\xi = -1$ and $\xi = +1$. If we also require that the ξ axis be attached to the bar and remain a lengthwise coordinate regardless of how the bar may be oriented in global coordinates, then ξ may be termed a natural coordinate.

Isoparametric elements were introduced in generally available literature by Irons (1). They are formulated using an intrinsic coordinate system $\xi\eta$, which is defined by element geometry and not by the element orientation in the global coordinate system. There is, of course, a relation between the two systems for each element of a specific structure, and this relation must be used in element formulation. Coordinates $\xi\eta$ (and ζ, if the element is three dimensional) are attached to the element and are scaled so that sides of a quadrilateral are defined by $\xi = -1$, $\xi = +1$, $\eta = -1$, and $\eta = +1$. As suggested in Section 4.6, uniformly integrated isoparametric elements are *invariant*. Later in this chapter we will demonstrate that isoparametric elements satisfy other requirements listed in Section 4.6.

5.2 A Plane Isoparametric Element

We consider a quadrilateral having eight d.o.f., namely u_i and v_i at each of four corner nodes i. The element has straight sides but is otherwise of arbitrary shape and may be considered as a distortion of a "parent" rectangular element (Fig. 5.2.1). Let us adopt the mapping function (2)

$$\begin{Bmatrix} x \\ y \end{Bmatrix} = \begin{bmatrix} N_1 & 0 & N_2 & 0 & N_3 & 0 & N_4 & 0 \\ 0 & N_1 & 0 & N_2 & 0 & N_3 & 0 & N_4 \end{bmatrix} \begin{Bmatrix} x_1 \\ y_1 \\ x_2 \\ \cdot \\ \cdot \\ \cdot \\ y_4 \end{Bmatrix} \qquad (5.2.1)$$

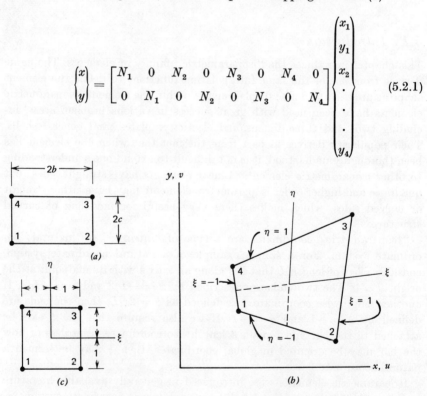

Figure 5.2.1. (a) Parent element. (b) Quadrilateral isoparametric element. (c) Mapping into square.

where

$$N_1 = \frac{(1 - \xi)(1 - \eta)}{4}, \qquad N_2 = \frac{(1 + \xi)(1 - \eta)}{4}$$

$$N_3 = \frac{(1 + \xi)(1 + \eta)}{4}, \qquad N_4 = \frac{(1 - \xi)(1 + \eta)}{4} \qquad (5.2.2)$$

This mapping relates a unit square in isoparametric coordinates $\xi\eta$ to the quadrilateral in xy coordinates whose size and shape are determined by the eight nodal coordinates $x_1, y_1, x_2, \ldots, y_4$. The mapping is also an interpolation scheme that yields the xy coordinates of any point in the element when the corresponding $\xi\eta$ coordinates are given.

The reader may recognize the "parent" element as that of Fig. 4.5.1. Note, however, that in Fig. 5.2.1b the xy axes need not be centroidal. Indeed, if we write

$$x = a_1 + a_2\xi + a_3\eta + a_4\xi\eta \qquad (5.2.3)$$

and similarly for y, and then replace a's by nodal coordinates using the substitutions $x = x_1$ at $\xi = \eta = -1$, etc., Eqs. 5.2.1 and 5.2.2 are produced (2).

Axes $\xi\eta$ are in general not orthogonal. They are orthogonal for a rectangular element, in which case they are merely dimensionless forms of rectangular centroidal coordinates. The 2 by 2 rectangular element, for which we may write $\xi = x$ and $\eta = y$, is a convenient special case to refer to when studying the following development.

Displacements within the element are defined by the same interpolation functions as used to define the element shape:

$$\{f\} = \{u \quad v\} = [N]\{u_1 \quad v_1 \quad u_2 \quad \ldots \quad v_4\} = [N]\{d\} \qquad (5.2.4)$$

where $[N]$ is the rectangular matrix of Eq. 5.2.1. Note that displacements u and v are directed parallel to x and y axes and *not* parallel to ξ and η axes.

It is worth noting that if the element is rectangular, the ξ-parallel normal strain ϵ_ξ varies directly with η but is independent of ξ; similarly $\epsilon_\eta =$ (const.)ξ, and $\gamma_{\xi\eta} = c_1 + c_2\xi + c_3\eta$ where the c's are constants.

The task is now that of formulating element matrices according to the "recipe" of Section 4.2. However, because it is impossibly tedious to write the shape functions in terms of x and y, we will carry out the formulation in terms of the isoparametric coordinates $\xi\eta$. This also may seem complicated on first reading but leads to a simple computer program.

Let us rewrite equations of this section in the forms

$$x = \sum_1^4 N_i x_i \qquad y = \sum_1^4 N_i y_i$$

$$u = \sum_1^4 N_i u_i \qquad v = \sum_1^4 N_i v_i \qquad (5.2.5)$$

Relations between derivatives in the two coordinate systems are established by the chain rule of differentiation (3)

$$\begin{Bmatrix} (u),_\xi \\ (u),_\eta \end{Bmatrix} = \begin{bmatrix} x,_\xi & y,_\xi \\ x,_\eta & y,_\eta \end{bmatrix} \begin{Bmatrix} (u),_x \\ (u),_y \end{Bmatrix} = [J] \begin{Bmatrix} (u),_x \\ (u),_y \end{Bmatrix} \tag{5.2.6}$$

where commas denote partial differentiation. From Eq. 5.2.5, $[J]$ is the Jacobian matrix

$$[J] = \begin{bmatrix} N_{1,\xi} & N_{2,\xi} & N_{3,\xi} & N_{4,\xi} \\ N_{1,\eta} & N_{2,\eta} & N_{3,\eta} & N_{4,\eta} \end{bmatrix} \begin{bmatrix} x_1 & y_1 \\ x_2 & y_2 \\ x_3 & y_3 \\ x_4 & y_4 \end{bmatrix} \tag{5.2.7}$$

Let $[J^*] = [J]^{-1}$. Thus, using the inverse relation from Eq. 5.2.6, we may write

$$\begin{Bmatrix} u,_x \\ u,_y \\ v,_x \\ v,_y \end{Bmatrix} = \begin{bmatrix} J_{11}^* & J_{12}^* & 0 & 0 \\ J_{21}^* & J_{22}^* & 0 & 0 \\ 0 & 0 & J_{11}^* & J_{12}^* \\ 0 & 0 & J_{21}^* & J_{22}^* \end{bmatrix} \begin{Bmatrix} u,_\xi \\ u,_\eta \\ v,_\xi \\ v,_\eta \end{Bmatrix} \tag{5.2.8}$$

The strain-displacement relation may be written

$$\{\epsilon\} = \begin{Bmatrix} \epsilon_x \\ \epsilon_y \\ \gamma_{xy} \end{Bmatrix} = \begin{bmatrix} 1 & 0 & 0 & 0 \\ 0 & 0 & 0 & 1 \\ 0 & 1 & 1 & 0 \end{bmatrix} \begin{Bmatrix} u,_x \\ u,_y \\ v,_x \\ v,_y \end{Bmatrix} \tag{5.2.9}$$

and, from Eq. 5.2.5,

$$\begin{Bmatrix} u,_\xi \\ u,_\eta \\ v,_\xi \\ v,_\eta \end{Bmatrix} = \begin{bmatrix} \begin{bmatrix} N_{i,\xi} & 0 \\ N_{i,\eta} & 0 \\ 0 & N_{i,\xi} \\ 0 & N_{i,\eta} \end{bmatrix}_{i=1} & \begin{bmatrix} \\ \\ \end{bmatrix}_{i=2} & \begin{bmatrix} \\ \\ \end{bmatrix}_{i=3} & \begin{bmatrix} \\ \\ \end{bmatrix}_{i=4} \end{bmatrix} \begin{Bmatrix} u_1 \\ v_1 \\ u_2 \\ \cdot \\ \cdot \\ \cdot \\ v_4 \end{Bmatrix} \tag{5.2.10}$$

Combination of Eqs. 5.2.8 through 5.2.10 yields the relation $\{\epsilon\} = [B]\{d\}$. It is left to the reader to write out the terms in matrix $[B]$.

Having established matrices $[N]$ and $[B]$ we may proceed to the integration of the expression $[B]^T[E][B]$. For a typical integral the change of coordinates is (3)

$$\iint (\ldots)\, dx\, dy = \int_{-1}^{1} \int_{-1}^{1} (\ldots) \det[J]\, d\xi\, d\eta \qquad (5.2.11)$$

If rows of $[J]$ are regarded as two vectors $\vec{A}$ and $\vec{B}$, $\det[J]$ is the magnitude of the vector product $\vec{A} \times \vec{B}$, which is the area of a parallelogram having $\vec{A}$ and $\vec{B}$ as adjacent sides. The determinant of $[J]$ is a magnification factor that yields area $dx\, dy$ from area $d\xi\, d\eta$, and which is a function of position within the element.

Having expressed all matrices in terms of ξ and η we are fully prepared to integrate in $\xi\eta$ coordinates. However, we cannot in general do the integration exactly because of the complexity of the expressions; polynomials in ξ and η appear in the denominator of $[J^*]$. Hence, integration is done approximately and numerically. We will outline the integration procedure in the following section and then complete formulation of the quadrilateral element.

5.3 Summary of Gauss Quadrature

There are many schemes for numerical evaluation of definite integrals. We will outline only the Gauss method (4), since it has proved most useful for finite element work.

To evaluate the integral

$$I = \int_{-1}^{1} y\, dx \qquad (5.3.1)$$

in the simplest and crudest way, we might sample (evaluate) y at the midpoint and multiply by the length of the interval, as shown in Fig. 5.3.1a. Thus we obtain $I = 2y_1$, a result that is exact if the curve happens to be a straight line.

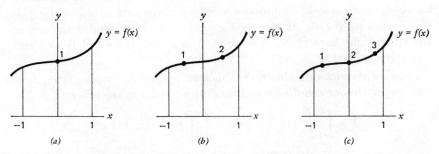

Figure 5.3.1. Gauss quadrature using (a) one, (b) two, and (c) three points.

Generalization of this simple formula leads to

$$I = \int_{-1}^{1} y \, dx \approx \sum_i W_i y_i \tag{5.3.2}$$

That is, to approximate the integral, evaluate the function at several sampling points, multiply each value y_i by the appropriate "weight" W_i and add. The Gauss method locates the sampling points so that for a given number of them, greatest accuracy is obtained. Sampling points are located symmetrically with respect to the center of the interval. Symmetrically paired points are given the same weight. Table 5.3.1 gives the appropriate

Table 5.3.1.   Gauss Quadrature Coefficients

No. of Points	Locations	Associated Weights W_i
1	$x_1 = 0.00000000000000000000000000$	$2.$
2	$x_1, x_2 = \pm 0.57735026918962576450914878$	$1.$
3	$x_1, x_3 = \pm 0.77459666924148337703585831$	$\dfrac{5}{9} \ (= 0.555\ldots)$
	$x_2 = 0.00000000000000000000000000$	$\dfrac{8}{9} \ (= 0.888\ldots)$

coefficients (5) for the first three orders. Thus, for example, using two points, we have simply $I \approx y_1 + y_2$, which is the exact result if $y = f(x)$ is a polynomial containing terms up to and including x^3. In general, Gauss quadrature using n points is exact if the integrand is a polynomial of degree $2n - 1$ or less. In using n points we effectively replace the given function $y = f(x)$ by a polynomial of degree $2n - 1$. Accuracy of numerical integration depends on how well the polynomial fits the given curve.

The derivation of a Gauss formula is illustrated as follows. Consider, for example, $y = c_0 + c_1 x + c_2 x^2 + c_3 x^3$. If integrated between -1 and $+1$ the area under this curve is $A = 2c_0 + 2c_2/3$. Using two symmetrically located points $x = \pm a$, we propose to calculate the area as $A_G = W y_{-a} + W y_{+a} = 2W(c_0 + c_2 a^2)$. If the error $e = A - A_G$ is to vanish for any c_0 and c_2, we must have $\partial e/\partial c_0 = \partial e/\partial c_2 = 0$, from which we find $W = 1$ and $a = 1/\sqrt{3} = 0.57735 \ldots$.

In two dimensions we obtain the quadrature formula by integrating first with respect to one coordinate and then with respect to the other:

$$I = \int_{-1}^{1} \int_{-1}^{1} f(\xi, \eta) \, d\xi \, d\eta = \int_{-1}^{1} \left[\sum_i W_i f(\xi_i, \eta) \right] d\eta$$

$$= \sum_j W_j \left[\sum_i W_i f(\xi_i, \eta_j) \right] = \sum_i \sum_j W_i W_j f(\xi_i, \eta_j) \tag{5.3.3}$$

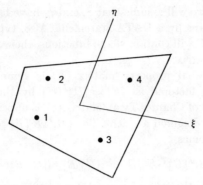

Figure 5.3.2.   Four-point Gauss rule.

Thus, for example, a four-point Gauss rule is shown in Fig. 5.3.2 and yields

$$I = (1.)(1.)f(\xi_1, \eta_1) + \cdots + (1.)(1.)f(\xi_4, \eta_4) \qquad (5.3.4)$$

where the four sampling points are at $\xi_i, \eta_i = \pm 0.57735 \cdots = \pm 1/\sqrt{3}$.

Extension to three dimensions is straightforward and yields (1)

$$I = \int_{-1}^{1} \int_{-1}^{1} \int_{-1}^{1} f(\xi, \eta, \zeta)\, d\xi\, d\eta\, d\zeta = \sum_i \sum_j \sum_k W_i W_j W_k f(\xi_i, \eta_j, \zeta_k) \quad (5.3.5)$$

We will usually use the same number of points in each coordinate direction, but it is not essential to do so.

5.4   Computer Subroutines for Plane Element

Formulation of the quadrilateral element of Section 5.2 is completed by writing computer routines to carry out the numerical work. A unit-thickness element is assumed. We must first introduce some of the notation and procedures used.

It is convenient to represent the shape functions and their derivatives in the form (2)

$$N_i = (1 + \xi\xi_i)(1 + \eta\eta_i)/4$$

$$N_{i,\xi} = \xi_i(1 + \eta\eta_i)/4 \qquad (5.4.1)$$

$$N_{i,\eta} = \eta_i(1 + \xi\xi_i)/4$$

where

$$\xi_i = -1., 1., 1., -1. \qquad \text{for} \qquad i = 1, 2, 3, 4$$

$$\eta_i = -1., -1., 1., 1. \qquad \text{for} \qquad i = 1, 2, 3, 4$$

$(5.4.2)$

In programming we will assume that ξ_i and η_i have been stored in arrays XII and ETI, perhaps by a DATA statement. Also, type REAL arrays N, NXI, NET, and JAC will contain shape functions, their ξ and η derivatives, and the Jacobian matrix.

Because the material property matrix $[E]$ is symmetric and positive definite, it may be factored as $[E] = [U]^T[U]$ by the Choleski method (Problem 5.15; 2, 10 of Chapter Two). Matrix $[U]$ is upper triangular, that is, all terms below the diagonal are *zero*. The $[B]^T[E][B]$ term of the element stiffness matrix becomes

$$[B]^T[U]^T[U][B] = ([U][B])^T([U][B]) \qquad (5.4.3)$$

This step is taken for the sake of computational efficiency: only the nonzero triangle in $[U]$ need be processed in forming $[U][B]$. If $[E]$ does not vary over the element, matrix $[U]$ is formed only once per element. Thus the expense of forming $[U]$ is overcome by the efficiency of the multiplication $[U][B]$, which must be done at every Gauss point. This and other economies are proposed in Refs. 1, 6, and 15.

Another convenience is that of storing element nodal loads as an extra column in the stiffness matrix array instead of as a separate matrix. For the present element we use a ninth column of array SE for this purpose. Illustrative purposes are served if we treat only those nodal loads arising from: (a) constant body forces BODYF(1) and BODYF(2) in x and y directions, respectively and (b) initial strains ϵ_{xo} and ϵ_{yo}, which have known nodal values. Nodal values may differ from one node to another and are stored in arrays EXO and EYO.

Array subscripting needed in handling body forces is handled by arrays NDXL and NDXM, which contain, respectively, the numbers 1,2,1,2,1,2,1,2 and 1,1,2,2,3,3,4,4.

The decomposition of Eq. 5.4.3 leads to the following form for terms in the integrand of the initial strain nodal load matrix:

$$[B]^T[E]\{\epsilon_o\} = ([U][B])^T([U]\{\epsilon_o\}) \qquad (5.4.4)$$

Equations 5.4.3 and 5.4.4 suggest that it would be convenient to store $\{\epsilon_o\}$ as an extra column of array $[B]$. This we will do.

For reasons to be explained later we elect to use four Gauss points. The Gauss weights W are unity, but we will carry them in general form for the sake of clarity. We will assume that TYPE statements, COMMON, DATA, etc., have already been established, that necessary data is passed between subroutines, and that information such as nodal coordinates XL and YL (preferably "localized" as suggested in Section 4.4), initial strains, etc., has been made available. Array AA locates the Gauss points; specifically AA(1) = $-0.57735\ldots$, AA(2) = $+0.57735\ldots$. At a later time we might wish to change these values; for example, to evaluate stress at node 4 we would wish to evaluate $[B]$ at that point and therefore would use $AA(1) = -1.$, $AA(2) = +1.$

Subroutine SHAPE (Fig. 5.4.1) calculates the shape functions, their derivatives, the Jacobian, its determinant and inverse, matrix $[B]$, and initial strains. These calculations are performed at the point in the element determined by XII(I), ETI(I), and the numbers in array AA. When forming the element stiffness this point is one of the Gauss points.

Subroutine INTEG (Fig. 5.4.2) forms matrix $[U]$ from $[E]$ and builds the element matrices by Gauss quadrature. This process involves as many calls

```
C       FIND SHAPE FUNCTIONS AND THEIR DERIVATIVES.
C       II   AND   JJ   ARE COMMUNICATED FROM SUBROUTINE INTEG.
        DO 10 I=1,4
        DUM1 = (1. + XII(I)*AA(II))*.25
        DUM2 = (1. + ETI(I)*AA(JJ))*.25
        N(I) = 4.*DUM1*DUM2
        NXI(I) = XII(I)*DUM2
     10 NET(I) = ETI(I)*DUM1
C       FIND JACOBIAN, ITS INVERSE AND ITS DETERMINANT.
        DO 15 I=1,2
        DO 15 J=1,2
     15 JAC(I,J) = 0.
        DO 20 I=1,4
        JAC(1,1) = JAC(1,1) + NXI(I)*XL(I)
        JAC(1,2) = JAC(1,2) + NXI(I)*YL(I)
        JAC(2,1) = JAC(2,1) + NET(I)*XL(I)
     20 JAC(2,2) = JAC(2,2) + NET(I)*YL(I)
        DETJAC = JAC(1,1)*JAC(2,2) - JAC(2,1)*JAC(1,2)
        DUM1 = JAC(1,1)/DETJAC
        JAC(1,1) =  JAC(2,2)/DETJAC
        JAC(1,2) = -JAC(1,2)/DETJAC
        JAC(2,1) = -JAC(2,1)/DETJAC
        JAC(2,2) =  DUM1
C       FORM THE STRAIN-DISPLACEMENT MATRIX B.
        DO 50 L=1,4
        J = 2*L
        I = J-1
        B(1,I) = JAC(1,1)*NXI(L) + JAC(1,2)*NET(L)
        B(1,J) = 0.
        B(2,I) = 0.
        B(2,J) = JAC(2,1)*NXI(L) + JAC(2,2)*NET(L)
        B(3,I) = B(2,J)
     50 B(3,J) = B(1,I)
C       INTERPOLATE INITIAL STRAINS FROM CORNER NODES.
C       STORE THEM IN AN EXTRA COLUMN OF ARRAY B.
        B(1,9) = 0.
        B(2,9) = 0.
        B(3,9) = 0.
        DO 60 I=1,4
        B(1,9) = B(1,9) + N(I)*EXO(I)
     60 B(2,9) = B(2,9) + N(I)*EYO(I)
        RETURN
        END
```

Figure 5.4.1. Statements in subroutine SHAPE.

```
C     CLEAR ARRAYS. OVERWRITE MATRIX  E  BY UPPER TRIANGLE  U.
      DO 3 K=1,8
      DO 3 L=K,9
    3 SE(K,L) = 0.
      E(1,1) = SQRT(E(1,1))
      E(1,2) = E(1,2)/E(1,1)
      E(1,3) = E(1,3)/E(1,1)
      E(2,2) = SQRT(E(2,2) - E(1,2)*E(1,2))
      E(2,3) = (E(2,3) - E(1,2)*E(1,3))/E(2,2)
      E(3,3) = SQRT(E(3,3) - E(1,3)*E(1,3) - E(2,3)*E(2,3))
C     START GAUSS QUADRATURE LOOP.
      DO 20 II=1,2
      DO 20 JJ=1,2
      CALL SHAPE
C     OVERWRITE MATRIX  B  BY  U*B  AND INITIAL
C     STRAINS BY   U   TIMES INITIAL STRAINS.
      DO 9 K=1,3
      DO 9 L=1,9
      DUM1 = 0.
      DO 8 M=K,3
    8 DUM1 = DUM1 + E(K,M)*B(M,L)
    9 B(K,L) = DUM1
C     ADD IN THE CONTRIBUTION TO ELEMENT MATRICES.
      DUM1 = W(II)*W(JJ)*DETJAC
      DO 20 NROW=1,8
C     FIRST, THE CONTRIBUTION OF BODY FORCES TO LOADS.
      L = NDXL(NROW)
      M = NDXM(NROW)
      SE(NROW,9) = SE(NROW,9) + N(M)*BODYF(L)*DUM1
C     NEXT, CONTRIBUTION TO INITIAL STRAIN EFFECT AND STIFFNESS MATRIX.
      DO 20 NCOL = NROW,9
      DUM2 = 0.
      DO 18 L=1,3
   18 DUM2 = DUM2 + B(L,NROW)*B(L,NCOL)
   20 SE(NROW,NCOL) = SE(NROW,NCOL) + DUM1*DUM2
C     QUADRATURE ENDED. COMPLETE STIFFNESS MATRIX BY SYMMETRY.
      DO 30 K=2,8
      DO 30 L=1,K
   30 SE(K,L) = SE(L,K)
      RETURN
      END
```

Figure 5.4.2. Statements in subroutine INTEG.

to SHAPE as there are Gauss points. Some effort is saved by forming only the upper triangle of the stiffness matrix in this way, then filling in the rest by symmetry.

A thorough understanding of the formulation of this particular element will be repaid by easier understanding of other isoparametric elements.

5.5   Some Additional Isoparametric Elements

The foregoing element is termed a "linear" element because its sides remain straight during deformation. A "quadratic" element is one for which edge displacements may vary as the second power of the edge coordinate. A plane quadratic element and its parent element are shown in Fig. 5.5.1. A clear advantage of the quadratic element is that it may have curved sides and

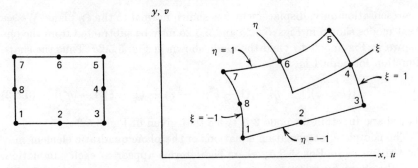

Figure 5.5.1. Quadratic parent element and isoparametric element.

therefore provides a better fit to curved sides of an actual structure. Straight sides may also be used; as the initial shape of an edge does not govern displacements along the edge, an initially straight edge may have quadratic displacements. Interpolation polynomials for shape and displacement in the quadratic element have the form (2)

$$a_1 + a_2\xi + a_3\eta + a_4\xi^2 + a_5\xi\eta + a_6\eta^2 + a_7\xi^2\eta + a_8\xi\eta^2 \qquad (5.5.1)$$

This expression may be cast in terms of nodal quantities by substituting the eight nodal values of x, y, u, or v for $\xi = \eta = -1$, $\xi = 0$ and $\eta = -1$, etc., and inverting the resulting 8 by 8 matrix. A more direct and physically appealing method (7) is to examine various displacement modes and combine them appropriately. These modes are shown in Fig. 5.5.2. For ease of

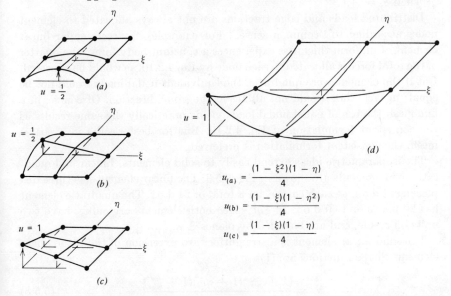

$$u_{(a)} = \frac{(1 - \xi^2)(1 - \eta)}{4}$$

$$u_{(b)} = \frac{(1 - \xi)(1 - \eta^2)}{4}$$

$$u_{(c)} = \frac{(1 - \xi)(1 - \eta)}{4}$$

Figure 5.5.2. Modes for quadratic element.

representation only, displacement u is shown normal to the $\xi\eta$ plane. We see that modes shown in Figs. 5.5.2a and 5.5.2b may be subtracted from the one shown in Fig. 5.5.2c to yield the mode shown in Fig. 5.5.2d. Thus the shape function for node 1 is

$$\frac{1}{4}[(1 - \xi)(1 - \eta) - (1 - \xi^2)(1 - \eta) - (1 - \xi)(1 - \eta^2)] \qquad (5.5.2)$$

The shape function for node 2 is twice that given in Fig. 5.5.2a.

The complete interpolation equations of the plane quadratic element may be written as in Eqs. 5.2.5, where eight terms appear in each summation, and the shape functions are

$$N_i = \frac{1}{4}(1 + \xi\xi_i)(1 + \eta\eta_i)(\xi\xi_i + \eta\eta_i - 1) \qquad \text{for} \qquad i = 1, 3, 5, 7$$

$$N_i = \frac{1}{2}(1 - \xi^2)(1 + \eta\eta_i) \qquad \text{for} \qquad i = 2, 6 \qquad (5.5.3)$$

$$N_i = \frac{1}{2}(1 + \xi\xi_i)(1 - \eta^2) \qquad \text{for} \qquad i = 4, 8$$

where for i running from 1 to 8 the consecutive values of ξ_i and η_i are

$$\xi_i = -1, 0, 1, 1, 1, 0, -1, -1 \qquad \eta_i = -1, -1, -1, 0, 1, 1, 1, 0$$
$$(5.5.4)$$

The first of Eqs. 5.5.3 follows from Eq. 5.5.2 because, for example, $(\xi\xi_i)^2 = \xi^2$ when $\xi_i = \pm 1$.

Distributed loads and edge tractions are not always allocated to element nodes according to "common sense." For example, if a rectangular linear element of uniform thickness experiences a constant body force, one quarter of the total force is allocated to each node by Eq. 4.2.10, as would be expected. But if the element were quadratic, the equivalent nodal loads would not be equal; in fact, they would not all act in the same direction. Of course, in a fine mesh the use of equal nodal loads yields practically the same results as the consistent formulation of Eq. 4.2.10. But for best results in a coarse mesh, the consistent formulation is preferred.

The isoparametric ideas extend easily to solid elements. Linear and quadratic solid elements are shown in Fig. 5.5.3. The linear element has three displacement d.o.f. at each corner, for a total of 24 d.o.f. The quadratic element has 20 nodes and 60 d.o.f. In $\xi\eta\zeta$ space both elements are cubes, have two units on a side, and are bounded by planes $\xi, \eta, \zeta = \pm 1$.

The solid linear element is a straightforward extension of the plane linear element. Shape functions are (1)

$$N_i = \frac{(1 + \xi\xi_i)(1 + \eta\eta_i)(1 + \zeta\zeta_i)}{8} \qquad (5.5.5)$$

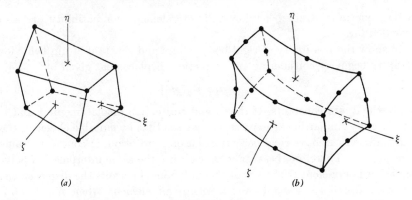

Figure 5.5.3. Linear and quadratic solid elements.

where ξ_i, η_i, $\zeta_i = \pm 1$. Relations between derivatives in $\xi\eta\zeta$ coordinates and those in xyz coordinates are, for linear and higher order elements,

$$
\begin{Bmatrix} (\ )_{,\xi} \\ (\ )_{,\eta} \\ (\ )_{,\zeta} \end{Bmatrix} = \begin{bmatrix} x_{,\xi} & y_{,\xi} & z_{,\xi} \\ x_{,\eta} & y_{,\eta} & z_{,\eta} \\ x_{,\zeta} & y_{,\zeta} & z_{,\zeta} \end{bmatrix} \begin{Bmatrix} (\ )_{,x} \\ (\ )_{,y} \\ (\ )_{,z} \end{Bmatrix} = [J] \begin{Bmatrix} (\ )_{,x} \\ (\ )_{,y} \\ (\ )_{,z} \end{Bmatrix} \qquad (5.5.6)
$$

If the three rows of $[J]$ are regarded as vectors $\bar{A}$, $\bar{B}$, and $\bar{C}$, det $[J]$ is equal to the scalar product $\bar{A} \cdot \bar{B} \times \bar{C}$, which is the volume of a parallelepiped having edge lengths A, B, and C.

All three displacements u, v, w are variables in the solid element, as are the six strains

$$
\{\epsilon\} = \{\epsilon_x \quad \epsilon_y \quad \epsilon_z \quad \gamma_{xy} \quad \gamma_{yz} \quad \gamma_{zx}\} \qquad (5.5.7)
$$

Strain-displacement relations $\epsilon_x = u_{,x}, \ldots, \gamma_{zx} = u_{,z} + w_{,x}$ are

$$
\{\epsilon\} = [SD]\{u_{,x} \quad u_{,y} \quad u_{,z} \quad v_{,x} \quad v_{,y} \quad v_{,z} \quad w_{,x} \quad w_{,y} \quad w_{,z}\} \qquad (5.5.8)
$$

where $[SD]$ is a matrix of 1's and 0's analogous to the array in Eq. 5.2.9. The square matrix corresponding to that in Eq. 5.2.8 is here 9 by 9. The remaining development is quite similar to that in Section 5.2.

More complicated isoparametric elements could be described, such as cubic elements, which have two nodes per edge, but linear and quadratic elements are sufficient for practical stress analysis.

5.6 *The Validity of Isoparametric Elements*

The validity of element formulations may be measured by the criteria of Section 4.6. Isoparametric elements satisfy the criteria of invariance and continuity of displacements within elements. We now show that rigid body modes, constant strain behavior, and interelement compatibility are also provided for.

To show the possibility of rigid body modes and constant strain behavior (2, 8) we begin by considering an x-direction displacement given by

$$u = a_0 + a_1 x + a_2 y + a_3 z \tag{5.6.1}$$

where the a_i are constants. If the y- and z-direction displacements v and w are given by similar linear expressions, we see that by suitable choice of the constants, we can have rigid body translation or rotation, or constant strains ϵ_x or γ_{xy}, etc. Therefore, because u, v, w all use the same interpolation polynomial, it is sufficient to show that this polynomial permits the displacement $u = a_0 + a_1 x + a_2 y + a_3 z$ to exist within an element when nodal d.o.f. are assigned values in accord with this displacement field. Thus, we prescribe the displacement of any node i as

$$u_i = a_0 + a_1 x_i + a_2 y_i + a_3 z_i \tag{5.6.2}$$

By definition $u = \sum N_i u_i$, so that

$$u = a_0 \sum N_i + a_1 \sum N_i x_i + a_2 \sum N_i y_i + a_3 \sum N_i z_i \tag{5.6.3}$$

But also by definition,

$$x = \sum N_i x_i, \qquad y = \sum N_i y_i, \qquad z = \sum N_i z_i \tag{5.6.4}$$

so that Eq. 5.6.3 reduces to Eq. 5.6.1 provided that $\sum N_i = 1$. To show that $\sum N_i = 1$, consider an element located by two different global coordinate systems (Fig. 5.6.1). We have

$$x = \sum N_i x_i$$
$$\bar{x} = \sum N_i \bar{x}_i \tag{5.6.5}$$

But $\bar{x} = d + x$, so that

$$d + x = \sum N_i (d + x_i) = d \sum N_i + \sum N_i x_i$$
$$d = d \sum N_i + \sum N_i x_i - x = d \sum N_i \tag{5.6.6}$$

from which we conclude that $\sum N_i = 1$.

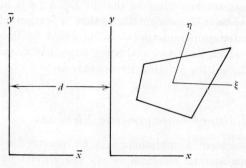

Figure 5.6.1. Two global coordinate systems.

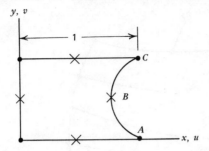

Figure 5.6.2. Superparametric element.

We have shown that rigid body and constant strain criteria are satisfied when element shape and element displacements are defined by the *same* shape functions. Zienkiewicz (8) has also mentioned "subparametric" and "superparametric" elements for which element shape is defined respectively by lower order and higher order polynomials than used to define element displacements.

The foregoing argument remains valid for "subparametric" elements, as we may argue that the functions that define displacements contain as a special case the functions that define shape. For a particular example we refer to the parent element of Fig. 5.5.1. If subparametric, element shape is defined by the four corner nodes. This linear shape is matched by linear edge displacements when Eq. 5.6.2 prevails; indeed in this circumstance the element is basically a linear one.

"Superparametric" elements are usually invalid; element shape is not contained within the functions that define displacements. Consider, for example, an element having a curved side but with displacements defined as for a linear element (Fig. 5.6.2). Impose nodal displacements in accord with the constant strain field $u = ax$, where a is a constant. Thus $u_A = u_C = a$. But the linear edge displacement requires that $u_B = u_A = u_C$. Strain ϵ_x across the element center is therefore greater than $u_{,x} = a$. We conclude that the element misrepresents constant strain conditions and is therefore invalid.

To show that interelement compatibility is preserved during deformation we note that displacements along any edge are uniquely determined by displacements of nodes on that edge and by no other nodes. This may be seen by substitution of ξ or η equal to -1 or $+1$ into the shape functions. Also, along a common edge ($\eta_1 = +1$ and $\eta_2 = -1$ in Fig. 5.6.3) the shape functions of the two elements are identical. Since $\xi_1 = \xi_2$ along this edge, it follows that compatibility is preserved. The same argument shows that isoparametric shape functions define element shapes so that before deformation, elements match exactly along common boundaries.

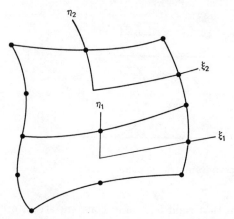

Figure 5.6.3.   Adjacent elements.

5.7   *Order of Quadrature Needed*

With other requirements satisfied, numerically integrated elements yield convergence toward correct results as the mesh is refined if numerical integration is adequate to evaluate the element volume exactly (1, 9, 10). This statement may be explained by noting that, as indicated by previous derivations, formation of a stiffness matrix is essentially the same as integration of a strain energy expression. As a mesh is refined and a constant strain condition comes to prevail in each element, the strain energy expression for an element assumes the form

$$\frac{1}{2} \int_{\text{vol}} (\{\epsilon\}^{T}[E]\{\epsilon\}) \, dx \, dy \, dz = \int_{\text{vol}} (\text{constants}) \det [J] \, d\xi \, d\eta \, d\zeta \quad (5.7.1)$$

Hence, in the limit, the strain energy of the structure is correctly assessed if the volume of each element is correctly assessed.

Examination of the Jacobian determinant yields the number of Gauss points needed to obtain the volume of particular elements. For a plane linear element of constant thickness (Fig. 5.2.1), $\det [J]$ is linear in ξ and η, hence only one Gauss point is needed. For a plane quadratic element of constant thickness (Fig. 5.5.1) $\det [J]$ contains ξ^3 and η^3, hence a 2 by 2 Gauss rule (four points) is the minimum that can be accepted. The linear solid element requires a 2 by 2 by 2 rule and the quadratic solid a 3 by 3 by 3 rule. It is possible for quadratic elements to have straight edges and midside nodes, in which case x, y, and z (but not u, v, and w) reduce to linear functions of ξ, η, and ζ, and the minimum permissible order of quadrature is also reduced.

For two-dimensional elements whose thickness t is variable, one must examine the product $t \det [J]$ in order to determine the minimum acceptable order of quadrature.

With most elements, use of a low order integration rule reduces element stiffness because the strain energy of higher-order deformation modes is not accounted for (8, 9). This effect may be beneficial because it tends to compensate for the overstiffness associated with use of assumed displacement fields. However, whether the finite-element structure is too stiff or too flexible now becomes problem-dependent, so that there is no guarantee that convergence of results will be monotonic. The "best" integration rule for an element is usually decided on by trial and experience.

A danger inherent in the use of low order integration rules is that *zero-energy deformation modes* may arise (11). Consider, for example, the use of a 2 by 2 Gauss rule with the quadratic plane element (Fig. 5.7.1). Let the element be square and let nodal d.o.f. be assigned the values $-u_1 = u_3 = u_5 = -u_7 = v_1 = v_3 = -v_5 = -v_7 = 1,$ $\quad -v_2 = -u_4 = v_6 = u_8 = 1/2,$ $u_2 = v_4 = u_6 = v_8 = 0$. At the Gauss points $\xi, \eta = \pm 1/\sqrt{3}$ we find that all strains are zero. Accordingly, because the Gauss points represent the element when its stiffness matrix is formed, the element offers no resistance at all to this particular deformation mode. Elements that are not square can have nodal displacements such that the same zero-energy mode is produced. A zero-energy mode, when and if it appears, is usually superposed on "legitimate" deformation modes.

The trouble could be removed, at some computational expense, by use of a 3 by 3 Gauss rule. Fortunately, however, this particular zero-energy mode is incompatible with the same mode in an adjacent element. The *assembly* of elements therefore cannot have a zero-energy mode (other than rigid body motion), hence the structure stiffness matrix will not be singular. In general, possible difficulty may be detected by computing eigenvalues of the element stiffness matrix; there will be as many zero eigenvalues as there are zero-strain-energy modes, including rigid body modes (see Section 16.3). Additional comments on zero-energy modes and their control appear in Section 9.5.

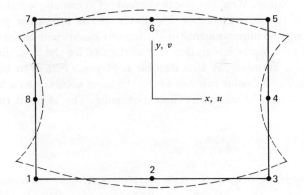

Figure 5.7.1. Possible zero-energy mode.

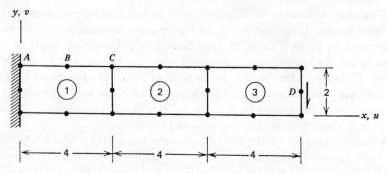

Figure 5.7.2.   Test problem. $E = 1.0$, $\nu = 1/3$, load $= 6$.

For a numerical example of the use of different integration rules, consider the end-loaded cantilever beam of Fig. 5.7.2. Stresses and deflections produced by use of four and nine Gauss points per element are shown in Table 5.7.1. Of the tip deflection of 0.5299 units, 0.0115 are due to shear deforma-

Table 5.7.1.   Stresses in Element 1 and End Deflection (Fig. 5.7.2)

	σ_x at A	σ_x at B	σ_x at C	v at D
Exact	108.0	90.0	72.0	-0.5299
2×2 rule	108.0	90.0	72.0	-0.5114
3×3 rule	105.5	90.0	74.5	-0.5084

tion and 0.5184 are due to bending. The same example problem has been solved using three *linear* elements, with apparently a 2 by 2 Gauss rule for each (2). Stress σ_x at A and deflection v at D were, respectively, 56 % and 65 % of the exact values. With the linear element, of course, σ_x is independent of x in each of the three elements. The latter results are rather poor, but in Chapter Six some improvements of the linear element will be described.

A 3 by 3 by 3 Gauss integration rule works well for the 20-node quadratic solid element. However, it appears that a 14-point rule is to be preferred (13, 14, 16). The 14-point rule has about the same accuracy as a 3 by 3 by 3 rule but is obviously much less time consuming. The 14-point rule has the form

$$\int_{-1}^{1} \int_{-1}^{1} \int_{-1}^{1} f(\xi, \eta, \zeta)\, d\xi\, d\eta\, d\zeta$$

$$= B_6[f(-b, 0, 0) + f(b, 0, 0) + f(0, -b, 0) + \cdots (6 \text{ terms})]$$

$$+ C_8[f(-c, -c, -c) + f(c, -c, -c) + \cdots (8 \text{ terms})] \qquad (5.7.2)$$

where

$$B_6 = 0.8864265927977839, \qquad b = 0.7958224257542215,$$
$$C_8 = 0.3351800554016621, \qquad c = 0.7587869106393281.$$

5.8 Concluding Remarks

Some of the versatility and convenience of numerically integrated iso-parametric elements may now be apparent. A variety of shapes and even curved sides are permitted. Material properties may vary over the element and are properly accounted for by use of the appropriate values at the Gauss points. These values may be interpolated from nodal values, in the same way as initial strains in Fig. 5.4.1 are interpolated. Similarly, the thickness of plane elements may vary and may be defined by nodal values.

If a program for the plane linear element is available, how must it be altered to use instead the plane quadratic element? There are few basic changes. A different set of shape functions must be used. The Jacobian is still 2 by 2 but is computed by use of more terms. Parameters of some DO-loops must be increased, because some arrays are larger.

Element shapes are not of unlimited variety. Regular shapes generally provide greatest accuracy. Thus, in a good mesh, most elements would be approximately square. Quite elongated or distorted elements can give good results if they happen to lie in a strain field that they can represent exactly; for example, a linear element could be used as a two-force member. Severe distortion can cause nonuniqueness of mapping; two different points in $\xi\eta$ coordinates may have the same xy coordinates, or xy coordinates of inter-polated points may lie outside the intended bounds of the element (8) (Fig. 5.8.1). Loosely stated, acceptable elements have mild curvature, corner angles well under 180 degrees, and evenly spaced side nodes. Apparently the chance for error increases as more side nodes are used in each element. Zienkiewicz (8) strongly recommends that the sign of the Jacobian be checked. If the Jacobian becomes negative at any location, error due to nonuniqueness of mapping is indicated. The Jacobian is always computed at

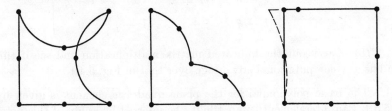

Figure 5.8.1. Unacceptable elements.

Gauss points during numerical integration. However, a negative det $[J]$ is more likely at corners, where stresses may be computed.

Markedly distorted elements, if used at all, may behave better if integrated using a Gauss rule of higher order than would be used for nearly rectangular elements.

While stresses may be computed at any node, stresses at corners are usually less accurate than stresses at side nodes and at Gauss points (8). If desired, one may compute stresses at Gauss points and then extrapolate to other points, using the element shape functions as the basis for an extrapolation scheme.

Finally, it is worth noting that isoparametric solid elements are considered preferable to hexahedral elements built as an assembly of tetrahedra (12) (tetrahedra are solid analogues of plane triangles). This conclusion was reached by a study of linear and quadratic elements of both types, considering both accuracy and computational expense.

PROBLEMS

5.1. Derive Eqs. 5.2.1 and 5.2.2, following the suggestion made in connection with Eq. 5.2.3.

5.2. Show that the Jacobian matrix provides the proper coordinate transformation in the case of a plane linear element, located and oriented such that $\xi = y/2$, $\eta = -x/2$.

5.3. (a) Let nodal coordinate arrays for the plane linear element be written as $\{x_n\} = \{x_1 \quad x_2 \quad x_3 \quad x_4\}$, $\{y_n\} = \{y_1 \quad y_2 \quad y_3 \quad y_4\}$. Show that the Jacobian determinant can be written in the form (2 of Chapter Six):

$$\det [J] = \frac{1}{8} \{x_n\}^T \begin{bmatrix} 0 & 1 - \eta & -\xi + \eta & -1 + \xi \\ -1 + \eta & 0 & 1 + \xi & -\xi - \eta \\ \xi - \eta & -1 - \xi & 0 & 1 + \eta \\ 1 - \xi & \xi + \eta & -1 - \eta & 0 \end{bmatrix} \{y_n\}$$

(b) Carry out the indicated matrix multiplication and see if this form has any computational advantage over that in Fig. 5.4.1.

5.4. The basic polynomial for the plane quadratic element is given by Eq. 5.5.1. Can interelement compatibility be inferred by study of this polynomial, without study of the shape functions derived from it? Explain.

5.5.   Write out each of the eight shape functions for the following elements and verify that they are suitable as interpolation functions:

(a)   Plane quadratic element.

(b)   Solid linear element.

5.6.   Devise suitable shape functions for the quadratic solid element (Fig. 5.5.3b).

5.7.   Write computer program statements, analogous to Figs. 5.4.1 and 5.4.2, which will formulate element matrices for the following elements:

(a)   Plane quadratic element.

(b)   Solid linear element.

(c)   Solid quadratic element.

(d)   How should Figs. 5.4.1 and 5.4.2 be revised to accommodate body forces that are functions of x and y?

5.8.   Consider an element two units on each side with edges parallel to xyz coordinate axes. Determine nodal loads resulting from:

(a)   A distributed traction in the $+x$ direction acting on the edge $\xi = 1$ of the plane linear element, varying from 0 at $\eta = -1$ to p_o units at $\eta = 1$.

(b)   Repeat Part a for the solid linear element. Assume that p_o does not vary with ζ.

(c)   A uniform traction of p_o units in the $+x$ direction on the edge $\xi = 1$ of the plane quadratic element.

(d)   Repeat Part c for the solid quadratic element.

(e)   Uniform body force of intensity q_o units, acting in the $+y$ direction in a plane linear element.

(f)   Repeat Part e for a plane quadratic element.

(g)   Repeat Part e for a solid linear element.

(h)   Repeat Part e for a solid quadratic element.

5.9.   A plane element having 16 external d.o.f. and 2 internal d.o.f. may be formed by combination of 8 constant-strain triangles, as shown. What advantages or disadvantages might this element have in comparison with the plane quadratic isoparametric element?

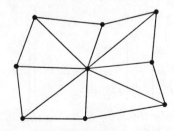

5.10.   Verify that the deformation mode of Fig. 5.7.1 produces zero strain at the Gauss points of a 2 by 2 integration rule.

5.11.   A cantilever beam problem is to be solved by use of a row of plane linear elements, as shown. If the minimum of one integration point per element is used, what results might be anticipated for:

    (a)   Loading by moment M only.

    (b)   Loading by force P only.

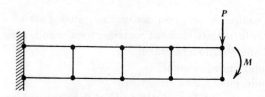

5.12.   How many Gauss quadrature points would be needed to produce the *exact* stiffness matrix for:

    (a)   A plane linear rectangular element?

    (b)   A plane quadratic rectangular element?

    (c)   A solid linear cubic element?

    (d)   A solid quadratic cubic element?

5.13.   Throughout this chapter, elements have been assumed to be of constant thickness. Now let the thickness be a function of ξ and η, and defined by nodal values.

    (a)   Revise Figs. 5.4.1 and 5.4.2 to accommodate a variable-thickness linear element.

    (b)   Will each element approach a state of constant strain as the mesh is refined?

    (c)   How many Gauss points are needed to evaluate the exact volume of a linear element?

    (d)   Repeat Part c for a quadratic element.

5.14.   Derive the sampling point locations and weights for the three-point Gauss rule of Table 5.3.1.

5.15.   Let $[E]$ be a symmetric positive definite matrix, which is decomposed by the Choleski method.

    (a)   If $[E]$ is 3 by 3, then

$$\begin{bmatrix} E_{11} & E_{12} & E_{13} \\ E_{12} & E_{22} & E_{23} \\ E_{13} & E_{23} & E_{33} \end{bmatrix} = \begin{bmatrix} U_{11} & 0 & 0 \\ U_{12} & U_{22} & 0 \\ U_{13} & U_{23} & U_{33} \end{bmatrix} \begin{bmatrix} U_{11} & U_{12} & U_{13} \\ 0 & U_{22} & U_{23} \\ 0 & 0 & U_{33} \end{bmatrix}$$

Hence, find U_{ij} and verify the Fortran statements in Fig. 5.4.2.

    (b)   If $[E]$ is n by n, determine general formulas for U_{ij}.

    (c)   Program the decomposition of Part b.

5.16. Reference 6 suggests the following economy for numerically integrated elements. Taking, for example, the plane linear element, we write the strain-displacement matrix $[B]$ as $[B] = [H][Q]$. Here $[H]$ is the rectangular matrix in Eq. 5.2.9 and $[Q]$ is formed from the rectangular matrices in Eqs. 5.2.8 and 5.2.10. The integrand term of the element stiffness matrix becomes

$$[B]^T[E][B] = [Q]^T[H]^T[E][H][Q] = [Q]^T[S][Q]$$

(a) Write out $[S]$ and $[Q]$ for the plane linear element.

(b) Can $[S]$ be decomposed, as was $[E]$ in Eq. 5.4.3? Explain.

(c) Program the multiplication $[Q]^T[S][Q]$, taking advantage of the special form of $[Q]$. What percentage saving is made in comparison with the multiplication $[B]^T[E][B]$?

(d) Repeat Parts a through c for the solid linear element (Fig. 5.5.3a).

CHAPTER SIX

Some Modifications of Elements

6.1 Introduction. Condensation of Internal D.O.F.

Isoparametric elements discussed in Chapter Five might be called "basic" for lack of a better term. This chapter introduces some modifications designed to improve accuracy and versatility. The principal modification is the introduction of additional d.o.f. These d.o.f. may be introduced to correct a specific defect of an element, or they may merely serve to improve accuracy by virtue of adding more terms to the assumed displacement field. The latter use is an alternative to prescribing more elements as input data and has the advantage that the additional d.o.f. can be supplied automatically by a computer program.

Added d.o.f. are "condensed" before elements are assembled to form the structure. To explain the procedure and its purpose, we will begin with a particular element in which the added d.o.f. are actual nodal displacements instead of simply generalized coordinates. This element is *not* isoparametric.

In modeling a plane structure it is often convenient to use quadrilateral elements of arbitrary shape. One way to form such a quadrilateral is to combine four of the constant strain triangles discussed in Section 4.4 (Fig. 6.1.1). This scheme was suggested in an early paper (1 of Chapter One). Input data to the program consists of quadrilaterals; the computer program divides the element into component triangular elements and properly assembles stiffnesses of the component triangles. In effect each quadrilateral becomes a small structure built of four elements. Customarily, the internal node is assigned coordinates

$$x_5 = \frac{(x_1 + x_2 + x_3 + x_4)}{4}$$

$$y_5 = \frac{(y_1 + y_2 + y_3 + y_4)}{4}$$

$$(6.1.1)$$

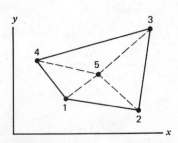 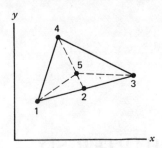

Figure 6.1.1. Elements with subelements.

The stiffness matrix of the quadrilateral "structure" is 10 by 10. It is both convenient and efficient to "reduce out" the two internal degrees of freedom before assembly of elements. The element stiffness matrix is thus reduced to size 8 by 8. This process is called *condensation* and amounts to beginning the solution of the complete system of equations on the element level prior to assembly.

Condensation is widely used, and in some elements there may be several internal d.o.f. Therefore we will describe the process in general terms. D.o.f. to be condensed may appear anywhere in the nodal displacement array and may be eliminated in any order. However, the most common and easily described case is treated in the following, where all d.o.f. to be condensed are grouped at the end of the array.

Let us partition the element stiffness equation so that $\{d_2\}$ represents the one or more internal d.o.f. to be condensed.

$$\begin{bmatrix} k_{11} & k_{12} \\ k_{21} & k_{22} \end{bmatrix} \begin{Bmatrix} d_1 \\ d_2 \end{Bmatrix} = \begin{Bmatrix} r_1 \\ r_2 \end{Bmatrix} \qquad (6.1.2)$$

As explained below, we regard Eq. 6.1.2 as a "fragment" of the structural equations, so that the right side represents loads applied to the nodes by the element. To condense, we first solve for $\{d_2\}$ in terms of $\{d_1\}$,

$$\{d_2\} = [k_{22}]^{-1}(\{r_2\} - [k_{21}]\{d_1\}) \qquad (6.1.3)$$

then substitute this result into the other of Eqs. 6.1.2 to get

$$([k_{11}] - [k_{12}][k_{22}]^{-1}[k_{21}])\{d_1\} = \{r_1\} - [k_{12}][k_{22}]^{-1}\{r_2\} \qquad (6.1.4)$$

The coefficient of $\{d_1\}$ is the condensed element stiffness matrix. Condensed element matrices are assembled and structural equations solved in the usual way, thus determining $\{d_1\}$ for each element. As a final step in solution of equations the internal d.o.f. $\{d_2\}$ may be found by application of Eq. 6.1.3 to each element.

It may now be seen that condensation and solution of equations are basically the same process. Both are processes in which unknowns are eliminated by substitution into the remaining equations. Indeed, if instead of using condensation we were to carry the internal d.o.f. into the structural system of equations, exactly the same results would be obtained. However, if condensation is avoided, the array of structural equations becomes longer and has a larger bandwidth, and the numerical effort of solution can be expected to increase. The effort would be no greater if the structural equations were ordered so that all internal d.o.f. are eliminated first—but this is exactly what is done by condensation before assembly of elements. The equations modified by condensation pertain only to nodes of the element being condensed; adjacent elements are not affected (why?).

When a d.o.f. is condensed, it is freed to displace as dictated by the remaining uncondensed d.o.f. Thus the remaining structure is made more flexible by condensation (see Problem 6.1). Usually, but not always, condensation is applied only to d.o.f. that do not affect element boundary displacements. If applied to (say) an element corner node, the condensed freedoms will not be tied to the corresponding structure node. The corner node will "float," with its motion dictated by the remaining element d.o.f.

The condensation indicated symbolically by Eq. 6.1.4 can be carried out by the following Gauss elimination routine (8). Let arrays SE and RE contain the element stiffness matrix and element nodal load vector, respectively. Let there be a total of NSIZE d.o.f., and suppose that the latter NUM of these d.o.f. are to be eliminated by condensation (Fig. 6.1.2). It is presumed that the matrix is symmetric. Accordingly, for economy, operations are carried out only on the lower triangle, that is, on terms below and including the main diagonal. Upon completion of Fig. 6.1.2, terms to the right of the main diagonal must be filled in by symmetry. The condensed matrices then reside in the first (NSIZE − NUM) rows and columns of arrays SE and RE. Assembly of elements then proceeds in the usual way, followed by solution of structural equations for element boundary d.o.f. Stress computation requires that internal d.o.f. also be known (but see also Problem 6.3).

If main or auxiliary storage for extra arrays is available, internal d.o.f. can be efficiently recovered as follows. First, immediately after condensation, store the last NUM rows of SE and RE in some arrays, say SM and RM (Fig. 6.1.3). Let M be the element number. Then, after other d.o.f. have been found by solution of the structural equations, recover the internal d.o.f.

```
      DO 20 K=1,NUM
      LL = NSIZE - K
      KK = LL + 1
      DO 20 L=1,LL
      DUM = SE(KK,L)/SE(KK,KK)
      DO 10 M=1,L
   10 SE(L,M) = SE(L,M)  - SE(KK,M)*DUM
   20 RE(L)   = RE(L)    - RE(KK)  *DUM
```

Figure 6.1.2.   Condensation routine.

```
    DO 40 I=1,NUM
    K = NSIZE - NUM + I
    DO 30 J=1,NSIZE
30  SM(M,I,J) = SE(K,J)        Figure 6.1.3.  Storage of condensation data for
40  RM(M,I)   = RE(K)          element M.
```

This is done by back substitution, thus completing the Gauss elimination solution begun during condensation. Let array D contain element d.o.f. (Fig. 6.1.4). If available storage does not permit use of this recovery method, the uncondensed element matrices must be reconstructed at some expense in computation time. Internal d.o.f. are then recovered by solution of the last NUM element equations, as shown by Eq. 6.1.3.

Condensation of "excessive" d.o.f. in dynamic problems is considered in Section 12.4.

6.2 Nonnodal Degrees of Freedom

In Chapter Three the Rayleigh-Ritz method was presented using generalized coordinates a_i as the primary unknowns to be determined. In the finite element form of the Rayleigh-Ritz method the primary unknowns became the nodal displacements. The two viewpoints may be combined. A review of Section 4.2 shows that the derivation may be repeated, except perhaps for convenient changes in notation, if the d.o.f. for *each element* consist of generalized coordinates $\{a\}$ as well as nodal freedoms $\{d\}$. Vector $\{a\}$ contains the d.o.f. to be condensed before assembly of elements. Specifically, let us write element displacements and strains as

$$\{f\} = [N_d \quad N_a]\{d \quad a\} \tag{6.2.1}$$

$$\{\epsilon\} = [B_d \quad B_a]\{d \quad a\}$$

Displacements $\{D\}$ of Eq. 4.2.6 may be augmented by arrays $\{a\}$ from each element, but only for purposes of derivation. In computation $\{D\}$ is in fact not augmented; each $\{a\}$ may be condensed, because it is common to but one element. Equations 4.2.9 and 4.2.10 are unchanged, and with Eq. 6.2.1

```
    DO 60 J=1,NUM
    JJ = NSIZE - NUM + J
    DUM = 0.
    K = JJ - 1
    DO 50 I=1,K
50  DUM = DUM + SM(M,J,I)*D(I)
60  D(JJ) = (RM(M,J) - DUM)/SM(M,J,JJ)
```

Figure 6.1.4. Recovery of internal d.o.f. for element M.

they yield

$$\begin{bmatrix} k_{dd} & k_{da} \\ k_{da}^T & k_{aa} \end{bmatrix} \begin{Bmatrix} d \\ a \end{Bmatrix} = \begin{Bmatrix} r_d \\ r_a \end{Bmatrix} \tag{6.2.2}$$

where

$$[k_{dd}] = \int_{\text{vol}} [B_d]^T [E][B_d] \, dV$$

$$[k_{da}] = \int_{\text{vol}} [B_d]^T [E][B_a] \, dV \tag{6.2.3}$$

$$[k_{aa}] = \int_{\text{vol}} [B_a]^T [E][B_a] \, dV$$

and $\{r_d\}$, $\{r_a\}$ are written by attaching subscripts d and a to matrices $[N]$ and $[B]$ in Eq. 4.2.10. Condensation of $\{a\}$ in Eq. 6.2.2 and subsequent recovery of these internal freedoms is done exactly as described in Section 6.1. Indeed, the element of Fig. 6.1.1 may be regarded as a special case of this section, in which it is possible to identify the additional freedom as an actual nodal displacement.

A slightly different viewpoint which leads to the same result is as follows (1). Form the expression for total potential Π_p of the structure as before, including element freedoms $\{d\}$ and $\{a\}$. With the notation of Eq. 6.2.3 we have

$$\Pi_p = \sum \left(\frac{1}{2} \{d\}^T [k_{dd}]\{d\} + \{d\}^T [k_{da}]\{a\} + \frac{1}{2} \{a\}^T [k_{aa}]\{a\} \right.$$

$$\left. - \{d\}^T \{r_d\} - \{a\}^T \{r_a\} \right) - \{D\}^T \{P\} \tag{6.2.4}$$

We now apply the Rayleigh-Ritz procedure under the assumption that all displacements $\{d\}$ are prescribed. Thus, only each element $\{a\}$ is permitted to vary, and because each element $\{a\}$ is independent of the others, we find for each element

$$\left\{ \frac{\partial \Pi_p}{\partial a} \right\} = 0 = [k_{da}]^T \{d\} + [k_{aa}]\{a\} - \{r_a\} \tag{6.2.5}$$

We may solve for $\{a\}$ from the latter equation and substitute it into Eq. 6.2.4. In doing so we generate the same condensation equations that were encountered in Section 6.1. Equation 6.2.4 is now a function of nodal d.o.f. $\{d\}$ only; arrays $\{d\}$ may be replaced by $\{D\}$ and the solution completed in the usual way.

The latter viewpoint clarifies the effect of the additional freedoms. As noted in Chapter Three, when assumed displacement fields are used, the equations of equilibrium are not satisfied at every point in a continuum but are more closely satisfied as more d.o.f. are used. Accordingly, introduction of element freedoms $\{a\}$ serves to more nearly satisfy equilibrium conditions within the element (1). We will see in Chapter Eight that some curved shell elements that do not remain strain-free under rigid body motion of the basic nodal freedoms are greatly improved in this regard by addition of internal freedoms. Reference 4 demonstrates improved accuracy in vibration analysis by adding one or two internal d.o.f. to simple elements.

Computer programming in the presence of nonnodal d.o.f. presents no new difficulties. For example, in adding d.o.f. to the plane linear isoparametric element, we must introduce additional shape functions, add more columns to array $[B]$, and enlarge the array that contains the uncondensed stiffness matrix. In these respects the changes resemble the alterations needed to change from a linear element to a quadratic one. However, the added d.o.f. are not used when computing the Jacobian matrix, because they do not affect original element shape; that is, the added d.o.f. do not affect the mapping between $\xi\eta\zeta$ and xyz coordinates. Also, of course, condensation and subsequent recovery of internal freedoms is required. Recovery may be omitted if stress calculations are omitted. Stress calculation involves the calculation of strains according to the second of Eqs. 6.2.1, where $\{d \quad a\}$ contains *all* element d.o.f.

The benefit of internal freedoms must be weighed against their computational cost. The cost may be high in nonlinear problems if the internal freedoms must be condensed and recovered in each cycle of an iterative solution.

6.3　Examples and Comments Regarding Internal D.O.F.

Two internal d.o.f. may be added to the plane linear element of Section 5.2, so that

$$u = \sum_1^5 N_i u_i, \qquad v = \sum_1^5 N_i v_i \qquad (6.3.1)$$

where N_1 through N_4 are given by Eq. 5.2.2 and (2)

$$N_5 = (1 - \xi^2)(1 - \eta^2) \qquad (6.3.2)$$

The element shape, which governs the Jacobian matrix, remains as stated in Eqs. 5.2.1 and 5.2.2. The added d.o.f.[1] are thus $\{a\} = \{u_5 \quad v_5\}$. The displacement modes associated with u_5 and v_5 vanish at all element edges, so that these d.o.f. are strictly internal and have no effect on interelement compatibility. Displacements u_{ca} and v_{ca} at the element center $\xi = \eta = 0$

---

[1] While additional d.o.f. are usually taken as polynomials, other choices are possible. For example, instead of Eq. 6.3.2, we might use (3) $N_5 = \cos(\pi\xi/2)\cos(\pi\eta/2)$.

are

$$u_{ca} = \frac{(u_1 + u_2 + u_3 + u_4)}{4} + u_5$$

$$v_{ca} = \frac{(v_1 + v_2 + v_3 + v_4)}{4} + v_5$$

(6.3.3)

so that u_5, v_5 are not actual displacements. Instead, they may be viewed simply as generalized coordinates, or as displacements *relative* to the displacements at $\xi = \eta = 0$ dictated by d.o.f. on the element boundary.

If element stresses are to be computed at any corner or at $\xi = \eta = 0$, it is not necessary to recover u_5 and v_5, as the mode of Eq. 6.3.2 produces no strain at these locations.

Unfortunately, the plane linear element is only slightly improved by use of Eqs. 6.3.1 and 6.3.2. Numerical results (2) are given subsequently in Table 6.4.1, where the basic plane linear element is designated $Q4$ and the modified element as $Q5$.

The condensation process alters element nodal loads as well as the stiffness matrix. Consider, for example, a plane square linear element, 2.0 units on a side, and carrying a uniform body force that totals 4.0 units of force. Use of Eqs. 6.3.1 and 6.3.2 leads to the nodal loads shown in Fig. 6.3.1. As the corner loads already add up to 4.0, the internal load may seem incorrect. However, condensation leads to the loads shown in Fig. 6.3.1b for a Poisson's ratio of 0.3. The redistributed loads of 0.1 are self-equilibrating, and with the internal d.o.f., they lead to improvement in computed stresses. The uncondensed element stiffness matrix in this example may also appear faulty because none of its columns sums to zero, as seems required for static equilibrium when a single d.o.f. is assigned a nonzero value. The condensed matrix, however, behaves as expected. The confusion arises because d.o.f. u_5 and v_5 are *relative*, not absolute displacements. Taking again the element of Fig. 6.3.1 for our

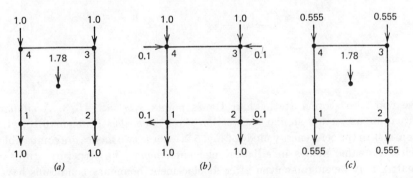

Figure 6.3.1. Element nodal loads for (a) uncondensed, internal "relative" d.o.f., (b) condensed, and (c) uncondensed, internal "absolute" d.o.f.

example, let v represent downward displacements and v_c, v_{ca} represent relative and absolute values of the internal d.o.f. They are related by the transformation

$$\begin{Bmatrix} v_1 \\ v_2 \\ v_3 \\ v_4 \\ v_{ca} \end{Bmatrix} = \begin{bmatrix} 1 & 0 & 0 & 0 & 0 \\ 0 & 1 & 0 & 0 & 0 \\ 0 & 0 & 1 & 0 & 0 \\ 0 & 0 & 0 & 1 & 0 \\ \dfrac{1}{4} & \dfrac{1}{4} & \dfrac{1}{4} & \dfrac{1}{4} & 1 \end{bmatrix} \begin{Bmatrix} v_1 \\ v_2 \\ v_3 \\ v_4 \\ v_c \end{Bmatrix} = [T] \begin{Bmatrix} v_1 \\ v_2 \\ v_3 \\ v_4 \\ v_c \end{Bmatrix} \qquad (6.3.4)$$

The same transformation matrix $[T]$ relates "relative" and actual d.o.f. u_c and u_{ca}. Coordinate transformation is discussed in Chapter Eleven. For the present, suffice it to say that the uncondensed stiffness matrix and load vector, formed using the "relative" d.o.f. of Eq. 6.3.2, may be converted by a standard process to correspond to use of actual displacements as internal d.o.f. When this is done each column of the transformed stiffness matrix represents a set of forces in static equilibrium. And, in particular, the forces $\{r\}$ of Fig. 6.3.1a are converted to "absolute" form by the operation $[T]^{-T}\{r\}$. The resulting forces are shown in Fig. 6.3.1c and sum to 4.0, as they should.

A danger of adding internal d.o.f. is that zero-energy deformation modes, first mentioned in Section 5.7, may arise in particular cases. Indeed this happens if the plane *quadratic* isoparametric element is augmented by the internal freedoms $(1 - \xi^2)(1 - \eta^2)u_9$ and $(1 - \xi^2)(1 - \eta^2)v_9$ and four-point Gauss quadrature is used (11 of Chapter Five). If, with nodes numbered as in Fig. 5.5.1, we consider a rectangular element that has nodal displacements

$$u_1 = u_3 = u_5 = u_7 = -\frac{2}{3}$$

$$u_2 = u_4 = u_6 = u_8 = +\frac{1}{3} \qquad (6.3.5)$$

$$u_9 = -\frac{3}{2}$$

we find zero strains at the four Gauss points ξ, $\eta = \pm 1/\sqrt{3}$. A similar situation arises for similar y-direction d.o.f. (see also Problem 6.10). In contrast to the zero-energy mode of Fig. 5.7.1, these two modes *are* compatible with the same modes in adjacent elements, hence the structure stiffness matrix may be singular even after displacement boundary conditions have been imposed, if these conditions happen not to suppress the zero-energy mode. The condensed element stiffness matrix $[k]$ will have order 16 but rank 10 because of three rigid body modes and a total of three zero-energy modes.

Zero-energy modes are often not obvious but may be detected by computing eigenvalues of $[k]$ (see Section 16.3). Often, they may be avoided by use of some simple tricks; this discussion is taken up in connection with elements for which use of internal d.o.f. is more profitable (see Section 9.5).

6.4 Avoiding Parasitic Shear in Linear Elements

Probably the major disadvantage of linear isoparametric elements is that they behave badly under pure bending. Suppose, for example, that a plane linear element is used to model a beam in pure bending. To model the beam exactly the deformation should be as shown in Fig. 6.4.1a. But this the element cannot do; its sides must remain straight, so it deforms as shown in Fig. 6.4.1b. For convenience let $x = \xi$ and $y = \eta$ in this discussion. Shear strain γ_{xy} should be zero throughout, but Eqs. 5.2.2 yield $\gamma_{xy} = u_1 x$ for nodal displacements $u_1 = -u_2 = u_3 = -u_4$. "Parasitic" shear exists at the Gauss points of all but a one-point rule, and the resulting element is too stiff in bending because deformation requires strain energy storage by shear strain as well as by normal strain.

Shear strain is correctly given as $\gamma_{xy} = 0$ in Fig. 6.4.1b at $\xi = 0$. In general, bending about either axis is possible, in which case γ_{xy} is correctly given only at $\xi = \eta = 0$. This observation led to the suggestion (2) that when forming element matrices by numerical integration, terms in matrix $[B]$ associated with shear strain should always be evaluated at $\xi = \eta = 0$, while other terms in $[B]$ are given the customary ξ and η values for the integration point at hand. This procedure greatly improves the performance of both the basic linear element and the augmented element of Eqs. 6.3.1. Unfortunately, both elements now fail the "patch test" unless they are rectangular. Also, neither element is invariant. Apparently, "nonuniform" integration must be used with great caution.

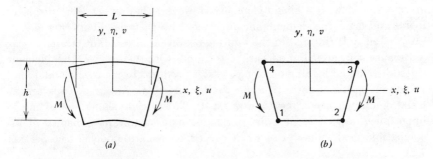

Figure 6.4.1. Pure bending.

A preferable way to improve linear elements was suggested by Wilson et al. (5; 5 of Chapter Two). With appropriate restraint of rigid body motion, the element of Fig. 6.4.1a has displacements

$$u = \frac{M}{EI} \xi\eta$$

$$v = \frac{ML^2}{8EI} (1 - \xi^2) + \nu \frac{Mh^2}{8EI} (1 - \eta^2) \tag{6.4.1}$$

where $EI =$ bending stiffness and $\nu =$ Poisson's ratio. But the best the linear element can do is to produce displacements of the form

$$u = c_1 \xi\eta$$

$$v = 0 \tag{6.4.2}$$

where c_1 is a constant. Similar consideration of bending about the y axis leads to the conclusion that, in general, where bending may be about either axis, the basic element errs by omitting displacements of the form

$$u = c_1(1 - \xi^2) + c_2(1 - \eta^2)$$

$$v = c_3(1 - \xi^2) + c_4(1 - \eta^2) \tag{6.4.3}$$

where c_1 through c_4 are constants. Accordingly, the element may be improved by adding these modes as internal freedoms. In this way pure bending, free of shear strain, may be represented exactly. Thus the complete displacement fields for the modified element are

$$u = \sum_1^4 N_i u_i + (1 - \xi^2)u_5 + (1 - \eta^2)u_6$$

$$v = \sum_1^4 N_i v_i + (1 - \xi^2)v_5 + (1 - \eta^2)v_6 \tag{6.4.4}$$

where N_1 through N_4 are given by Eq. 5.2.2. The added modes $\{a\} = \{u_5 \quad u_6 \quad v_5 \quad v_6\}$ are condensed before assembly of elements. As in Section 6.3 the added freedoms may be visualized as displacements relative to displacements defined by the corner nodes. These d.o.f. are internal in that they are associated with a single element. However, they affect edge displacements and make the elements incompatible with their neighbors (Fig. 6.4.2). Compatibility is restored in the limiting condition of constant strain.

Because the added d.o.f. are associated with motion of the element sides, distributed loading on element sides produces terms in the nodal load vector in positions corresponding to the added d.o.f. These loads are as usual computed according to Eq. 4.2.10. These loads should be of minor importance, especially in a fine mesh. A more significant difficulty of incompatible modes may arise when curved bodies are modeled (see Section 7.2).

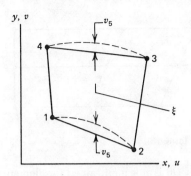
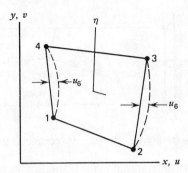

Figure 6.4.2. Incompatible modes.

The element of Eqs. 6.4.4 is designated as $Q6$. Element $Q4$ is that of Section 5.2, and element $Q5$ is described by Eqs. 6.3.1 and 6.3.2. These elements are compared in Table 6.4.1 (2, 5). The test problems are shown in Fig. 6.4.3.

Table 6.4.1. Behavior of Elements (Fig. 6.4.3)

Element	Total d.o.f. per Element	External d.o.f. per Element	Compat-ible Element?	Displacement v at C Load A	Load B	Stress σ_x at D Load A	Load B
$Q4$	8	8	yes	70.6	72.3	−2188.	−2954.
$Q5$	10	8	yes	75.7	77.7	−2270.[a]	−3056.[a]
$Q6$	12	8	no	100.0	101.3	−3000.	−4050.
Beam theory	—	—	—	100.0	103.0	−3000.	−4050.

[a] Extrapolated from values at element centers.

Because elements are rectangular, stress σ_x is constant along edges $y =$ constant in each element of Table 6.4.1. However, nodal values of σ_x from element $Q6$ are also exact if stresses from adjacent elements are averaged. Element $Q6$ is invariant and under four-point Gauss quadrature has no zero energy modes other than rigid body motion.

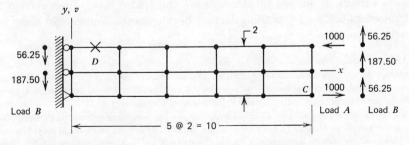

Figure 6.4.3. Test problems for plane elements. $E = 1500$, $v = 0.25$.

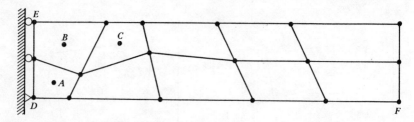

Figure 6.4.4. Beam with nonrectangular elements.

Unfortunately, the accuracy of element $Q6$ is not maintained for elements of arbitrary shape. Figure 6.4.4 is a scale drawing of a test problem having nonrectangular elements. Table 6.4.2 shows results obtained under the same

Table 6.4.2. Stresses and Displacements of Beam of Fig. 6.4.4 Under Pure Moment Loading

Point	Q6 Elements	Q4 Elements	Exact
	Axial stress σ_x		
A	1931.	1356.	1800.
B	−1290.	−1013.	−1200.
C	−1209.	−643.	−1350.
D	1851.	2156.	3000.
E	−3249.	−2107.	−3000.
	Vertical displacement		
F	96.41	59.16	100.00

moment loading as given in Fig. 6.4.3 (load A). Stresses at element centers A, B, C in $Q6$ elements are acceptable. However, it is more disturbing that under purely axial load on this structure, $Q6$ elements give nodal stresses σ_x in error by $\pm 40\%$, while $Q4$ elements model the structure exactly. In other words, arbitrarily shaped $Q6$ elements fail the "patch test," Requirement 2 of Section 4.6. Clearly, pending further developments, element $Q6$ must be used with caution.[1]

---

[1] An assumed-stress *hybrid* element having five stress modes behaves as well as element Q6 when elements are rectangular. In addition, nonrectangular hybrid elements pass the patch test. Invariance is achieved by formulation in a local coordinate system whose orientation is governed by the orientation of the element.

Parasitic shear may also be avoided by using the least-squares scheme of Problem 6.3 to define the shear strain. The resulting element does not appear to be invariant. Studies of the possible merits and pitfalls of this approach are not presently available. The student is invited to explore the matter, perhaps as a term project.

Incompatible modes have been added to the linear solid isoparametric elements, and it also is greatly improved (5; 5 of Chapter Two). As might be expected in view of the behavior of element Q6, these elements pass the patch test only if they are rectangular parallelepipeds. Nevertheless, it is claimed (5) that a single layer of such elements is adequate for analysis of arch dams and thick pipe joints. Displacement fields for the uncondensed element are

$$u = \sum_1^8 N_i u_i + (1 - \xi^2)u_9 + (1 - \eta^2)u_{10} + (1 - \zeta^2)u_{11}$$

$$v = \sum_1^8 N_i v_i + (1 - \xi^2)v_9 + (1 - \eta^2)v_{10} + (1 - \zeta^2)v_{11} \qquad (6.4.5)$$

$$w = \sum_1^8 N_i w_i + (1 - \xi^2)w_9 + (1 - \eta^2)w_{10} + (1 - \zeta^2)w_{11}$$

where shape functions N_i are given by Eq. 5.5.5. A 2 by 2 by 2 rule, having eight Gauss points, is used for integration of rectangular elements, while a 3 by 3 by 3 rule is recommended for distorted elements (5 of Chapter Two). Condensation of the 9 incompatible d.o.f. yields a condensed (and incompatible) element having 24 d.o.f.

6.5   Degrading of Elements

It is sometimes convenient and desirable to use a mixture of elements in a given mesh. We then must have a way to join different elements, for example, a linear element must be joined to a quadratic one. And in some places triangular (or tetrahedral) elements are convenient. The joining problem and triangular elements may both be approached by degrading of elements.

Suppose, for example, that a plane quadratic element is to be joined to a plane linear element (Fig. 6.5.1). Edge 3–5 of the quadratic element must be straight originally and must remain straight if interelement compatibility

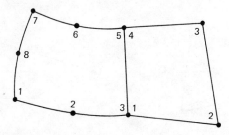

Figure 6.5.1.   Mixing of elements.

is to be maintained during deformation. To generate such an element we begin with the basic quadratic element whose shape functions are given by Eqs. 5.5.3 and 5.5.4. Let α represent x, y, u, or v. Shape functions for the degraded element are obtained by enforcing linearity; we substitute

$$\alpha_4 = \frac{(\alpha_3 + \alpha_5)}{2} \tag{6.5.1}$$

into the expression

$$\alpha = \sum_1^8 N_i \alpha_i \tag{6.5.2}$$

and gather coefficients of the seven remaining α_i. Thus we obtain new shape functions N_i' for the degraded element:

$$\alpha = \sum_1^3 N_i' \alpha_i + \sum_5^8 N_i' \alpha_i \tag{6.5.3}$$

In practice it would be convenient to use seven consecutive numbers for nodes of the new element. Clearly any edge may be treated in this way and up to three edges may be treated. If all four edges are treated the element becomes a linear element.

As another example of degrading, consider the formation of a triangular element from a linear quadrilateral (Fig. 6.5.2). In this case we may begin with shape functions of the plane linear isoparametric element (Eqs. 5.2.2). By replacing, say, x_4 by x_1 and gathering terms, we obtain $x = N_1' x_1 + N_2' x_2 + N_3' x_3$ where N_i' are shape functions of the degraded element. These new shape functions define y, u, and v as well as x, so that the superposed nodes are forced to have the same displacements as well as the same co-ordinates. The new element properly represents constant strain states and rigid body motion and is compatible. It is, in short, a constant strain triangle (Fig. 4.4.1).

In similar fashion a linear-strain triangle may be produced by enforcing equality of coordinates and displacements of nodes 1, 7, and 8 (Fig. 6.5.3). Unfortunately the shape function associated with node 4 now gives infinite

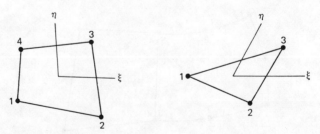

Figure 6.5.2. A triangular element.

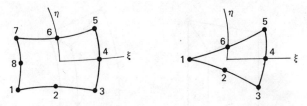

Figure 6.5.3.   Another triangular element.

strains at node 1. However, the trouble may be removed by multiplying the
offending shape function by $(1 + \xi)/2$ (8 of Chapter Five).

Application of Gaussian quadrature to triangular elements with $\xi\eta$ coordinates is not entirely satisfactory. As seen in Fig. 6.5.4, a different choice
of axes ξ and η produces a different set of Gauss points. Accordingly, unless
there are so many Gauss points that the computed stiffness matrix is almost
exact, there is no assurance that the two choices of axes will produce the
same stiffness matrix. Special rules adapted to triangular regions are to be
preferred (Section 17.6).

Triangular elements are useful but not essential in grading a mesh from
coarse to fine. When a few triangular elements are used for this purpose,
their shortcomings, such as lack of invariance, are not very important.
Invariance may be achieved by use of area coordinates (Section 17.6).
Readers who have more than occasional use for triangular elements will
wish to study other references, for example, Refs. 6 and 7.

It should be apparent that degrading is a technique not limited to the
examples presented. It may be applied to solid elements, or to produce one
linear and one quadratic edge on a cubic element, etc. Useful solid wedge
elements (Fig. 6.5.5) may be formed not by degrading but by use of area coordinates in triangular cross sections and an isoparametric coordinate along
the prism axis (7).

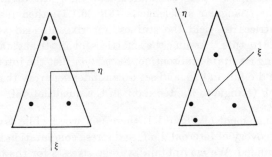

Figure 6.5.4.   Gauss quadrature points in a triangle.

Figure 6.5.5. Solid wedge elements.

One should be aware that it is not acceptable to produce a triangular element by merely assigning the same coordinates to adjacent nodes, say by input data to a program based on quadrilateral elements. When two nodes of an element are superposed in this way, very large coefficients appear in the element stiffness matrix because relative motion between the two nodes has not been suppressed. Numerical difficulties are therefore to be expected (Section 16.5), and computed stresses are likely to be quite inaccurate.

PROBLEMS

6.1.   Rearrange Eqs. 2.8.1 so that the order of unknowns is $\{u_3 \quad u_2 \quad u_1\}$. Condense u_2 and u_1 by use of Eq. 6.1.5. The condensed equation involves only u_3 and should agree with the equation for u_3 appearing in Eq. 2.8.3.

6.2.   Write computer program statements that will divide a quadrilateral into four triangular subelements (Fig. 6.1.1); then reassemble the component matrices to yield the stiffness matrix and load vector of the quadrilateral "structure." Assume that matrices for any triangle are generated merely by calling a suitable subroutine. *Suggestion.* Let the internal node be associated with d.o.f. 9 and 10, and set up an indexing array that relates the six d.o.f. of each triangle to the ten d.o.f. of the quadrilateral.

6.3.   For the element of Fig. 6.1.1, there is a worthwhile saving in computer time if recovery of internal d.o.f. and stress computation in individual triangles are avoided. We may obtain average stresses for the quadrilateral by fitting a linear field $u = a_1 + a_2 x + a_3 y$, $v = a_4 + a_5 x + a_6 y$ to the corner node displacements. We write

$$
\begin{Bmatrix} u_1 \\ u_2 \\ u_3 \\ u_4 \end{Bmatrix} - \begin{bmatrix} 1 & x_1 & y_1 \\ 1 & x_2 & y_2 \\ 1 & x_3 & y_3 \\ 1 & x_4 & y_4 \end{bmatrix} \begin{Bmatrix} a_1 \\ a_2 \\ a_3 \end{Bmatrix} = \begin{Bmatrix} e_1 \\ e_2 \\ e_3 \\ e_4 \end{Bmatrix}
$$

or $\{u\} - [Q]\{a\} = \{e\}$. There is a similar equation for v. Residuals $\{e\}$ are present because three constants a_i cannot exactly satisfy four equations. Show that minimization of $\{e\}^T\{e\}$ with respect to $\{a\}$ (least squares fit) yields

$$
\{a\} = ([Q]^T[Q])^{-1}[Q]^T\{u\}
$$

Hence, derive expressions for strains ϵ_x, ϵ_y, and γ_{xy} in terms of x_i, y_i, u_i, and v_i, where $i = 1, 2, 3, 4$.

This stress computation has been used successfully for solids of revolution (9; 2 of Chapter Seven).

6.4.   Derive the "condensation equations" by substitution of Eq. 6.2.5 into Eq. 6.2.4 and verify that the result agrees with the equations of Section 6.1.

6.5.   Suppose that internal d.o.f. are used in an element. Suppose also that recovery of internal d.o.f. is to be preceded by removal of rigid body motion, as suggested in Section 4.7. Will the same values of internal d.o.f. be computed? And the same element stresses? Explain.

6.6.   Consider the constant-strain triangle of Fig. 4.4.1. According to the discussion of Section 6.2, can introduction of internal d.o.f. be of any benefit to this element? Explain.

6.7.   Write computer program statements that will generate the stiffness matrix of the element of Eqs. 6.3.1 and 6.3.2. Include condensation.

6.8.   (a)   Is there any computational advantage to associating internal d.o.f. with the shape function $(1 - \xi^2)(1 - \eta^2)$ instead of $\cos(\pi\xi/2) \cos(\pi\eta/2)$? Explain.

(b)   Consider a square element, of constant thickness and two units on a side, carrying a constant body force in the $+y$ direction. Determine the force associated with the internal d.o.f. using each of the two functions of Part a.

6.9.   The uniform two-force bar has cross-sectional area A and elastic modulus E. It is subjected to a uniform body force that totals F force units in the $+x$ direction. Let displacements be given by

$$u = [(1 - x/2L) \quad x/2L \quad x(2L - x)]\{u_1 \; u_2 \; u_c\},$$

where u_c is an internal d.o.f.

    (a) Derive the 3 by 3 uncondensed stiffness matrix and the nodal load vector. Do these results appear to satisfy static equilibrium? (See Fig. 6.3.1.)

    (b) If coordinate transformation has been studied, convert the result of Part a to correspond with an actual internal displacement. Compare the results with those of Problem 4.5a.

    (c) Remove u_c by condensation of the results of Part a. Compare the results with those of Problem 4.5b.

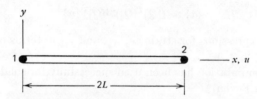

6.10.   (a) Verify that Eqs. 6.3.5 produce zero strain at the Gauss points of a 2 by 2 integration rule.

    (b) Let boundary node displacements of the quadratic isoparametric element be $u_1 = u_7 = -v_1 = -v_3 = 4, u_8 = -v_2 = -2, u_2 = u_6 = -v_4 = -v_8 = -1, u_3 = u_4 = u_5 = v_5 = v_6 = v_7 = 0$. What absolute displacements at $\xi = \eta = 0$ render this a zero-energy mode? The element is square, has two internal d.o.f. associated with $(1 - \xi^2)(1 - \eta^2)$, and is integrated by a 2 by 2 Gauss rule. *Suggestion.* Sketch and u and v displacement modes separately.

    (c) Hence, what relative displacements at $\xi = \eta = 0$ are associated with $(1 - \xi^2)(1 - \eta^2)$ in the zero energy mode of Part b?

    (d) Suppose that instead of using N_5 as given by Eq. 6.3.2, the trigonometric form is used (see footnote on page 128). Is it now possible for an assembly of quadratic elements to have a zero-energy mode?

6.11.   (a) Derive Eqs. 6.4.1.

    (b) What would be the objection to omitting $(1 - \xi^2)u_5$ and $(1 - \eta^2)v_6$ from Eqs. 6.4.4?

    (c) After computation of displacements using element $Q6$, what would be the probable effect of omitting internal d.o.f. during stress computation, say for the two problems of Fig. 6.4.3?

6.12.   Write computer program statements that will generate element matrices for the following elements. Include condensation.

    (a) Element $Q6$, Eqs. 6.4.4.

    (b) Corresponding solid element, Eqs. 6.4.5.

6.13.   Discuss the desirability and the validity of joining plane linear and quadratic elements as shown in the sketch. (See Section 11.7 for a rational solution to this problem.)

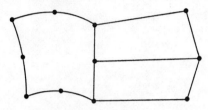

6.14.   Show that a plane linear element is produced by degrading all four edges of a plane quadratic element.

6.15.   Reference 1 of Chapter Five suggests using side nodes as "departures from linearity." Thus, for example, with $\alpha = x$, y, u, or v, we write $\alpha_2 = (\alpha_1 + \alpha_3)/2 + \alpha_2'$ where α_2' is the departure from linearity.

(a)   Write the shape functions for the plane quadratic element using departures from linearity as nodal d.o.f. of the four side nodes.

(b)   Hence, how is linearity of an edge enforced?

(c)   What are possible advantages and disadvantages of this scheme?

6.16.   Consider the triangular element discussed in connection with Fig. 6.5.2.

(a)   Write the shape functions for this element.

(b)   Let nodes 1, 2, 3 have xy coordinates $(-1, 0)$, $(1, -1)$, and $(1, 1)$, respectively. Write matrix $[B]$ for this element.

(c)   Consider the displacement field $u = a + bx + cy$, $v = d + ex + fy$ where a through f are constants. Show that nodal displacements consistent with this field cause matrix $[B]$ of Part b to produce the correct strains $\epsilon_x = b$, $\epsilon_y = f$, $\gamma_{xy} = c + e$.

CHAPTER SEVEN

Solids of Revolution

7.1 Introduction

Solids of revolution under axisymmetric loads are two-dimensional insofar as analysis for stress and deformation is concerned. Displacements are confined to the r (radial) and z (axial) directions. Accordingly, little change is needed to adapt the plane elements of Chapter Five to the axisymmetric problem, and the improvements and modifications discussed in Chapter Six may also be used.

In cases where the geometry and elastic properties of the solid remain independent of the circumferential coordinate θ, arbitrary loadings may be accommodated by the Fourier series method. Thus, the given loading is expressed as a sum of several separate loadings, and an analysis is carried out for each. According to the principle of superposition the original problem is solved by adding the solutions of the component problems. Each separate analysis remains essentially two dimensional, so that the original three-dimensional problem is exchanged for several two-dimensional problems. The exchange is worthwhile because a single three-dimensional solution is usually much more expensive than several two-dimensional solutions. A similar treatment of plate problems is mentioned in Section 9.1.

Finite element analysis of solids of revolution, including both axisymmetric and general loading, was described by Wilson (1). Various computer programs are widely available, but most are based on the classic but comparatively inefficient constant-strain triangle of Fig. 4.4.1 (2, 3). The similarity of plane and axisymmetric analyses makes it convenient to incorporate both in a single program (2).

7.2   Formulation for Axisymmetric Loading

The finite element is a ring of constant cross section (Fig. 7.2.1). A quantity of these elements suffices to model a solid of revolution, such as that suggested by dashed lines in Fig. 7.2.1. The solid and each of its elements are solids of revolution about the z axis. Nodal points become nodal circles whose centers lie on the z axis.

Under axisymmetric loading no point has any θ direction displacement. Accordingly, the element displacement field has only r and z components:

$$\{f\} = \{u \; v\} = [N]\{d\} \tag{7.2.1}$$

If, for example, the element is a linear isoparametric element, $[N]$ is the same shape function matrix used in Section 5.2, and $\{d\}$ is the same array of eight nodal d.o.f. Or, if element $Q6$ of Section 6.4 is used, $\{d\}$ contains 12 d.o.f. prior to condensation.

The axisymmetric problem differs from the plane problem in that the hoop normal strain ϵ_θ is neither zero nor uniquely determined by the other strains, and therefore must be present in explicit form. For the case of isotropy, the stress-strain relation $\{\sigma\} = [E](\{\epsilon\} - \{\epsilon_o\})$ is therefore

$$
\begin{Bmatrix} \sigma_r \\ \sigma_z \\ \sigma_\theta \\ \gamma_{rz} \end{Bmatrix}
=
\frac{E}{(1+\nu)(1-2\nu)}
\begin{bmatrix}
1-\nu & \nu & \nu & 0 \\
\nu & 1-\nu & \nu & 0 \\
\nu & \nu & 1-\nu & 0 \\
0 & 0 & 0 & \dfrac{1-2\nu}{2}
\end{bmatrix}
$$

$$
\times
\left(
\begin{Bmatrix} \epsilon_r \\ \epsilon_z \\ \epsilon_\theta \\ \gamma_{rz} \end{Bmatrix}
-
\begin{Bmatrix} \alpha T \\ \alpha T \\ \alpha T \\ 0 \end{Bmatrix}
\right)
\tag{7.2.2}
$$

where E = elastic modulus, ν = Poisson's ratio, α = thermal expansion coefficient, and T = temperature change. The strain-displacement relation is

$$
\begin{Bmatrix} \epsilon_r \\ \epsilon_z \\ \epsilon_\theta \\ \gamma_{rz} \end{Bmatrix}
=
\begin{bmatrix}
1 & 0 & 0 & 0 & 0 \\
0 & 0 & 0 & 1 & 0 \\
0 & 0 & 0 & 0 & \dfrac{1}{r} \\
0 & 1 & 1 & 0 & 0
\end{bmatrix}
\begin{Bmatrix} u_{,r} \\ u_{,z} \\ v_{,r} \\ v_{,z} \\ u \end{Bmatrix}
\tag{7.2.3}
$$

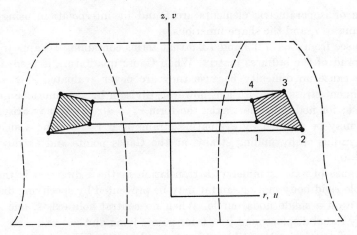

Figure 7.2.1. Axisymmetric element.

which replaces Eq. 5.2.9. For simple element types such as the triangle of Fig. 4.4.1 and rectangle of Fig. 4.5.1, u and v may be written directly in terms of r, z, and nodal d.o.f. For isoparametric elements the relations between displacement quantities in coordinates rz and $\xi\eta$ are also needed and are

$$\begin{Bmatrix} u_{,r} \\ u_{,z} \\ v_{,r} \\ v_{,z} \\ u \end{Bmatrix} = \begin{bmatrix} J_{11}^* & J_{12}^* & 0 & 0 & 0 \\ J_{21}^* & J_{22}^* & 0 & 0 & 0 \\ 0 & 0 & J_{11}^* & J_{12}^* & 0 \\ 0 & 0 & J_{21}^* & J_{22}^* & 0 \\ 0 & 0 & 0 & 0 & 1 \end{bmatrix} \begin{Bmatrix} u_{,\xi} \\ u_{,\eta} \\ v_{,\xi} \\ v_{,\eta} \\ u \end{Bmatrix} \tag{7.2.4}$$

where $[J^*]$ is the inverse of the Jacobian matrix $[J]$.

All ingredients are now at hand and the usual recipe for formulating element properties may be applied, just as was done in Section 5.2. We see that matrix $[B]$ now has four rows as compared with three for the plane problem; however, matrix $[E]$ is now 4 by 4 as well. For a given element type, both plane and axisymmetric elements have the same number and type of nodal d.o.f. Isoparametric elements involve the usual coordinate transformation prior to numerical integration:

$$\int_{\text{vol}} (---)\, r\, d\theta\, dr\, dz = \int_{\text{vol}} (---)\, r\, d\theta \det[J]\, d\xi\, d\eta \tag{7.2.5}$$

where $\det[J]$ is the Jacobian determinant. For convenience we may use only a one-radian segment of the axisymmetric solid; thus all integrands are reduced to $(---)\, r\, dr\, dz$ or $(---)\, r \det[J]\, d\xi\, d\eta$. Values of r at the Gauss points of isoparametric elements are found by interpolation, using nodal coordinates r_i and the shape functions.

We see from Eqs. 7.2.3 and 7.2.5 that terms containing $1/r$ appear in the integrand of the stiffness matrix. When Gauss quadrature is used, the $1/r$ terms cause no difficulty because they are never evaluated at $r = 0$. In subsequent stress computation, however, division by zero must be guarded against; for nodes on the z axis, the form $\epsilon_\theta = u/r = 0/0$ may arise. Difficulty may be avoided by evaluating strains near $r = 0$ using a small but finite radius or by finding strains at the Gauss points and extrapolating to $r = 0$.

Because of axial symmetry, a translation in the z direction is the only possible rigid body motion, and it may be prevented by specifying displacement v on a single nodal circle. When no central hole exists, one should specify radial displacement $u = 0$ for points lying on the z axis.

Use of certain incompatible elements, such as $Q6$ of Eqs. 6.4.4, involves a modest difficulty (7). Radial displacement of the corner nodes produces hoop strain ϵ_θ. As shown in Fig. 7.2.2, resistance to this strain activates the mode $u = (1 - \eta^2)u_6$, which is associated with nonzero shear strain γ_{rz} at all points except those along $\eta = 0$. This shear strain is small in comparison with other strains if $r \gg a$, but it should not be present at all. A treatment that removes the difficulty is that of setting to zero the coefficient in the ϵ_θ row of matrix $[B]$, which multiplies u_6. Thus, the improved bending response of the element remains intact. But this treatment is not completely satisfactory unless all ξ

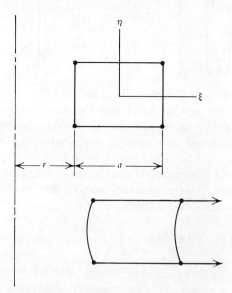

Figure 7.2.2.   Distortion of axisymmetric $Q6$ element.

axes and all element faces $\eta = \pm 1$ are in planes perpendicular to the axis of revolution. We can, of course, apply the treatment in the ϵ_θ row to the u_5 coefficient as well, but now there is an appreciable loss in element quality unless $r \gg a$ (7). Probably the best scheme for general use is to retain all d.o.f. in all strain terms, but in final stress calculation evaluate shear strain γ_{rz} at only the element center $\xi = \eta = 0$.

It may be noted that the unwanted deformation shown in Fig. 7.2.2 may occur in a quadratic (eight-node) element if nodal forces caused by pressure loading are not consistently applied. But fully consistent nodal loads will not prevent $Q6$ elements on the interior of a mesh from behaving as described above. Thus, axisymmetric $Q6$ elements violate the "patch test" of Section 4.6, but only slightly if $r \gg a$.

7.3 Nonsymmetric Loading: Introduction

To introduce the analysis method and show its validity, we begin by presuming that displacements of an arbitrary point in the body are

$$\text{radial disp.} = u = \bar{u} \cos n\theta$$
$$\text{axial disp.} = v = \bar{v} \cos n\theta \qquad (7.3.1)$$
$$\text{circumferential disp.} = w = \bar{w} \sin n\theta$$

where $\bar{u}, \bar{v}, \bar{w}$ are functions of r and z only, and n is an *integer*. Thus for given r and z each displacement forms n complete waves around the circumference. All three displacements must be retained because the problem is fully three dimensional. We now seek the circumstances under which this particular displacement field will prevail.

The strain-displacement relations in cylindrical coordinates are (4):

$$
\begin{Bmatrix} \epsilon_r \\ \epsilon_z \\ \epsilon_\theta \\ \gamma_{rz} \\ \gamma_{z\theta} \\ \gamma_{\theta r} \end{Bmatrix}
=
\begin{bmatrix}
1 & 0 & 0 & 0 & 0 & 0 & 0 & 0 & 0 & 0 & 0 \\
0 & 0 & 0 & 0 & 1 & 0 & 0 & 0 & 0 & 0 & 0 \\
0 & 0 & 0 & 0 & 0 & 0 & 0 & 0 & \dfrac{1}{r} & \dfrac{1}{r} & 0 \\
0 & 1 & 0 & 1 & 0 & 0 & 0 & 0 & 0 & 0 & 0 \\
0 & 0 & 0 & 0 & 0 & \dfrac{1}{r} & 0 & 1 & 0 & 0 & 0 \\
0 & 0 & \dfrac{1}{r} & 0 & 0 & 0 & 1 & 0 & 0 & 0 & -\dfrac{1}{r}
\end{bmatrix}
\begin{Bmatrix} u_{,r} \\ u_{,z} \\ u_{,\theta} \\ v_{,r} \\ v_{,z} \\ v_{,\theta} \\ w_{,r} \\ w, \\ w_{,\theta} \\ u \\ w \end{Bmatrix}
$$

$$(7.3.2)$$

For purposes of explanation we will consider a particular form of stress-strain relation. The most general form is considered in Section 7.5. For the time being let the stress-strain relation have the form:

$$
\begin{Bmatrix} \sigma_r \\ \sigma_z \\ \sigma_\theta \\ \tau_{rz} \\ \tau_{z\theta} \\ \tau_{\theta r} \end{Bmatrix} = \begin{bmatrix} E_{11} & E_{12} & E_{13} & E_{14} & 0 & 0 \\ E_{21} & E_{22} & E_{23} & E_{24} & 0 & 0 \\ E_{31} & E_{32} & E_{33} & E_{34} & 0 & 0 \\ E_{41} & E_{42} & E_{43} & E_{44} & 0 & 0 \\ 0 & 0 & 0 & 0 & E_{55} & E_{56} \\ 0 & 0 & 0 & 0 & E_{65} & E_{66} \end{bmatrix} \begin{Bmatrix} \epsilon_r \\ \epsilon_z \\ \epsilon_\theta \\ \gamma_{rz} \\ \gamma_{z\theta} \\ \gamma_{\theta r} \end{Bmatrix} \tag{7.3.3}
$$

An example of a solid described by this relation is shown in Fig. 7.3.1. Here a prismatic bar having principal directions of orthotropy $x'y'z'$ has been bent into a toroidal shape. Apparently this form of stress-strain relation will suffice for the great majority of problems.

Substitution of the displacements of Eq. 7.3.1 into Eq. 7.3.2 and the resulting strains into Eq. 7.3.3 shows that stresses have the form:

$$
\begin{aligned}
\sigma_r &= \bar{\sigma}_r \cos n\theta & \tau_{rz} &= \bar{\tau}_{rz} \cos n\theta \\
\sigma_z &= \bar{\sigma}_z \cos n\theta & \tau_{z\theta} &= \bar{\tau}_{z\theta} \sin n\theta \\
\sigma_\theta &= \bar{\sigma}_\theta \cos n\theta & \tau_{\theta r} &= \bar{\tau}_{\theta r} \sin n\theta
\end{aligned} \tag{7.3.4}
$$

where the barred quantities are functions of elastic constants, r, z, and n but not θ. The equilibrium equations are (4)

$$
\sigma_{r,r} + \frac{1}{r}\tau_{\theta r,\theta} + \tau_{rz,z} + \frac{\sigma_r - \sigma_\theta}{r} + F_r = 0
$$

$$
\sigma_{z,z} + \tau_{rz,r} + \frac{1}{r}\tau_{z\theta,\theta} + \frac{\tau_{rz}}{r} + F_z = 0 \tag{7.3.5}
$$

$$
\frac{1}{r}\sigma_{\theta,\theta} + \tau_{\theta r,r} + \tau_{z\theta,z} + \frac{2\tau_{\theta r}}{r} + F_\theta = 0
$$

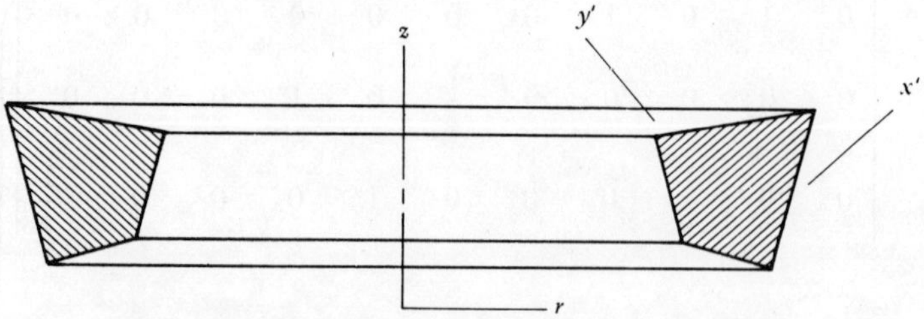

Figure 7.3.1. Orthotropic material properties.

in which F_r, F_z, F_θ are body forces per unit volume in the r, z, θ directions, respectively. Let these forces and temperature change T have the form

$$F_r = \bar{F}_r \cos n\theta \qquad F_\theta = \bar{F}_\theta \sin n\theta$$
$$F_z = \bar{F}_z \cos n\theta \qquad T = \bar{T} \cos n\theta \tag{7.3.6}$$

where the barred quantities are functions of r and z only. With Eq. 7.3.3, this type of temperature field yields initial stresses of the form shown in Eq. 7.3.4 but with $\tau_{z\theta} = \tau_{\theta r} = 0$. Externally applied surface loads in r, z, and θ directions must have the same θ variation as F_r, F_z, and F_θ, respectively.

Upon substitution of Eqs. 7.3.4 and 7.3.6, Eqs. 7.3.5 assume the form

$$(---)\cos n\theta = 0, \qquad (---)\cos n\theta = 0, \qquad (---)\sin n\theta = 0$$
$$\tag{7.3.7}$$

where $(---)$ represents terms containing r, z, and n but independent of θ. These equilibrium equations must be satisfied for all values of θ, hence the expressions $(---)$ must each vanish, thereby producing a set of three partial differential equations with r and z as the independent variables and $\bar{u}$, $\bar{v}$, $\bar{w}$ as the dependent variables. Thus we see that use of Eqs. 7.3.1 and 7.3.3 has rendered the problem two dimensional in a mathematical sense. Note also that loading that has n circumferential waves is associated with displacements and stresses that also have n circumferential waves.

We therefore conclude that under body and surface loading of the form given by Eqs. 7.3.6 and with the stress-strain relation of Eq. 7.3.3, displacements and stresses will be as given by Eqs. 7.3.1 and 7.3.4. Thus the finite element problem may be solved by use of assumed displacement fields in the form of Eqs. 7.3.1. The problem is essentially two dimensional because we must find $\bar{u}$, $\bar{v}$, and $\bar{w}$, which are functions of r and z but not θ. For a given n the solution is unique, so that use of a different n constitutes an entirely separate problem, with different loading and a different solution. In general, the different loadings corresponding to $n = 0, 1, 2, \ldots$ are each components of the Fourier series representation of some specified load. For a linear problem the principle of superposition allows us to obtain the solution for the specified load by simply adding the separate solutions that are obtained from the separate load components. In most practical problems only a few terms of the series are needed.

In the most general circumstances, displacements and the body and surface loads are represented by the series

$$(u, v, F_r, F_z, T) = \sum_0^\infty (---)_n \cos n\theta + \sum_1^\infty (\overline{\overline{---}})_n \sin n\theta$$

$$(w, F_\theta) = \sum_1^\infty (---)_n \sin n\theta + \sum_0^\infty (\overline{\overline{---}})_n \cos n\theta \tag{7.3.8}$$

where $(\overline{---})$ and $(\overline{\overline{---}})$ represent functions of r and z only. For example, in the series for radial displacement u we have $(\overline{---})_n = \bar{u}_n(r, z)$ and $(\overline{\overline{---}})_n = \bar{\bar{u}}_n(r, z)$. Preceding arguments involved only the single barred series, with the explanation based on a single representative harmonic of the series. Clearly the arguments could be repeated with "sin" and "cos" interchanged to justify use of the double barred series. If the loading has symmetry with respect to some rz plane, this plane may be defined as the $\theta = 0$ position, and only the single barred series need be retained (Fig. 7.3.2). The axially symmetric case is represented by use of only the $n = 0$ term of the single barred series. Antisymmetric loading is represented by the double barred series (Fig. 7.3.2). Use of only the $n = 0$ term of the double barred series corresponds to pure torque. Thus we may study shafts of variable diameter under pure torsion—but not with great efficiency, because displacements u and v are everywhere zero (4) but are retained as d.o.f. that must be computed or otherwise processed in a standard computer program. A more compact finite element solution is based on a stress function (11, 12).

Use of only the even terms $n = 0, 2, 4, \ldots$ of the single barred series corresponds to loading and deformation having both $\theta = 0$ and $\theta = 90°$ as planes of symmetry. In such cases the radial and circumferential displacements u and w are zero at $r = 0$ in all harmonics.

The simplest displacement boundary condition, that of zero displacement along a given nodal circle, requires that the displacement coefficients be zero in each harmonic. Nonzero and nonsymmetric displacement conditions may also be used by representing the specified displacements as Fourier series and imposing the appropriate Fourier coefficient as a displacement in each harmonic.

Specified displacements that should be imposed at $r = 0$ are deduced as follows. Equations 7.3.8 are substituted into Eq. 7.3.2. For strains to remain finite at $r = 0$, each coefficient of $1/r$ must vanish. Accordingly, for all z at $r = 0$ (10),

$$\text{for } n = 0 \qquad \bar{u}_0 = \bar{w}_0 = 0 \tag{7.3.9}$$

$$\text{for } n = 1 \qquad \bar{u}_1 = \bar{w}_1$$

$$\bar{\bar{u}}_1 = -\bar{\bar{w}}_1 \tag{7.3.10}$$

$$\bar{v}_1 = \bar{\bar{v}}_1 = 0$$

$$\text{for } n > 1 \qquad \bar{u}_n = \bar{v}_n = \bar{w}_n = 0$$

$$\bar{\bar{u}}_n = \bar{\bar{v}}_n = \bar{\bar{w}}_n = 0 \tag{7.3.11}$$

The conditions of Eqs. 7.3.10 do not eliminate all singular terms for $n = 1$, and a special displacement field is recommended (10). If Eqs. 7.3.9 through 7.3.11 are not enforced, certain terms in the element stiffness matrix are an order of magnitude larger than others. These terms will probably have

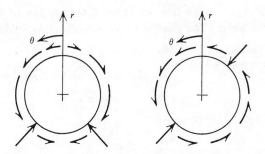

Figure 7.3.2.   Examples of symmetric and antisymmetric loading.

negligible effect in a static analysis but may require an extremely small time
step in a dynamic response analysis (10).

7.4   *Nonsymmetric Loading: Miscellaneous*

To form matrices for finite elements we begin with displacement assumptions $\{f\} = [N_n]\{d\}$, which for harmonic n assumes the form

$$
\begin{Bmatrix} u \\ v \\ w \end{Bmatrix} =
\begin{bmatrix}
N_{u1} \cos n\theta & 0 & 0 & \cdots \\
0 & N_{v1} \cos n\theta & 0 & \cdots \\
0 & 0 & N_{w1} \sin n\theta & \cdots
\end{bmatrix}
\begin{Bmatrix} u_1 \\ v_1 \\ w_1 \\ u_2 \\ \cdot \\ \cdot \\ \cdot \end{Bmatrix}
\tag{7.4.1}
$$

If, for example, the element is of rectangular cross section with four nodal
circles, there are 12 nodal d.o.f. We may use $N_u = N_v = N_w$, where these
shape functions are defined by Eqs. 4.4.6. The same shape functions are used
for all harmonics. In general, if shape functions are written directly in terms
of r, z, and θ, we may establish the relation between displacement derivatives
and nodal d.o.f. as

$$
\{u_{,r} \quad u_{,z} \quad u_{,\theta} \quad v_{,r} \quad \cdots \quad w_{,z} \quad w_{,\theta} \quad u \quad w\} = [Q]\{d\} \tag{7.4.2}
$$

where matrix $[Q]$ contains terms involving r, z, n, $\sin n\theta$ and $\cos n\theta$. Substitution of Eq. 7.4.2 into Fig. 7.3.2 yields the strain-displacement matrix
$[B]$ as a function of the same quantities.

   Continuing the development, we find that $\sin^2 n\theta$ or $\cos^2 n\theta$ appears in
every term of the element stiffness matrix and load vector and must be

integrated from $\theta = -\pi$ to $\theta = \pi$. These integrals are of a well-known type in Fourier analysis. The general forms are stated as follows:

$$\int_{-\pi}^{\pi} \sin m\theta \sin n\theta \, d\theta = \begin{cases} \pi & \text{for } m = n \neq 0 \\ 0 & \text{for } m \neq n \text{ and for } m = n = 0 \end{cases}$$

$$\int_{-\pi}^{\pi} \cos m\theta \cos n\theta \, d\theta = \begin{cases} 2\pi & \text{if } m = n = 0 \\ \pi & \text{if } m = n \neq 0 \\ 0 & \text{if } m \neq n \end{cases} \qquad (7.4.3)$$

$$\int_{-\pi}^{\pi} \sin m\theta \cos n\theta \, d\theta = 0 \quad \text{for all } m \text{ and } n$$

where m and n are integers. Because π or 2π is a common factor in each stiffness equation it may be cancelled out. Integration involving dr and dz is conveniently done by Gauss quadrature. The completed stiffness matrix is found to be a function of n.

If the isoparametric formulation is used, Eq. 7.4.2 is replaced by a similar equation where differentiation is with respect to ξ, η, and θ, as seen in Eq. 5.2.10. Equation 7.4.2 is recovered by use of a transformation relation quite similar to Eq. 7.2.4:

$$\begin{Bmatrix} u_{,r} \\ u_{,z} \\ u_{,\theta} \\ \cdot \\ \cdot \\ \cdot \\ \text{---} \\ u \\ w \end{Bmatrix} = \begin{bmatrix} J_{11}^* & J_{12}^* & 0 & \cdots & & \\ J_{21}^* & J_{22}^* & 0 & \cdots & \bigcirc & \\ 0 & 0 & 1 & \cdots & & \\ \cdot & \cdot & \cdot & & & \\ \cdot & \cdot & \cdot & & & \\ \cdot & \cdot & \cdot & & & \\ \text{---} & \text{---} & \text{---} & & \text{---} & \\ & \bigcirc & & & 1 & 0 \\ & & & & 0 & 1 \end{bmatrix} \begin{Bmatrix} u_{,\xi} \\ u_{,\eta} \\ u_{,\theta} \\ \cdot \\ \cdot \\ \cdot \\ \text{---} \\ u \\ w \end{Bmatrix} \qquad (7.4.4)$$

$$\quad 1 \times 11 \qquad\qquad\qquad 11 \times 11 \qquad\qquad\qquad 1 \times 11$$

As compared with plane analysis, the foregoing analysis is more tedious to carry out, largely because of the many terms involved. However, there are few fundamental differences.

A computer program may treat all harmonics specified by the user in a single computer run. Analysis results for each harmonic may be both printed and stored on tape or disc files. A final subroutine recalls the component analyses and superposes them to form the resultant (total) solution. Values of final displacements and stress may be printed at each of several user-specified values of θ (3).

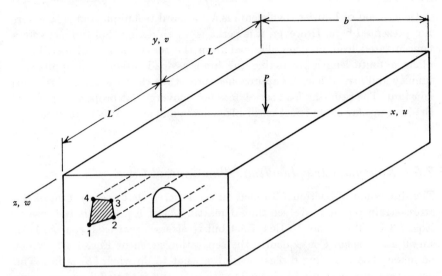

Figure 7.4.1. Prismatic solid.

The Fourier series treatment has also been applied to prismatic solids (6, 9). Such a solid is shown in Fig. 7.4.1, with a typical element. Each element is also a prismatic solid but has the usual plane displacement field in the xy plane. Displacements are expanded as stated in Eq. 7.3.8 except that $\pi z/L$ replaces θ. If only the single barred series are used, we have deformation and loading symmetric about the xy plane, with $w = 0$ at $z = -L$, 0, L. In a physical sense what has been done is to take a toroidal solid extending from $-\pi$ to π and straighten it out to form a prismatic solid extending from $-L$ to L. Again arbitrary loads and displacements may be treated by finding their Fourier coefficients and making a separate analysis for each.

The particular case shown in Fig. 7.4.1 represents a concentrated load acting over a tunnel, with the prismatic solid being a suitably large portion of the surrounding earth or rock. If dimensions L and b are large in comparison with the tunnel dimensions, displacements along the bottom and at $x = 0$, b could be set to zero with little error. Similar problems (6, 9) have been satisfactorily solved by use of 9 to 19 Fourier coefficients. The advantage of this approach over a fully three-dimensional analysis lies in the probable saving of much computer time.

Problems suited for analysis as prismatic solids under Fourier series loading may be solved approximately as solids of revolution under non-symmetric loading. For example, the problem of Fig. 7.4.1 may be approximated as toroidal solid by moving the y axis far to the left and making it an axis of revolution. The geometry is only slightly distorted if dimension b is small in comparison with the radius of the toroid.

Generation of Fourier coefficients is a standard technique and is therefore not presented here. However, it is perhaps worth noting that the series for a concentrated load may be obtained from the series for a load distributed along a finite length by taking the limit as load intensity p approaches infinity and loaded length l approaches zero, while the total force pl remains constant. The resulting load series does not converge but properly represents the load insofar as its effect on displacements and stresses is concerned.

7.5 Nonsymmetric Loading: The General Case

The discussion of Section 7.3 relied on a restricted (but useful) form of the stress-strain relation. We see that if matrix $[E]$ of Eq. 7.3.3 is full, use of Eqs. 7.3.1 will lead not to Eqs. 7.3.4 but to stresses that each contain both $\sin n\theta$ and $\cos n\theta$. Consequently the separation given in Eqs. 7.3.7 cannot be made. The following discussion shows what to do when $[E]$ is full. The derivation presumes use of elements with u, v, and w as nodal displacement d.o.f. but otherwise is general and complements the special case described in Section 7.3.

We begin with the displacements of Eqs. 7.3.8, including all terms. In matrix notation the element displacement field and strain field derived from it are

$$\{f\} = [N]\{d\} = [N_0 \quad \tilde{N}_1 \quad \tilde{\tilde{N}}_1 \quad \tilde{N}_2 \quad \tilde{\tilde{N}}_2 \quad \ldots]\{d_0 \quad d_1 \quad \bar{d}_1 \quad d_2 \quad \bar{d}_2 \quad \ldots\}$$
$$3 \times 1$$
$$(7.5.1)$$

$$\{\epsilon\} = [B]\{d\} = [B_0 \quad \tilde{B}_1 \quad \tilde{\tilde{B}}_1 \quad \tilde{B}_2 \quad \tilde{\tilde{B}}_2 \quad \ldots]\{d_0 \quad d_1 \quad \bar{d}_1 \quad d_2 \quad \bar{d}_2 \quad \ldots\}$$
$$6 \times 1$$
$$(7.5.2)$$

where, if m is the number of element d.o.f., each $[N_n]$ is 3 by m, each $[B_n]$ is 6 by m, and each d_n in the column vector is m by 1. The source of submatrices $[N_n]$ and $[B_n]$ in these equations, from the single or double barred series of Eqs. 7.3.8, is denoted by tildes instead of bars to indicate that these terms contain $\sin n\theta$ and $\cos n\theta$ as well as r and z. Arrays $[N_0]$ and $[B_0]$ describe the zeroth (axisymmetric) mode and contain no trigonometric functions.

For the purpose of explanation let us suppose that matrix $\lfloor E \rfloor$ is decomposed by the Choleski process as described in Section 5.4. Thus the integrand of the element stiffness matrix may be written

$$[B]^T[E][B] = [B]^T[U]^T[U][B] = [b]^T[b] \qquad (7.5.3)$$

If $[E]$ is as given by Eq. 7.3.3, each $[\tilde{b}_n]$ has only $\cos n\theta$ terms in its first four rows and only $\sin n\theta$ terms in its last two rows. Similarly, each $[\tilde{\tilde{b}}_n]$

has only $\sin n\theta$ terms in its first four rows and only $\cos n\theta$ terms in its last two rows. If $[E]$ is full, $\sin n\theta$ and $\cos n\theta$ terms are added in every coefficient of $[\breve{b}_n]$ and $[\tilde{b}_n]$. We also take note of Eqs. 7.4.3. With these observations in mind we examine the integrand $[b]^T[b]$ of the element stiffness matrix:

$$
\begin{bmatrix}
b_0{}^T b_0 & b_0{}^T \breve{b}_1 & b_0{}^T \tilde{b}_1 & b_0{}^T \breve{b}_2 & b_0{}^T \tilde{b}_2 & \cdots \\
 & \breve{b}_1{}^T \breve{b}_1 & \breve{b}_1{}^T \tilde{b}_1 & \breve{b}_1{}^T \breve{b}_2 & \breve{b}_1{}^T \tilde{b}_2 & \cdots \\
 & & \tilde{b}_1{}^T \tilde{b}_1 & \tilde{b}_1{}^T \breve{b}_2 & \tilde{b}_1{}^T \tilde{b}_2 & \cdots \\
 & & & \breve{b}_2{}^T \breve{b}_2 & \breve{b}_2{}^T \tilde{b}_2 & \cdots \\
\text{symmetric} & & & & \tilde{b}_2{}^T \tilde{b}_2 & \cdots
\end{bmatrix}
\qquad (7.5.4)
$$

The diagonal submatrix $[b_0]^T[b_0]$ does not contain θ. In all other submatrices integration with respect to θ is carried out from $-\pi$ to π. Thus, each integral of $[b_c]^T[\breve{b}_n]$ and $[b_0]^T[\tilde{b}_n]$ vanishes for $n > 0$, because we integrate simply $\sin n\theta$ or $\cos n\theta$ in every term. Indeed, *all* terms with unlike subscripts vanish because of Eqs. 7.4.3. In the diagonal submatrices $[\breve{b}_n]^T[\breve{b}_n]$ and $[\tilde{b}_n]^T[\tilde{b}_n]$, $\sin^2 n\theta$ and $\cos^2 n\theta$ terms integrate to π, while any $\sin n\theta \cos n\theta$ terms present if $[E]$ is full integrate to zero according to Eqs. 7.4.3. We are now left with only the $[\breve{b}_n]^T[\breve{b}_n]$ submatrices. If $[E]$ is restricted as in Eqs. 7.3.3, we encounter only $\sin n\theta \cos n\theta$ terms, which vanish upon integration. But if $[E]$ is more general, $\sin^2 n\theta$ and $\cos^2 n\theta$ terms are present and do not vanish when integrated.

Integration with respect to θ therefore leads to an element stiffness matrix of the form

$$
\begin{bmatrix}
k_{00} & 0 & 0 & \cdots \\
0 & k_{11} & 0 & \cdots \\
0 & 0 & k_{22} & \cdots \\
\cdot & \cdot & \cdot & \\
\cdot & \cdot & \cdot & \\
\cdot & \cdot & \cdot &
\end{bmatrix}
\qquad (7.5.5)
$$

where, if m is the number of element d.o.f., $[k_{00}]$ is m by m. Each remaining $[k_{nn}]$ is $2m$ by $2m$. However, if $[E]$ is as given by Eq. 7.3.3, each $[k_{nn}]$ is composed of two m by m matrices on-diagonal and two m by m null matrices off-diagonal. By conceptually expanding element matrices to "structure size" we see that the assembled structural stiffness matrix has the same form as Eq. 7.5.5.

Thus we again conclude that the unsymmetric loading problem uncouples into a series of entirely separate problems, one for each harmonic. There is no coupling between harmonics, but in the general case there is coupling between the single barred and double barred series component of a given harmonic (9 of Chapter Eight).

Treatment of applied loads presents no special problems and is left to the reader as an exercise (Problem 7.7).

Perhaps the most general case of all occurs when loads *and* elastic properties (but not physical shape) vary with θ. This happens when the temperature field is unsymmetric and material properties are temperature dependent. In this case *all* harmonics are coupled, leading to a large set of equations whose length and bandwidth are directly proportional to the number of harmonics used. Details of the analysis are given by Crose (8).

7.6 Pure Moment Loading

Pure moment loading of the type shown in Fig. 7.6.1 would not usually be considered as within the province of axisymmetric finite element analysis, despite the fact that the structure and its stresses are axisymmetric. A way to treat such problems by an axisymmetric analysis (5) may be explained with the aid of Fig. 7.6.2. The structure (curved beam, toroidal shell, etc.) is first shown with a gap in the axisymmetric geometry. Analysis under moments M_o is desired. When these moments are applied the gap closes, moment M_o prevails for all θ, and the reference radius a changes by an amount Δa (here Δa is negative). Under the assumption that plane sections $\theta = $ constant remain plane, the curved beam theory of strength of materials gives

$$d\theta' = \frac{a}{a + \Delta a}\, d\theta \qquad (7.6.1)$$

as the new inclination of cross sections that were previously separated by an angle $d\theta$. Each point in a cross section has a rigid-body radial displacement of amount Δa, and also a radial displacement u because of straining under the stress field produced by moment M_o. Hence the hoop strain at any point is

$$\epsilon_\theta = \frac{ds' - ds}{ds} = \frac{(r + \Delta a + u)\, d\theta' - r\, d\theta}{r\, d\theta} \qquad (7.6.2)$$

With $\Delta a \ll a$, Eqs. 7.6.1 and 7.6.2 yield

$$\epsilon_\theta = \left(\frac{1}{r} - \frac{1}{a}\right)\Delta a + \frac{u}{r} \qquad (7.6.3)$$

M_o M_o *Figure 7.6.1.*   Bending of a plane curved beam.

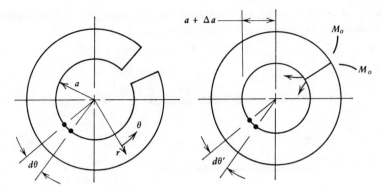

Figure 7.6.2.   Pure moment loading.

Equation 7.6.3 is the usual equation for axisymmetric conditions, namely $\epsilon_\theta = \alpha T + u/r$, except that the thermal strain αT is replaced by $(1/r - 1/a)\,\Delta a$.

The device, then, is to specify a and Δa as input data, and load the structure by using $(1/r - 1/a)\,\Delta a$ as a "thermal" load. Displacements and stresses are calculated in the usual manner. The moment M_o associated with these stresses is computed from the stresses themselves:

$$M_o = \iint \sigma_\theta r\, dr\, dz, \qquad \text{since} \qquad \iint \sigma_\theta\, dr\, dz = 0 \qquad (7.6.4)$$

Satisfaction of the latter integral forms a partial check on solution accuracy. Sample problems (5) show very good agreement with theory.

If the structure is a thin shell whose cross section flattens appreciably under load, the problem is no longer linear.

PROBLEMS

7.1. Revise the computer routines of Figs. 5.4.1 and 5.4.2 to use an axisymmetric isoparametric element of the type
    (a) $Q6$ (see Eqs. 6.4.4).
    (b)   Quadratic (4 corner and 4 side nodes).

7.2. Suppose that a flat, washer-shaped element is to be constructed, for use in plane-stress problems of axisymmetric circular plates under axisymmetric loading. Thus, only radial edge loads, thermal and body forces are permitted. Nodal d.o.f. are u_1 and u_2. Construct the stiffness matrix and nodal load vectors, using a linear displacement assumption. For simplicity set Poisson's ratio $\nu = 0$ and assume isotropy.

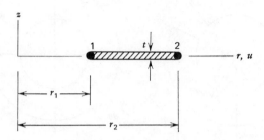

7.3. Repeat Problem 7.2 but with the generalization that there may be both radial and circumferential loads which vary as $\cos n\theta$ and $\sin n\theta$, respectively. Add displacement quantities and nodal d.o.f. as needed, but continue to assume isotropy and $\nu = 0$.

7.4. For an element of rectangular cross section whose displacement function is given by Eqs. 4.5.4, write out matrix $[B]$ and compute nodal loads due to heating a fully restrained element an amount $T = T \cos n\theta$ where T is a constant. Assume isotropy and for convenience set Poisson's ratio $\nu = 0$.

7.5. A solid circular cylinder has a flat end ($z =$ constant). An axisymmetric, z-direction traction p is applied ($p =$ force per unit area). If this traction varies linearly with r as shown in the sketches below, determine the nodal loads for

    (a)   A linear isoparametric element (Fig. 5.2.1).
    (b)   A quadratic isoparametric element (Fig. 5.5.1).

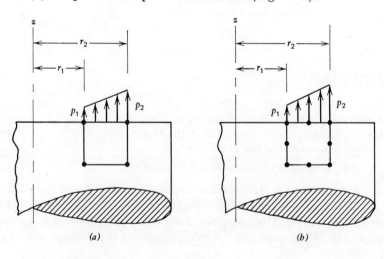

(a)   (b)

7.6. What is the minimum acceptable number of Gauss points that can be used for an axisymmetric element, if the element is:

    (a) A linear isoparametric element (4 nodes).

    (b) A quadratic isoparametric element (8 nodes).

    (c) How are your answers changed if the regular-shaped elements of Problem 7.5 are used?

Suggestion. The Jacobian determinant *and* radius r must be considered (see Eq. 7.2.5).

7.7. Suppose that unsymmetric body forces are given by the series

$$\{F\}_{3\times 1} = \{F_0 + \tilde{F}_1 + \tilde{\tilde{F}}_1 + \tilde{F}_2 + \tilde{\tilde{F}}_2 + \cdots\}$$

where $\sim$ and $\approx$ indicate from which series in Eqs. 7.3.8 the terms originate. Each $\{\tilde{F}_n\}$ and $\{\tilde{\tilde{F}}_n\}$ contains r, z, and $\sin n\theta$ or $\cos n\theta$. Pursue the development of Section 7.5 to discover how the components of $\{F\}$ are allocated in the resulting element nodal load vector.

7.8. Repeat Problem 7.3, assuming that radial and circumferential loads and displacements each contain *both* $\cos n\theta$ and $\sin n\theta$ terms.

7.9. Verify Eqs. 7.3.9, 7.3.10, and 7.3.11.

CHAPTER EIGHT

Thin Shells of Revolution

8.1 Introduction

In this chapter we will consider only "thin" shells, that is, shells for which transverse shear deformation is negligible. Elements that account for transverse shear deformation are considered in Chapter Ten.

Elements for shells of revolution may be singly curved (conical frusta) or doubly curved, as shown in Fig. 8.1.1. There are nodal circles instead of nodal points. Loads may be axisymmetric or asymmetric. Clearly the situation is similar to that of solids of revolution (Chapter Seven), hence the specifics of element formulation will not be presented in great detail.

We will begin by discussing three desirable properties that in large measure determine the quality of a shell element. They are:

- Use of doubly curved elements for doubly curved shells.
- Freedom from strain under rigid body motion.
- Similar competence for all assumed displacement fields.

These properties are also desirable in elements for shells of general shape, but they are more easily discussed in connection with elements for shells of revolution.

8.2 Idealization of Geometry and Displacement

The earliest shell of revolution elements were conical frusta. A hemisphere modeled by such elements is shown in Fig. 8.2.1. Under internal pressure

161

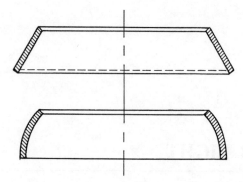

Figure 8.1.1. Axisymmetric shell elements.

only membrane stresses should exist in the actual shell. However, the conical element model displays bending stress as well. This effect is characteristic of shells; bending stresses arise where curvature changes abruptly, because membrane stresses alone cannot provide static equilibrium. Use of a great many conical elements gives good results. But for good results with few elements, there is ample evidence that shells of revolution having double curvature should be modeled by elements having double curvature (1, 2).

The foregoing conclusion is also reached by noting that meridional membrane and bending actions are independent of one another within conical elements. In reality these actions interact with one another throughout a doubly curved shell. Accordingly, elements that model the interaction should be better than singly curved elements, which do not.

We now consider rigid body motion. Figure 8.2.2 shows the meridian of a typical element. Displacements w and u are normal and tangential to the meridian. Angle ϕ denotes the inclination of the meridian with respect to the z axis. Nodal d.o.f. may be taken as displacements w_i and u_i and rotation

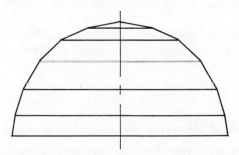

Figure 8.2.1. Use of conical frusta.

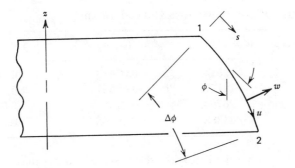

Figure 8.2.2. Element meridian.

$\beta_i = (dw/ds + u\, d\phi/ds)_i$, where $i = 1, 2$ (9). An acceptable assumed displacement field for the axisymmetric problem is

$$w = a_1 + a_2 s + a_3 s^2 + a_4 s^3$$
$$u = a_5 + a_6 s$$
$$(8.2.1)$$

where s is distance along the arc of the meridian. Such an assumption does not provide for rigid body motion if the element meridian is curved. That is, the element is not strain-free if the nodal d.o.f. are assigned values consistent with a z-direction translation. Haisler and Stricklin (3) have shown that the fallacious strain vanishes as the angle $\Delta\phi$ subtended by the meridional arc approaches zero. Thus, rigid body capability is recovered, and increasing mesh refinement produces convergence to correct results.

Although rigid body motion need not be represented explicitly and exactly, various authors (4, 5, 6) have found that implicit introduction of rigid body motion leads to a much improved element. What is done is to retain w as given in Eqs. 8.2.1, but replace u by either

$$u = a_5 + a_6 s + a_7 s(s - L)$$

or

$$u = a_5 + a_6 s + a_7 s(s - L) + a_8 s^2(s - L)$$
$$(8.2.2)$$

where L is the element meridian arc length and a_7 and a_8 are internal d.o.f. to be condensed before assembly of elements. Freedoms a_7 and a_8 assume numerical values that greatly reduce, but do not eliminate, strain under rigid body motion of the nodal d.o.f. The benefit of the internal freedoms apparently far exceeds their computational expense, as good accuracy now becomes possible with a much smaller number of elements. Some numerical evidence is shown in Table 8.2.1, taken from Ref. 4. The element is a curved beam rather than a shell (further remarks about beams appear in Section 17.3). The eigenvalue quoted, which should be zero, is proportional to the

Table 8.2.1.   Eigenvalues of a Curved Beam

Meridian $\Delta\phi$	$u = f(a_5, a_6)$	$u = f(a_5, a_6, a_7)$	$u = f(a_5, a_6, a_7, a_8)$
2°	2013.	0.05688	0.00005
6°	12177.	0.02387	0.00019
12°	26794.	0.21930	0.00618
20°	44388.	2.46182	0.08052

strain energy produced when nodal d.o.f. are assigned values corresponding to rigid body motion. The lowest eigenvalue associated with elastic deformation is on the order of 10^6.

Finally, with regard to displacement assumptions, analyses (7, 13) indicate that for best results the polynomials representing in-plane and normal displacements should be of the same degree, as in Eqs. 8.2.1 and 8.2.2 when $u = f(a_5, a_6, a_7, a_8)$ is used. Limited numerical evidence (8) suggests that doubly curved shells are more successfully modeled by flat elements having displacement fields of equal competence than by doubly curved elements having in-plane displacement fields of appreciably lower degree than the lateral displacement field.

In summary, doubly curved shells will usually be most effectively treated by elements that are curved instead of flat, contain rigid body motion either explicitly or implicitly, and use equal competence of interpolation for all displacement components. These remarks apply to general shells as well as to shells of revolution. However, it must be admitted that doubly curved elements for general shells are often complicated, and that flat elements serve well enough in modeling general shells that they are preferred by some users.

8.3   Strain-Displacement Relations

It is necessary to use shell theory only to the extent of obtaining strain-displacement relations. These we will write without derivation. The necessary notation is identified in Fig. 8.3.1, where it is assumed that loads, displacements, and stresses may lack axial symmetry.

Coordinate s denotes distance along the arc of a meridian. The inclination of the shell midsurface at a point P with respect to the axis of geometric symmetry is ϕ, and r measures the distance between P and the axis. Both ϕ and r are functions of s. Displacements u, v, w form an orthogonal triad with u and w in the s and shell-normal directions, respectively. The conventional notation for shell displacements is used here, despite the fact that displacement w is circumferential in the conventional notation for solids of

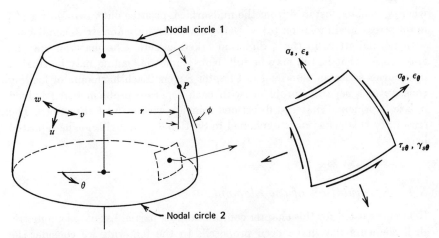

Figure 8.3.1. Shell element geometry and internal forces.

revolution (Chapter Seven). Positive senses for stresses and strains are as shown; these quantities may vary with s, θ, and distance from the shell midsurface.

The shell is assumed to be "thin," so that lines initially normal to the midsurface remain straight and normal during deformation. According to the Novozhilov theory, midsurface strains $(\epsilon)_o$ and curvatures κ are expressed in terms of displacements as follows (9). A comma denotes partial differentiation.

$$(\epsilon_s)_o = u,_s - w\phi,_s$$

$$(\epsilon_\theta)_o = \frac{1}{r}(v,_\theta + u\sin\phi + w\cos\phi)$$

$$(\gamma_{s\theta})_o = \frac{1}{r}(rv,_s - v\sin\phi + u,_\theta)$$

$$\kappa_s = -w,_{ss} - u\phi,_{ss} - u,_s\phi,_s \qquad\qquad (8.3.1)$$

$$\kappa_\theta = -\frac{1}{r}\left[\frac{1}{r}w,_{\theta\theta} - \frac{\cos\phi}{r}v,_\theta + (w,_s + u\phi,_s)\sin\phi\right]$$

$$\kappa_{s\theta} = \frac{2}{r}\left[-w,_{s\theta} + \frac{\sin\phi}{r}w,_\theta + (\cos\phi)v,_s - \frac{\sin\phi\cos\phi}{r}v - \phi,_su,_\theta\right]$$

If meridians are straight, $\phi,_s = 0$. If loads and displacements are axisymmetric, $v = 0$ and all derivatives with respect to θ vanish.

Strains at an arbitrary point in the shell are (9)

$$\{\epsilon\} = \{\epsilon_s \quad \epsilon_\theta \quad \gamma_{s\theta}\} = \{(\epsilon_s)_o \quad (\epsilon_\theta)_o \quad (\gamma_{s\theta})_o\} + \zeta\{\kappa_s \quad \kappa_\theta \quad \kappa_{s\theta}\} \quad (8.3.2)$$

where ζ denotes distance from the midsurface, positive outward. Matrix $[E]$ in the stress-strain relation $\{\sigma\} = [E]\{\epsilon\}$ is written for plane stress conditions, with normal stress in the ζ direction taken as zero. The material may be anisotropic; that is, $[E]$ may be full. Nonsymmetric loading may be treated by Fourier series as described in Chapter Seven. Each harmonic of loading constitutes a separate problem, with each separate problem being halved in size if principal material directions coincide with s and θ directions. The reason for this behavior is explained in Section 7.5 and will not be repeated here.

8.4 Formulation of an Element

References listed for this chapter comprise only a partial list of axisymmetric shell elements that have been proposed. In the following we consider the element described in Refs. 4, 9, 10, and 11. This element appears to be as straightforward and effective as any other. The choice may rest largely on personal preference.

The first task is to represent the element geometry. Formulas are needed because the quantities r, $\sin\phi$, $\cos\phi$, and $\phi_{,s}$ appear in the strain-displacement expressions, Eqs. 8.3.1. Figure 8.4.1 shows the meridian of a typical element.

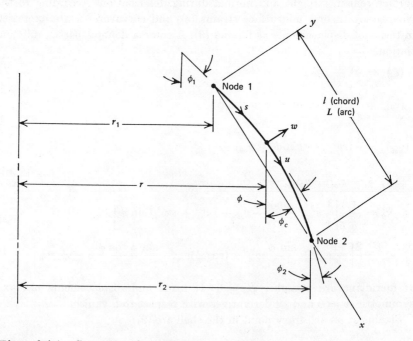

Figure 8.4.1. Geometry of meridian.

As usual the nodal labels 1 and 2 are arbitrary and serve only as convenient labels. The following formulation of meridian geometry is usually adequate, but more exact (and more complicated) expressions have been developed (5, 6). We may write (9, 11)

$$\phi = \alpha_0 + \alpha_1 s + \alpha_2 s^2 \tag{8.4.1}$$

which permits adjacent elements to have the same slope where they meet. Constants α_0, α_1, α_2 are evaluated from the conditions

$$\phi_{s=0} = \phi_1, \qquad \phi_{s=L} = \phi_2,$$
$$\int_0^L \sin(\phi - \phi_c)\, ds \approx \int_0^L (\phi - \phi_c)\, ds = 0 \tag{8.4.2}$$

where the slope $\tan(\phi - \phi_c) = dy/dx$ is presumed small. The results are

$$\alpha_0 = \phi_1$$
$$\alpha_1 = \frac{(6\phi_c - 4\phi_1 - 2\phi_2)}{L} \tag{8.4.3}$$
$$\alpha_2 = \frac{(3\phi_1 + 3\phi_2 - 6\phi_c)}{L^2}$$

A first approximation for arc length L is simply $L = l$, where the chord length l is easily found from the nodal r and z coordinates. A better approximation may be obtained as follows (see also Problem 8.4). In accordance with Eq. 8.4.1, the distance y between the arc and its chord may be taken as a cubic in x (the same function was used for beam deflection, Section 3.9). The cubic that passes through the nodes with appropriate slopes is

$$y = \left[x\left(1 - \frac{x}{l}\right)^2 \quad \frac{x^2}{l}\left(-1 + \frac{x}{l}\right) \right] \begin{Bmatrix} \phi_1 - \phi_c \\ \phi_2 - \phi_c \end{Bmatrix} \tag{8.4.4}$$

Again assuming small slope dy/dx so that $\phi - \phi_c \approx dy/dx$, we write (2 of Chapter Fourteen)

$$L - l = \int_0^l [1 - \cos(\phi - \phi_c)]\, ds \approx \frac{1}{2}\int_0^l \left(\frac{dy}{dx}\right)^2 dx \tag{8.4.5}$$

Hence,

$$L \approx l + \frac{l}{60}\begin{Bmatrix} \phi_1 - \phi_c \\ \phi_2 - \phi_c \end{Bmatrix}^T \begin{bmatrix} 4 & -1 \\ -1 & 4 \end{bmatrix}\begin{Bmatrix} \phi_1 - \phi_c \\ \phi_2 - \phi_c \end{Bmatrix} \tag{8.4.6}$$

Distance r may be written as

$$r = r_1 + \int_0^s \sin\phi\, ds \tag{8.4.7}$$

where ϕ might be approximated as $\phi \approx \phi_c + dy/dx$, followed by numerical integration to evaluate the integral. Thickness variation of the shell may be included, most easily by linear interpolation between given nodal values.

In preceding chapters it has been our custom in deriving an element stiffness matrix to begin with a displacement field expressed in terms of nodal displacements. An alternate approach is to formulate element properties in terms of generalized coordinates, saving the transformation to nodal d.o.f. until last. The latter approach was described near the end of Section 4.2 and has been commonly used with axisymmetric shells. We therefore pursue that approach in the following.

For brevity and simplicity of notation we will write the displacement field for one component of a full Fourier series loading symmetric about the plane $\theta = 0$. The "full problem," involving both symmetric and antisymmetric terms, may be treated as described in Section 7.5. For the displacement components shown in Fig. 8.3.1 we write (4)

$$w = [a_1 + a_2 s + a_3 s^2 + a_4 s^3] \cos n\theta$$

$$u = [a_5 + a_6 s + a_9 s(s - L) + a_{10} s^2(s - L)] \cos n\theta \qquad (8.4.8)$$

$$v = [a_7 + a_8 s + a_{11} s(s - L) + a_{12} s^2(s - L)] \sin n\theta$$

where the a_i are generalized coordinates. Displacements u and v are clearly independent of a_9 through a_{12} at $s = 0$ and at $s = L$. Accordingly, a_9 through a_{12} are freedoms to be condensed before assembly of elements. For axisymmetric loading, $v = 0$, and w and u are independent of θ. Stricklin et al. suggest that use of internal d.o.f. in high harmonics may yield a too-flexible stiffness matrix (see references for Chapter Fourteen).

Equations 8.4.8, and strains derived from them as in Eq. 8.3.2, may be written in matrix notation as

$$\{f\} = \{w \quad u \quad v\} = [N_a]\{a\}$$

$$\{\epsilon\} = \{\epsilon_s \quad \epsilon_\theta \quad \epsilon_{s\theta}\} = [B_a]\{a\} \qquad (8.4.9)$$

where $[N_a]$ and $[B_a]$ are 3 by 12. Note that $[B_a]$ contains the variables ζ, s, r, $\sin \phi$, $\cos \phi$ and in the unsymmetric case also n, $\sin n\theta$, and $\cos n\theta$. The general formulas of Chapter Four still apply and may be used to generate the stiffness matrix and nodal loads due to applied pressure, body force, etc., all referred to generalized coordinates $\{a\}$. The next task is to evaluate the integrals appearing in these expressions. Integration with respect to θ is carried out as described in Chapter Seven. Integration with respect to the shell-normal coordinate ζ may be performed explicitly. Note that for linearly elastic materials, with ζ measured from the neutral surface of bending, integrals of terms containing ζ to the first power are zero. Further details appear in Chapters Nine and Ten, where the same integration is encountered. Integration with respect to s involves r, $\sin \phi$, and $\cos \phi$, all of which are functions of s. This integration may be done numerically within the computer program.

Condensation is now applied to remove the generalized coordinates a_9 through a_{12}. Thus the element stiffness matrix is reduced to size 8 by 8, or to 6 by 6 in the axisymmetric case. Note that this simple condensation is in consequence of the fact that a_9 through a_{12} have no effect on displacements at the nodes. If instead of Eq. 8.4.8 we had written

$$u = [a_5 + a_6 s + a_9 s^2 + a_{10} s^3] \cos n\theta \qquad (8.4.10)$$

and similarly for v, the relation among the a_i is changed and condensation must be done *after* the a_i have been replaced by nodal d.o.f. In this case the d.o.f. to be eliminated could be chosen as u and v at $x = l/3$ and at $x = 2l/3$ (5). The transformation from $\{a\}$ to nodal d.o.f. then involves a 12 by 12 matrix instead of an 8 by 8 matrix.

The final step before assembly of elements is the replacement of generalized coordinates $\{a\}$ by nodal displacements and rotations. The nodal d.o.f. in Fig. 8.4.2a are

$$\{d\} = \{w_1 \quad u_1 \quad v_1 \quad \beta_1 \quad w_2 \quad u_2 \quad v_2 \quad \beta_2\} \qquad (8.4.11)$$

in which $\beta_i = (w_{,s} + u\phi_{,s})_i$ where $i = 1, 2$. Transformation between generalized coordinates $\{a\}$ and the nodal d.o.f. $\{d\}$ of Fig. 8.4.2a is given by

$$\{d\} = [DA]\{a\} \qquad (8.4.12)$$

where $[DA]$ is 6 by 6 in the symmetric case and 8 by 8 in the unsymmetric case. Matrix $[DA]$ is formed by evaluating the bracketed expressions of Eqs. 8.4.8 at the nodal points $s = 0$ and $s = L$, thus automatically excluding a_9 through a_{12}. Thus $[DA]$ contains L and, from the equation for ϕ, also ϕ_1, ϕ_2, and ϕ_c. Conversion of element stiffness and nodal load matrices $[k_a]$ and $\{r_a\}$ from the $\{a\}$-basis to the $\{d\}$-basis is given by

$$[k] = [DA]^{-T}[k_a][DA]^{-1}, \qquad \{r\} = [DA]^{-T}\{r_a\} \qquad (8.4.13)$$

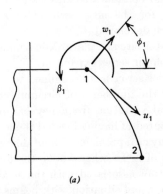
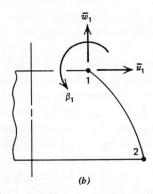

(a)   (b)

Figure 8.4.2. Nodal degrees of freedom.

This transformation is satisfactory if there is no abrupt change in ϕ across nodal circles. If there is an abrupt change in slope as, for example, where a cylinder meets a cone, nodal displacements w and u in the adjacent elements no longer refer to common directions. In such cases the nodal d.o.f. of Fig. 8.4.2b may be used. The new nodal d.o.f. are related to the old by the equations

$$\{\bar{d}_1\} = \{\bar{w}_1 \quad \bar{u}_1 \quad \bar{v}_1 \quad \bar{\beta}_1\} = [\Lambda_1]\{d_1\}$$

$$[\Lambda_1] = \begin{bmatrix} \sin\phi_1 & -\cos\phi_1 & 0 & 0 \\ \cos\phi_1 & \sin\phi_1 & 0 & 0 \\ 0 & 0 & 1 & 0 \\ 0 & 0 & 0 & 1 \end{bmatrix} \tag{8.4.14}$$

Similar expressions apply at node 2. Thus

$$\{\bar{d}\} = [\Lambda]\{d\} = [\Lambda][DA]\{a\} = [T_{DA}]\{a\}$$

$$\underset{8\times 8}{[\Lambda]} = \begin{bmatrix} \Lambda_1 & 0 \\ 0 & \Lambda_2 \end{bmatrix}, \qquad \{d\} = \{d_1 \quad d_2\} \tag{8.4.15}$$

Matrix $[DA]$ in Eq. 8.4.13 is thus replaced by $[T_{DA}]$. The latter transformation is applicable whether or not ϕ changes smoothly with s. It is desirable to obtain an algebraic expression for the inverse of the transformation matrix, so that repeated numerical inversion may be avoided.

8.5   Concluding Remarks

At a pole, $r = 0$, and infinite terms appear in the strain-displacement relations, Eqs. 8.3.1. This was also the case in Chapter Seven when elements touched the z axis. Again we may simply avoid use of $r = 0$. However, accuracy may be improved by use of special boundary conditions for $r = 0$ and special strain-displacement relations for the neighborhood of $r = 0$ (9, 11). In this way special cap elements are developed (6).

If $\phi = 90°$ for all r, we have the case of a circular plate. Nonsymmetric situations may be treated by use of the Fourier series expansion.

It might be mentioned that use of Fourier series is not confined to static problems. For example, one might specify the number n of circumferential waves and then solve for the fundamental vibration frequency or lowest buckling load. In general one cannot be assured of having located the lowest frequency or buckling load for the actual physical problem without doing the eigenvalue analysis for several trial values of n. In other words, the lowest frequency or lowest buckling load is often not associated with the smallest number of waves; this is in contrast to the usual situation with beams and plates.

A program with capability for treatment of nonsymmetric loads on shells of revolution may be used to analyze cylindrical shells that lack axial symmetry and that have discrete axial stiffening members and varying thickness, so long as the cylindrical shell has simply supported ends (12). The technique is similar to that mentioned in the latter part of Section 7.4. What is actually analyzed is a toroidal shell; if its cross-sectional dimensions are small in comparison with its distance from the axis of revolution, a short piece of the toroid behaves almost as if it were straight. Effectively, independent variables are switched; the axial coordinate of the cylinder becomes the circumferential coordinate of the toroid. With many Fourier harmonics in the circumferential direction, the torus is effectively built of many simply supported pieces, each of which models the unsymmetric cylindrical shell in question.

Internal d.o.f. beyond those used in Eqs. 8.4.8 may be added and applied to the shell-normal displacement w as well as to u and v. Element properties are further improved, but at some computational expense. The success of this approach is shown in Ref. 6, with a somewhat different element than described above.

The finite-difference method is well-suited to shells of revolution, and some versatile programs are based on this approach (14). Finite-difference analyses based on an energy formulation should be comprehensible to students of finite elements.

PROBLEMS

8.1. Verify that Eqs. 8.3.1 reduce to those that describe axisymmetric bending and stretching of
    (a)   A flat circular thin plate.
    (b)   A circular cylindrical thin shell.

8.2. (a) Verify Eq. 8.4.3.
      (b) Verify Eq. 8.4.6.

8.3. Let the meridian of an element be a circular arc with $\phi_1 = 60°$ and $\phi_2 = 30°$. Compare the exact arc length L with L as computed by Eq. 8.4.6.

8.4. If the element meridian is taken as a circular arc subtending the angle (in radians) $\Delta\phi = |\phi_1 - \phi_2|$, arc length may be computed as $L = (l/2)(\Delta\phi)/\sin(\Delta\phi/2)$. Compare this length with that computed by Eq. 8.4.6 if
    (a) $\phi_c = 30°$, $\phi_1 = 45°$, $\phi_2 = 25°$.
    (b) $\phi_c = 30°$, $\phi_1 = 36°$, $\phi_2 = 36°$.

CHAPTER NINE

Bending of Flat Plates

9.1 Introduction. Plate Elements in General

A great many plate elements have been devised, and here no attempt is made even to list them all. Those mentioned are illustrative of the displacement models available and also show some difficulties peculiar to plate elements. The latter part of this chapter considers a particular element in some detail. This element is reasonably accurate, includes transverse shear deformation effects, and leads directly to a general shell element discussed in Chapter Ten.

In the "thin plate" problem transverse shear deformation is neglected, and deformation is completely described by a function $w = w(x, y)$, where lateral displacement w and coordinate z are perpendicular to the plate (1). According to Section 4.6, a polynomial for w must contain the constant curvature states $w_{,xx}$ and $w_{,yy}$ and constant twist $w_{,xy}$. Also, it should have no preferred directions and should be compatible with its neighbors. A single function w having all these properties is not easy to devise, especially for triangular and general quadrilateral elements. In surmounting the difficulties, investigators have produced many elements, some of them quite complicated.

An early yet quite successful element is based on the 12-term polynomial (1 of Chapter Four)

$$w = a_1 + a_2 x + a_3 y + a_4 x^2 + a_5 xy + a_6 y^2 + a_7 x^3$$
$$+ a_8 x^2 y + a_9 xy^2 + a_{10} y^3 + a_{11} x^3 y + a_{12} xy^3 \quad (9.1.1)$$

which is an incomplete quartic because the terms x^4, $x^2 y^2$, and y^4 are missing. The element is restricted to rectangular shape and has three d.o.f. at each corner i, namely displacement w_i and rotations $(w_{,x})_i$ and $(w_{,y})_i$. Element sides are parallel to x and y coordinate directions. Reference 3 gives numerical examples of the behavior of this and several other elements and states element stiffness matrices explicitly. Convergence of the 12-term element is not monotonic; thus, a mesh of these elements may be too stiff in some problems and too flexible in others. This happens because normal slopes—for example, $w_{,x}$ along an edge $x =$ constant—are not compatible between elements. The reader may demonstrate the incompatibility as an exercise (Problem 4.16). It should be mentioned that another 12 d.o.f. rectangular element was previously derived by the same author, using physical arguments instead of a variational principle (2). This element also does not provide monotonic convergence but is very effective (3).

Attempts to achieve interelement compatibility in thin plate elements encounter a fundamental difficulty (4). Let A and B represent two adjacent edges of a *compatible* element. Twist $(w_{,xy})_A$ along edge A is uniquely determined by the nodal d.o.f. on edge A. A similar statement applies to twist $(w_{,xy})_B$ along edge B. Hence at the corner common to edges A and B we will not in general have $(w_{,xy})_A = (w_{,xy})_B$. The conclusion is that twist $w_{,xy}$ cannot be uniquely defined at corners of compatible thin-plate elements if nodal d.o.f. consist only of displacements and rotations.

The "uniqueness-compatibility" difficulty may be overcome by use of more d.o.f. per node. For example, Ref. 5 adds twist $w_{,xy}$ to the list of nodal d.o.f. to produce a 16 d.o.f. compatible rectangular element, which has excellent convergence properties. Other examples include triangular elements, based on a complete fifth degree polynomial for w, and having curvatures $w_{,xx}$ and $w_{,yy}$ as nodal d.o.f. as well as w, $w_{,x}$, $w_{,y}$, and $w_{,xy}$ (6, 7). Such "higher order" elements are usually quite accurate, but have two significant disadvantages. First, where adjacent elements have different thicknesses or different elastic properties, or where stiffeners are attached, curvatures and twist change abruptly across interelement boundaries. Therefore, continuity of these higher order nodal d.o.f. must not be enforced; instead, the extra freedoms should be viewed as internal d.o.f. and should be removed by condensation before assembly of elements. Second, boundary conditions involving the higher order d.o.f. may not be obvious and are often cumbersome as, for example, along curved boundaries, in shell problems where in-plane displacements and their derivatives may also be used as nodal freedoms, and if adjacent elements are not coplanar. In summary, we may conclude that higher order d.o.f. make the finite element method awkward in application to problems for which it is usually best suited—to problems involving thickness changes, stiffeners, mixtures of different element types, and members meeting at angles.

Abel and Desai (21) have compared several plate elements on the basis of accuracy versus computational expense. It is worth noting that when judged in this way, the higher order elements mentioned above may be surpassed by simpler elements.

If the effects of transverse shear deformation are to be included, it is important that the nature of nodal rotation d.o.f. be recognized. Let θ_x and

θ_y be the rotation components, about x and y axes, respectively, of a line that was perpendicular to the midsurface of the undeformed plate. Unless transverse shear deformation is taken as zero, θ_x and θ_y are *not* equal in magnitude to the surface slopes $w_{,y}$ and $w_{,x}$ (Fig. 9.1.1). Indeed, surface slopes may be discontinuous across interelement boundaries, as when adjacent elements have different shear moduli. Accordingly, it is the rotations θ_x and θ_y of the midsurface normal that must be matched between elements and used as nodal d.o.f.

Thin-plate elements can be constructed by suppressing transverse shear deformation. For example, a "discrete Kirchhoff" element (8) is generated by use of *independent* polynomials for w, θ_x, and θ_y, and by suppressing shear deformation at certain points. Thus in Fig. 9.1.1 we set $\theta_y = -w_{,x}$ and $\theta_x = w_{,y}$ at each node, thereby coupling the polynomials and producing a thin-plate element.

Other elements worth noting include the "finite strip" elements. References 9 to 12 are sample references. These elements are best suited to flat plates and folded plates where two opposite edges are simply supported, although other support conditions can also be treated. In the simply supported case the deflection of a strip is given by

$$w = \sum_{n=1,2,3,\ldots} f(x) \sin \frac{n\pi y}{L} \qquad (9.1.2)$$

where L = longitudinal span between simple supports and $f(x)$ = polynomial beam function in the transverse direction. Equation 9.1.2 automatically gives zero deflection and zero moment at the simple supports. If $f(x)$ is cubic, element freedoms consist of two displacements and two slopes that define $f(x)$ for a given n. The load is broken into its Fourier series components, and a separate analysis is carried out for each component. Thus, instead of using general plate elements and many d.o.f. in a single solution, we may use finite strips, solve several problems with few d.o.f. in each, and superpose results. Clearly the method resembles the Fourier series treatment of solids of revolution (Chapter Seven).

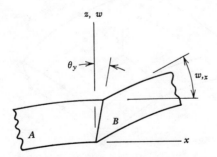

Figure 9.1.1. Shear deformation in adjacent elements.

Finally it should be mentioned that not all good elements are based on assumed displacement fields. In particular, elements formulated as assumed-stress hybrids are among the best plate elements (see Section 17.1).

9.2   A Plate Element. The Displacement Field

The element to be described has been presented in Refs. 13 through 15. In brief, the approach is as follows. We note that all physical problems exist in three-dimensional space; it is only because of special circumstances that such idealizations as plane stress and plate theory provide good approximations of reality. Accordingly, we begin with a finite element for elastic solids and impose certain simplifications and specializations so as to transform it into an effective and efficient plate element. Less extensive modification of the solid element leads to a shell element; however, for convenience of explanation, we will postpone consideration of general shells until Chapter Ten.

We begin with the 20-node isoparametric solid element, shown in Fig. 5.5.3b and again in Fig. 9.2.1. This element could be used as is to model plates or shells, but there are two reasons for not doing so. First, if the element is made quite thin relative to its span, the stiffness in the thickness direction becomes much higher than other element stiffness coefficients, thus producing ill-conditioned equations and inaccurate results. Second, the element is uneconomical because the large number of d.o.f. along lines normal to the midsurface is not necessary. These two drawbacks are overcome by specializing the element so that strain energy of stresses normal to the midsurface is ignored and by constraining lines initially normal to the midsurface to remain straight so that fewer d.o.f. are needed to define the displacement field. Note that the midsurface normals are not to be kept normal during

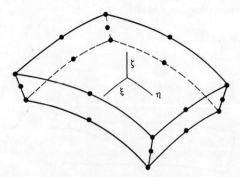

Figure 9.2.1.   Sixty d.o.f. isoparametric solid element.

deformation; they may make angles of other than 90 degrees with the *deformed* midsurface. In this way the ability to model transverse shear deformation is retained.

When used to model a flat plate, the element shape is defined by nodal values of x and y coordinates and by thicknesses at the nodes. Thus, as regards shape, the element is identical to the plane element of Fig. 5.5.1, and its shape functions are Eqs. 5.5.3.

With regard to deformations, we will use as nodal d.o.f. the displacement w and rotations θ_x and θ_y of thickness-direction lines at each of the eight nodes of Fig. 9.2.2, thus obtaining a 24 d.o.f. element (Fig. 9.2.3a). This step is justified as follows. In consequence of this assumption that midsurface-normals remain straight, displacements u and v in x and y directions are completely specified by rotations θ_x and θ_y, as shown in Fig. 9.2.3b (plate bending and not stretching is considered here, so that $u = v = 0$ on the midsurface). And, because stress $\sigma_z = 0$ and strain ϵ_z is assumed to make no contribution to strain energy, more than one node in the thickness direction is superfluous, as the extra nodes would serve only to define ϵ_z. Accordingly, the assumed displacement field for the eight-node plate element is

$$\begin{Bmatrix} u \\ v \\ w \end{Bmatrix} = \sum_{i=1}^{8} \begin{bmatrix} 0 & zN_i & 0 \\ 0 & 0 & -zN_i \\ N_i & 0 & 0 \end{bmatrix} \begin{Bmatrix} w_i \\ \theta_{yi} \\ \theta_{xi} \end{Bmatrix} \tag{9.2.1}$$

where shape functions N_i are again given by Eqs. 5.5.3. Coordinate z originates at the midsurface. There is no need for isoparametric coordinate ζ, because z and ζ have identical directions.

Use of the foregoing displacement field preserves interelement compatibility. The element might be classed as "superparametric" (Section 5.6)

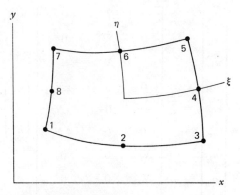

Figure 9.2.2. Plan view of plate element.

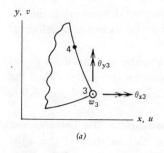

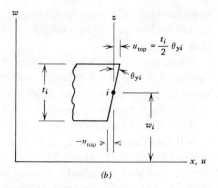

Figure 9.2.3. (a) Typical nodal d.o.f. (b) Displacements at arbitrary node.

because in the normal direction the thickness may vary but nodal d.o.f do not define thickness changes arising from strain. However, the element remains valid because transverse normal strain plays no role in the formulation of element properties.

To construct the strain-displacement matrix $[B]$ we proceed as in Section 5.2. The essential relations are as follows.

$$
\begin{Bmatrix} \epsilon_x \\ \epsilon_y \\ \gamma_{xy} \\ \gamma_{yz} \\ \gamma_{zx} \end{Bmatrix} =
\begin{bmatrix}
1 & 0 & 0 & 0 & 0 & 0 & 0 & 0 & 0 \\
0 & 0 & 0 & 0 & 1 & 0 & 0 & 0 & 0 \\
0 & 1 & 0 & 1 & 0 & 0 & 0 & 0 & 0 \\
0 & 0 & 0 & 0 & 0 & 1 & 0 & 1 & 0 \\
0 & 0 & 1 & 0 & 0 & 0 & 1 & 0 & 0
\end{bmatrix}
\begin{Bmatrix} u_{,x} \\ u_{,y} \\ u_{,z} \\ v_{,x} \\ v_{,y} \\ v_{,z} \\ w_{,x} \\ w_{,y} \\ w_{,z} \end{Bmatrix}
\tag{9.2.2}
$$

$$
\begin{Bmatrix} u_{,\xi} \\ u_{,\eta} \\ u_{,z} \\ v_{,\xi} \\ v_{,\eta} \\ v_{,z} \\ w_{,\xi} \\ w_{,\eta} \\ w_{,z} \end{Bmatrix} = \sum_{i=1}^{8}
\begin{bmatrix}
0 & zN_{i,\xi} & 0 \\
0 & zN_{i,\eta} & 0 \\
0 & N_i & 0 \\
0 & 0 & -zN_{i,\xi} \\
0 & 0 & -zN_{i,\eta} \\
0 & 0 & -N_i \\
N_{i,\xi} & 0 & 0 \\
N_{i,\eta} & 0 & 0 \\
0 & 0 & 0
\end{bmatrix}
\begin{Bmatrix} w_i \\ \theta_{yi} \\ \theta_{xi} \end{Bmatrix}
\tag{9.2.3}
$$

$$\{u_{,x} \quad u_{,y} \quad \cdots \quad w_{,z}\}_{9\times1} = \begin{bmatrix} J^* & 0 & 0 \\ 0 & J^* & 0 \\ 0 & 0 & J^* \end{bmatrix}_{9\times9} \{u_{,\xi} \quad u_{,\eta} \quad \cdots \quad w_{,z}\}_{9\times1} \tag{9.2.4}$$

where $[J^*]$ is the inverse of the Jacobian matrix $[J]$, and

$$[J] = \begin{bmatrix} x_{,\xi} & y_{,\xi} & 0 \\ x_{,\eta} & y_{,\eta} & 0 \\ 0 & 0 & 1 \end{bmatrix} \tag{9.2.5}$$

We introduce the notation

$$a_i = J_{11}^* N_{i,\xi} + J_{12}^* N_{i,\eta}$$
$$b_i = J_{21}^* N_{i,\xi} + J_{22}^* N_{i,\eta} \tag{9.2.6}$$

Hence, Eqs. 9.2.2 through 9.2.6 combine to yield $\{\epsilon\} = [B]\{d\}$:

$$\begin{Bmatrix} \epsilon_x \\ \epsilon_y \\ \gamma_{xy} \\ \gamma_{yz} \\ \gamma_{zx} \end{Bmatrix} = \sum_{i=1}^{8} \left(\begin{bmatrix} 0 & 0 & 0 \\ 0 & 0 & 0 \\ 0 & 0 & 0 \\ b_i & 0 & -N_i \\ a_i & N_i & 0 \end{bmatrix} + \begin{bmatrix} 0 & za_i & 0 \\ 0 & 0 & -zb_i \\ 0 & zb_i & -za_i \\ 0 & 0 & 0 \\ 0 & 0 & 0 \end{bmatrix} \right) \begin{Bmatrix} w_i \\ \theta_{yi} \\ \theta_{xi} \end{Bmatrix} \tag{9.2.7}$$

The complete matrix $[B]$ is 5 by 24. The reason for arbitrarily splitting $[B]$ so as to separate terms linear in z will be seen in the next section.

9.3 Elastic Properties and Integration Through the Thickness

The stress-strain relation $\{\sigma\} = [E](\{\epsilon\} - \{\epsilon_o\})$ may be written in the form (1)

$$\begin{Bmatrix} \sigma_x \\ \sigma_y \\ \tau_{xy} \\ \tau_{yz} \\ \tau_{zx} \end{Bmatrix} = \begin{bmatrix} E_x' & E'' & 0 & 0 & 0 \\ E'' & E_y' & 0 & 0 & 0 \\ 0 & 0 & G & 0 & 0 \\ 0 & 0 & 0 & G_{yz} & 0 \\ 0 & 0 & 0 & 0 & G_{zx} \end{bmatrix} \left(\begin{Bmatrix} \epsilon_x \\ \epsilon_y \\ \gamma_{xy} \\ \gamma_{yz} \\ \gamma_{zx} \end{Bmatrix} - \begin{Bmatrix} \alpha_x T \\ \alpha_y T \\ 0 \\ 0 \\ 0 \end{Bmatrix} \right) \tag{9.3.1}$$

Where x and y axes are presumed to coincide with principal directions of an orthotropic material. Initial strains are presumed due to heating an amount T, where α_x and α_y are principal coefficients of thermal expansion. In the case of isotropy $G_{zx} = G_{yz}$ and (1)

$$E'_x = E'_y = \frac{E''}{\nu} = \frac{E}{1 - \nu^2}, \qquad G = \frac{E}{2(1 + \nu)} \tag{9.3.2}$$

where $E =$ elastic modulus and $\nu =$ Poisson's ratio.

Equation 9.3.1 is more specialized than necessary. Actually, axes x and y need not coincide with principal material directions. However, it is assumed that whatever the nature of the material, in-plane stresses remain independent of γ'_{yz} and γ'_{zx}; that is, $E_{ij} = E_{ji} = 0$ for $i = 1, 2, 3$ and $j = 4, 5$. Thus $[E]$ consists of what might be termed a 3 by 3 "in-plane" matrix and a 2 by 2 "transverse shear" matrix.

As suggested by Eq. 9.2.7, the *entire* matrix $[B]$ may be split into a part $[B_o]$ independent of z and a part $z[B_1]$ linear in z. The element stiffness matrix is therefore given by

$$[k] = \int_{\text{vol}} [B]^T[E][B] \, dx \, dy \, dz = \int_{\text{vol}} [B_o + zB_1]^T[E][B_o + zB_1] \, dx \, dy \, dz \tag{9.3.3}$$

Integration with respect to z is between the limits $\pm t/2$. Because of the distribution of zeros in $[B_o]$, $[B_1]$, and $[E]$, products $z[B_o]^T[E][B_1]$ and $z[B_1]^T[E][B_o]$ vanish, $[B_o]^T[E][B_o]$ involves only the "transverse shear" portion of $[E]$, and $z^2[B_1]^T[E][B_1]$ involves only the "in-plane" portion of $[E]$ with z^2 in every term. The net result of these circumstances and integration through the thickness is that the element stiffness matrix of Eq. 9.3.3 becomes

$$[k] = \int_{\text{area}} [\bar{B}]^T[D][\bar{B}] \, dx \, dy = \int_{\text{area}} [\bar{B}]^T[D][\bar{B}] \det [J] \, d\xi \, d\eta \tag{9.3.4}$$

where

$$[\bar{B}] = \begin{bmatrix} 0 & a_1 & 0 & \dots & 0 & a_8 & 0 \\ 0 & 0 & -b_1 & \dots & 0 & 0 & -b_8 \\ 0 & b_1 & -a_1 & \dots & 0 & b_8 & -a_8 \\ b_1 & 0 & -N_1 & \dots & b_8 & 0 & -N_8 \\ a_1 & N_1 & 0 & \dots & a_8 & N_8 & 0 \end{bmatrix} \tag{9.3.5}$$

and, using $[E]$ as given by Eq. 9.3.1, the matrix $[D]$ of flexural rigidities is (1)

$$D_{11} = \frac{E'_x t^3}{12} \qquad D_{22} = \frac{E'_y t^3}{12}$$

$$D_{12} = D_{21} = \frac{E'' t^3}{12} \qquad D_{33} = \frac{G t^3}{12} \tag{9.3.6}$$

$$D_{44} = G_{yz} t \qquad D_{55} = G_{zx} t$$

In performing the integration of Eq. 9.3.4 one must recognize that $[D]$ may be a function of x and y. This may happen because of moduli being functions of x and y or because of thickness changes. It is in the calculation of $[D]$ that the "thickness component" of element volume is accounted for.

Use of $D_{44} = G_{yz}t/1.2$ and $D_{55} = G_{zx}t/1.2$ has been suggested in order to account for the fact that transverse shear stress varies quadratically with z (13). Reference 16 finds that the denominator should be 1.0 instead of 1.2 to yield answers for thick plates in agreement with the Reissner theory (1).

Note that the shear stiffness terms D_{44} and D_{55} are independent of the remaining flexural stiffnesses in $[D]$. Accordingly, our analysis need not be restricted to homogeneous plates; layered plates may also be treated if the stiffnesses are available from calculations or physical tests. For example, an isotropic sandwich plate (17) has stiffnesses

$$D_{11} = D_{22} = \frac{D_{12}}{\nu} = \frac{D_{21}}{\nu} = \frac{Eh(c+h)^2}{2(1-\nu^2)}$$

$$D_{33} = \frac{Eh(c+h)^2}{4(1+\nu)} \tag{9.3.7}$$

$$D_{44} = D_{55} = \frac{G(c+h)^2}{c}$$

where E, ν = elastic modulus and Poisson's ratio of facings, h = thickness of each facing, c = core thickness, and G = shear modulus of the core. The validity of this formulation has been confirmed by computational test (16, 18).

Because the last two rows of $\{\epsilon_o\}$ in Eq. 9.3.1 are zero, portion $[B_o]$ of matrix $[B]$ has no connection with initial strains. Let us assume that $T = 2zT_o/t$, where T_o is the temperature of the upper surface $z = t/2$. Use of the usual equation for calculation of nodal loads due to $\{\epsilon_o\}$ leads to the same conversion seen in going from Eq. 9.3.3 to 9.3.4, and for the same reasons. We obtain

$$\{r\} = \int_{\text{vol}} [B]^T[E]\{\epsilon_o\}\, dx\, dy\, dz = \int_{\text{area}} [\bar{B}]^T[D]\{\bar{\epsilon}_o\}\, dx\, dy \tag{9.3.8}$$

where $[\bar{B}]$ is given by Eq. 9.3.5, $[D]$ by Eq. 9.3.6 or 9.3.7, and

$$\{\bar{\epsilon}_o\} = \frac{2T_o}{t}\{\alpha_x \quad \alpha_y \quad 0 \quad 0 \quad 0\} \tag{9.3.9}$$

The third term of $\{\bar{\epsilon}_o\}$ need not be zero if x and y are not principal axes. If $[D]$ contains stiffness coefficients of a sandwich plate, use of $[D]\{\bar{\epsilon}_o\}$ in Eq. 9.3.8 still yields the correct thermal moments, provided that t represents the effective thickness $c + h$.

When all nodal d.o.f. have been determined we may evaluate strains and stresses by use of Eqs. 9.2.7 and 9.3.1. Alternatively, curvatures, moments, and shear forces may be determined as follows. In thin-plate theory (1), strains and curvatures are related by the equations

$$\epsilon_x = -zw_{,xx}, \qquad \epsilon_y = -zw_{,yy}, \qquad \epsilon_{xy} = -2zw_{,xy} \qquad (9.3.10)$$

where $w_{,xx}$ and $w_{,yy}$ are positive when the center of curvature is on the $+z$ side. A study of how these relations are derived (1) shows that $w_{,x}$ and $w_{,y}$ represent rotations of a midsurface-normal line, even when these rotations may differ from slopes of the plate surface because of transverse shear deformation. Accordingly, substitution of Eqs. 9.3.11 below makes use of the proper rotation quantities even in the presence of transverse shear deformation. Therefore, from Eq. 9.2.7, the present plate element predicts the "effective" curvatures

$$w_{,xx} = -\sum_{i=1}^{8} a_i \theta_{yi}, \qquad w_{,yy} = \sum_{i=1}^{8} b_i \theta_{xi}$$

$$2w_{,xy} = \sum_{i=1}^{8} (-b_i \theta_{yi} + a_i \theta_{xi})$$

$$(9.3.11)$$

Stresses from Eq. 9.3.1 may be substituted into the usual expressions (1):

$$M_x = \int_{-t/2}^{t/2} \sigma_x z \, dz, \qquad M_{xy} = \int_{-t/2}^{t/2} \tau_{xy} z \, dz, \qquad Q_x = \int_{-t/2}^{t/2} \tau_{zx} \, dz, \quad \text{etc.} \quad (9.3.12)$$

Hence, for the case $T = 2zT_o/t$, we obtain from Eqs. 9.3.1, 9.3.6, 9.3.10, and 9.3.12,

$$M_x = -\left[D_{11}\left(w_{,xx} + \frac{2\alpha_x T}{t}\right) + D_{12}\left(w_{,yy} + \frac{2\alpha_y T}{t}\right) \right]$$

$$M_y = -\left[D_{21}\left(w_{,xx} + \frac{2\alpha_x T}{t}\right) + D_{22}\left(w_{,yy} + \frac{2\alpha_y T}{t}\right) \right]$$

$$M_{xy} = -2D_{33}w_{,xy}$$

$$(9.3.13)$$

$$Q_x = D_{44}\gamma_{zx}, \qquad Q_y = D_{55}\gamma_{yz}$$

As noted below Eqs. 9.3.10, these equations are valid in the presence of transverse shear deformation if $w_{,xx}$, $w_{,yy}$, and $w_{,xy}$ are obtained from Eqs. 9.3.11.

In Eqs. 9.3.13 coordinates xy are taken as principal material directions. Again Eqs. 9.3.7 may be used for stiffnesses in matrix $[D]$ if a sandwich plate is being treated, but as in $\{\bar{\epsilon}_o\}$, t must be regarded as the effective thickness $c + h$. Moments and forces are per unit length in the xy plane. The sign convention differs from that in Ref. 1 and is shown in Fig. 9.3.1.

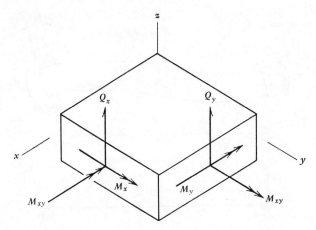

Figure 9.3.1. Bending moments and shear forces.

9.4 Location of Gauss Quadrature Points

If the eight-node element is to be used only for thick plates, the order of
Gauss quadrature is not critical. But if four Gauss points are used, element
behavior is improved; indeed, as elements become thinner, convergence
toward correct results will happen *only* when four Gauss points are used (14).
The reason for this behavior (15) will now be explained with reference to a
uniform beam. The conclusions remain qualitatively valid for plate elements
of varying thickness and arbitrary shape, as has been shown by the good
behavior of such elements.

Let a uniform beam be subjected to end rotations consistent with linearly
varying bending moment. Rotations of the ends, originally normal to the
midsurface, consist of a portion θ_B because of bending and a portion θ_S
because of the constant shear force, as shown in Fig. 9.4.1.

Let us now subject the plate element to nodal displacements that correspond to displacements of the beam element. The shape functions permit
rotation θ to vary only quadratically with the edge coordinate s. Displacement w vanishes, as it also is quadratic in s and must vanish at $s = -a, 0, a$.
The deformed shape and variation of θ with s are shown in Fig. 9.4.2, which
is an edge view of a plate element.

The trouble with the finite element is that because $w = 0$ for all s in Fig.
9.4.2, rotation θ is interpreted as shear strain all along the length. As θ is
much greater than the shear strain θ_S that should exist (Fig. 9.4.1), a great
excess of shear strain energy is stored by the element. The element is stiffened
to the extent that very poor results are obtained if $t \ll a$. The remedy is to
base numerical integration on sampling points that are located at places

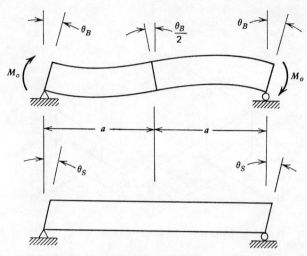

Figure 9.4.1. Beam deformation under linearly varying moment, shown separated into bending and pure shear components.

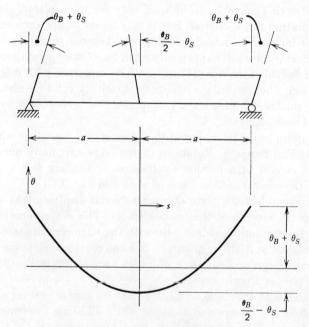

Figure 9.4.2. Plate element deformation under linearly varying moment.

where shear strain is correctly evaluated. Rotation in Fig. 9.4.2 is

$$\theta = -\left(\frac{\theta_B}{2} - \theta_S\right) + \frac{3\theta_B}{2}\left(\frac{s}{a}\right)^2 \tag{9.4.1}$$

Setting $\theta = \theta_S$ we obtain

$$\frac{s}{a} = \pm\frac{1}{\sqrt{3}} \tag{9.4.2}$$

which are precisely the Gauss point locations of a 2 by 2 quadrature rule.

During final strain computation, in-plane strains ϵ_x, ϵ_y, and γ_{xy} may be evaluated at any point in the element. But transverse strains γ_{yz} and γ_{zx} are greatly in error except at the four Gauss points. The fact that γ_{yz} and γ_{zx} are practically zero at the Gauss points of a very thin element raises the possibility of reducing the number of d.o.f. from 24 to 16 by explicitly enforcing this condition (see Problem 10.9).

The 2 by 2 quadrature rule introduces the possibility of zero-energy modes; rotations θ_x and θ_y may have nodal values such that the four Gauss points remain unstrained. These zero-energy modes are of the same type as shown in Fig. 5.7.1; to see this we need only replace u by $z\theta_y$ and v by $-z\theta_x$. Therefore, as argued in Section 5.7, no difficulty will arise if more than one element is used.

As noted in Section 5.7, the 2 by 2 quadrature rule is of too low an order to integrate all stiffness terms exactly even if the element is rectangular. Accordingly, we cannot say in advance whether displacements in a given problem will be overestimated or underestimated. Indeed, it appears that a 2 by 2 quadrature rule may not even evaluate element volume exactly. To reach this conclusion, consider how element volume $dV = \det[J]\, d\xi\, d\eta\, dz$ varies with an isoparametric coordinate, say ξ. For the most general shape of element, the ξ-, η-, and z-parallel dimensions of dV are respectively functions of ξ, ξ^2, and ξ^2. Accordingly, dV is a function of ξ^5, and two integration points in the ξ direction are not sufficient. But judging by the success of the element in practice, there is no cause for alarm. In practice, as a mesh is refined, elements are usually made simpler in shape, and errors in volume calculation tend to vanish. The 2 by 2 rule is, of course, adequate for elements of constant thickness.

9.5 Internal D.O.F. and Numerical Examples

Let us consider the merits of adding three internal d.o.f. to the eight-node plate element. Thus, summations in Eqs. 9.2.1, 9.2.3, and 9.2.7 run from 1 to 9 and three columns are added to matrix $[B]$ of Eq. 9.3.5. Shape function N_9, which appears in these equations and in a_9 and b_9 of Eq. 9.2.6, is taken as $N_9 = (1 - \xi^2)(1 - \eta^2)$. No change is made in the Jacobian $[J]$, as element shape is still defined by the eight nodal values of x and y. As N_9 vanishes at $\xi, \eta = \pm 1$, interelement compatibility is unaffected. The added d.o.f. are $\{w_9 \quad \theta_{y9} \quad \theta_{x9}\}$, which may be viewed as displacement and rotations at $\xi = \eta = 0$ *relative* to the values of w, θ_y, and θ_x at $\xi = \eta = 0$ dictated by d.o.f. of the eight boundary nodes.

The three added d.o.f. are condensed prior to assembly of elements. Condensation alters element stiffness and also has a beneficial effect in distributing loads to the eight boundary nodes. For example, under lateral pressure the nodal force associated with w_9 causes moments to appear at boundary nodes. In contrast, forces but not moments appear when only the basic d.o.f. are used.

Let us consider use of these internal d.o.f. with a 2 by 2 Gauss quadrature rule. There may be a *zero-energy mode* (see Sections 5.7 and 6.3) in which in-plane displacement u (or v) for $z \neq 0$, in a square element two units on each side, is proportional to (19)

$$1 - 3x^2 - 3y^2 + 9x^2y^2 \qquad (9.5.1)$$

An element deformed in this way remains compatible with surrounding elements having the same deformation. Therefore the *assembly* of elements has a zero-energy mode, and the structure stiffness matrix may be singular if the displacement boundary conditions happen not to restrain this particular mode. Singularity may be suppressed by a device described below; however, numerical evidence indicates that use of w_9 *only* is preferable to use of all three internal freedoms (numerical examples follow).

Even when using only w_9 as an added d.o.f., the zero-energy mode of w proportional to

$$1 + x^2 + y^2 - 3x^2y^2 \qquad (9.5.2)$$

may arise under 2 by 2 Gauss quadrature. Again, surrounding elements may have the same mode, rendering the structure stiffness matrix singular. This spurious mode is superposed on other element deformations. According to Eqs. 9.3.11 and 9.3.13 it has no effect on computed moments (unless it happens to be so large as to overwhelm the significant data and cause numerical error in the computed nodal rotations). Moreover, the mode will arise only if displacement w is unrestricted at *all* side nodes of the mesh—a circumstance not likely to be met in practice (20).

Fortunately, unwanted modes can be kept under control by the following trick. Let k_c be the diagonal term of the uncondensed element stiffness matrix associated with an internal d.o.f., which leads to singularity (in the present case k_c is the 25th diagonal term, associated with w_9). Singularity under *any* support condition is avoided by the simple device of multiplying k_c by $(1 + e)$, where e is a small number. Physically this amounts to adding a soft spring of stiffness ek_c that resists the relative displacement. The ability of the element to assume modes dictated by the basic boundary d.o.f. is not impaired. Legitimate activity of the internal freedom is only slightly inhibited by the soft spring. The element may be "tuned" by selecting the value of e that is most generally beneficial. If e is made large the element is reduced to the basic one, which has no internal d.o.f. (however, use of large internal stiffnesses is not recommended; see Section 16.5).

This "e device" fully suppresses a spurious mode if the loading has no tendency to create it. But in general the unwanted modes are only under control, not eliminated. Certain loadings and support conditions—not present in the problems of Table 9.5.1—may activate the unwanted modes to a considerable degree. In such cases stress computation must be done at the four Gauss points, where the extraneous strains are zero.

It is worth mentioning that an attempt was made to control spurious modes by combining a fraction of the uncondensed element stiffness matrix with the condensed matrix. Computed results for stress and deflection were greatly in error.

Figure 9.5.1 shows quadrants of plates used as test cases. Symmetry exists about x and y axes. Numerical results appear in Table 9.5.1. Span to thickness ratios for the clamped plates are from 320 to 400. Along supported edges of the sandwich plate, transverse shear deformation parallel to the edge was prohibited. Transverse shear deformation accounts for more than

Table 9.5.1.   Data from Test Cases, Eight-Node Isoparametric Plate Element. Moments Calculated Directly at Nodes, Not Extrapolated from Gauss Points

Quantity	Number of Internal d.o.f. Used Per Element				
	None	$1(w)$	$2(\theta_x, \theta_y)$	$3(w, \theta_x, \theta_y)$	Theory
Clamped circular orthotropic plate, uniform load					
w at A	0.5237	0.6821	0.6713	0.6796	0.678
M_x at A	0.38	1.29	1.15	1.18	1.12
M_y at A	11.44	15.28	13.78	14.60	13.85
Clamped square isotropic plate, uniform load					
w at A	0.0840	0.2422	0.2377	0.2441	0.2405
M_x at A	2.975	6.318	5.272	6.486	5.91
M_y at B	-7.665	-10.77	-10.78	-11.35	-13.12
Simply supported sandwich plate, aspect ratio = 2, uniform load					
w at A	0.6737	0.6741	0.6737	0.6741	0.674
M_x at A	10.584	10.581	10.515	10.527	10.71
M_y at A	4.647	4.640	4.652	4.648	4.64
M_y at B	0.771	0.770	0.860	0.861	0.00

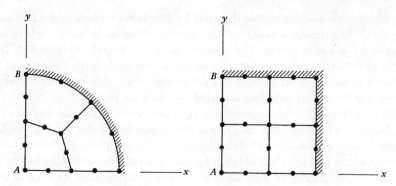

Figure 9.5.1.   Test cases.

half the deflection of the sandwich plate. Theoretical results are taken from Refs. 1 and 17. If the "*e* device" is introduced, multiplying $k_{25,25}$ by 1.004 before condensation in the above clamped square plate case produces the results

$$w \text{ at } A = 0.2233$$
$$M_x \text{ at } A = 5.919 \tag{9.5.3}$$
$$M_y \text{ at } B = -10.40$$

The clamped-plate examples of Table 9.5.1 are particularly favorable to the use of an internal freedom. In most problems the improvement is more modest, and in some problems an internal d.o.f. is slightly detrimental. On balance, however, the internal d.o.f. seems well worth the slight additional effort it requires.

PROBLEMS

9.1.   Suppose that lateral displacement w of a triangular thin-plate element is taken as a complete fifth-degree polynomial (21 terms). Without deriving element matrices, devise an arrangement of nodes and nodal d.o.f. that should be acceptable.

9.2.   Verify Eq. 9.2.7.

9.3.   Verify that Eq. 9.3.4 follows from Eq. 9.3.3 according to the arguments given.

9.4.   Let constant z-direction pressure p act on the surface of the element described in Sections 9.2 and 9.3. If the element is rectangular, evaluate the nodal loads which result.

9.5. Verify Eq. 9.3.8 and verify that either of Eqs. 9.3.6 or 9.3.7 can be used for matrix $[D]$.

9.6. Verify Eqs. 9.3.13.

9.7. (a) For an eight-node rectangular plate element, determine the displacement field resulting from imposition of zero displacement w and nonzero rotation $\theta_B + \theta_S$ along edges $s = -a$ and $s = a$ in Fig. 9.4.2.

(b) Hence, if w and θ are assumed independent of the in-plane coordinate perpendicular to s, determine from the stationary value of total potential Π_p the resulting displacement and rotation at $s = 0$. Use exact integration instead of Gauss quadrature in forming Π_p.

9.8. Let internal freedom $(1 - \xi^2)(1 - \eta^2)w_9$ be added to the eight-node plate element. Hence, repeat Problem 9.4, but do not carry out the condensation.

9.9. (a) Let u be given by Eq. 9.5.1. Show that no strain exists at the Gauss points of a 2 by 2 quadrature rule. Evaluate the nodal displacements of a square element two units on a side and sketch the deformed element.

(b) Hence, what is the magnitude of internal d.o.f. θ_{y9} in the displacement mode $(1 - \xi^2)(1 - \eta^2)\theta_{y9}$?

9.10. (a, b) Repeat Problem 9.9 with reference to Eq. 9.5.2 and internal d.o.f. w_9.

(c) Explain why the arguments regarding zero-energy modes remain valid if Eqs. 9.5.1 and 9.5.2 are interchanged.

9.11. Suppose that stresses in the eight-node plate element have been computed at the Gauss points of a 2 by 2 quadrature rule. Describe in detail (e.g., by writing computer program statements) how to extrapolate these stresses to the eight boundary nodes. Base the argument on the linear isoparametric function $(1 + \xi\xi_i)(1 + \eta\eta_i)$.

CHAPTER TEN

General Shells

10.1   Introduction. Shell Elements in General

Some desirable properties of shell elements were listed in Section 8.1 and expanded on in Section 8.2. Many remarks in Section 9.1 about difficulties in plate elements also apply to shells, since bending action occurs in both types of elements. If shell elements are made flat, they should be usable for plate bending problems under transverse loading or for plane-stress problems under in-plane loading. From all this we conclude that good shell elements are among the most difficult elements to devise because of the considerable demands for correctness of response and versatility of behavior.

In latter sections of this chapter a specific general shell element is considered in some detail. We begin with some general comments regarding the application of finite elements to general shells.

Flat elements, such as plane triangles, were used in early efforts to model shells (1). This approach has the merit of simplicity; plane stress and plate bending element matrices are expanded in size to accommodate the total number of shell element d.o.f. and are then directly added. The element thus formed is properly oriented in space by applying a coordinate transformation to its stiffness matrix and load vector. The element is capable of rigid body motion without strain. This capability, which is lacking in some doubly curved elements, tends to permit good accuracy without having to use a great many elements. However, the opposite tendency is present because the element is flat. Nevertheless, this approach works well enough that flat elements are still preferred by some users.

Let n represent a direction perpendicular to the actual shell. In a model composed of flat elements, n at a node represents a direction that is generally not normal to any of the surrounding elements. To facilitate transformation of flat elements to global coordinates, the rotation θ_n about direction n is

191

introduced as a nodal d.o.f., thus increasing the freedoms per node from (say) 5 to 6. This expansion is accomplished merely by adding to the element stiffness matrix a row and a column of zeros corresponding to each θ_n to be added. The assembly of elements does not have a singular structure stiffness matrix, provided that adjacent elements at a node are not all coplanar. If a θ_n d.o.f. is then eliminated (not condensed, but discarded) from a nonsingular set of structure stiffness equations, the constraint $\theta_n = 0$ is imposed.

Freedom θ_n does not appear as a variable in shell theory. Also, on physical grounds, θ_n might be expected to be negligibly small in comparison with the other nodal rotations. This appears to be the case; whether the θ_n's are carried or eliminated from the list of structural d.o.f., the computed solution to a practical shell problem remains substantially the same (2). Even a cantilever shell, some six times longer than its depth and built of flat elements, was little affected (2); here one might have expected that restraints $\theta_n = 0$ would have a significant effect. Appreciable effect would occur where, for some reason, a rigid body rotation of the shell is permitted, or possibly where the shell is extremely flexible. To eliminate a freedom θ_n from the list of structural d.o.f. we may use one of the schemes described in Section 2.7, thus imposing the boundary condition $\theta_n = 0$. Alternatively, θ_n's could be retained, but at some computational expense in solving equations. Specifically, with 6 d.o.f. per node, the cost is increased by a factor of $(6/5)^3 = 1.728$ (2). If θ_n is kept at a node where surrounding elements are coplanar, the structure stiffness matrix will be singular unless θ_n is restrained by, say, a soft torsional spring connected to ground. In curved elements where θ_n is not present in the original element nor introduced for coordinate transformation, the absence of θ_n in the structural equations does not imply restraint of rotations about a normal to the shell.

A possible incompatibility between shell elements is easily described with reference to flat elements. If in-plane and lateral displacement polynomials are not well matched, we may have, for example, linear in-plane displacement normal to an element edge while the same edge has cubic lateral displacement. Such edge displacements match those of an adjacent element only if the two elements are coplanar. Any discrepancy tends toward zero as the mesh is refined and adjacent elements become more nearly coplanar. The same discrepancy may appear when certain doubly curved elements are used and a "kink" in the actual shell must be modeled.

Membrane and bending strains are coupled in shell action but are independent of one another *within* flat elements. The necessary coupling for shell action arises because flat elements are inclined to one another where they meet. Use of shallow shell theory makes elements curved and introduces coupling within elements but avoids the intricacies of differential geometry that prevail in general shell theory. Representative shallow shell elements appear in Refs. 3 and 4. It might be argued that even in a deep shell each individual element is likely to be small enough to be adequately treated as shallow. Test cases in the literature show good accuracy for deep shells modeled by elements based on shallow shell theory.

Elements specially devised for cylindrical shells need not be shallow but are often restricted to circular cylinders, 90-degree corner angles and must have their two straight sides parallel to the axis of revolution (5, 6). Reference 7 begins with assumptions regarding the strain field instead of the displacement field and obtains a very effective element for thin circular cylindrical shells. The element has 5 d.o.f. at each of its four corners. This element satisfies rigid body motion and constant-strain criteria and succeeds in problems where the more complicated elements of Refs. 5 and 6 fail. However, it is doubtful that the derivation method could be adapted to elements of general shape and curvature.

Cantin (8) has shown that some theories of cylindrical shells are defective in that they produce strain when rigid body motion is imposed. More generally, there are many different shell theories, each incorporating approximations in an attempt to represent reality without great complexity, and it is by no means obvious which one the analyst should choose as the basis for deriving a shell element. However, if because of the shell theory used, an element happens not to be strain-free under rigid body motion, this defect can be corrected (9; 13 of Chapter Eight).

Use of general shell theory to derive doubly curved elements (10) involves the geometric complexities that make shell theory difficult. Fortunately we may avoid these difficulties and obtain a good doubly curved element by appropriately specializing a three-dimensional solid element. The same solid element used in Chapter Nine is used in the following to yield an eight-node general shell element having 40 d.o.f. References for the theory are 10 of Chapter Four, and 13 through 15 of Chapter Nine.

10.2 A General Shell Element. Shape and Displacement Fields

The 60 d.o.f. solid element is shown again in Fig. 10.2.1. Boundaries of the element are ξ, η, $\zeta = \pm 1$. In contrast to the plate element of Section 9.2, directions z and ζ do not coincide, so that here ζ must be retained. In subsequent expressions, subscripts p and q are appended to *nodal* quantities to identify, respectively, the bottom $\zeta = -1$ and the top $\zeta = 1$ of the element. No subscript on nodal quantities indicates reference to the midsurface $\zeta = 0$.

If lines joining nodes p and q are constrained to be straight, the shape of

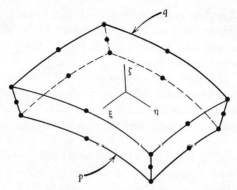

Figure 10.2.1. Sixty d.o.f. isoparametric solid element.

the element may be defined as

$$\begin{Bmatrix} x \\ y \\ z \end{Bmatrix} = \sum_{i=1}^{8} N_i \frac{1+\zeta}{2} \begin{Bmatrix} x_{iq} \\ y_{iq} \\ z_{iq} \end{Bmatrix} + \sum_{i=1}^{8} N_i \frac{1-\zeta}{2} \begin{Bmatrix} x_{ip} \\ y_{ip} \\ z_{ip} \end{Bmatrix} \qquad (10.2.1)$$

where x_{iq}, x_{ip}, etc., are global cartesian coordinates of the 16 nodes on $\zeta = -1$ and $\zeta = 1$, and N_i are shape functions of the eight-node element (Eqs. 5.5.3). Equation 10.2.1 locates coordinates xyz on $\zeta = 0$ as the average of coordinates xyz on $\zeta = -1$ and $\zeta = +1$, thus removing the need for three nodes along midsurface-normal edges. We define

$$2\{x_i\, y_i\, z_i\} = \{x_{iq}\, y_{iq}\, z_{iq}\} + \{x_{ip}\, y_{ip}\, z_{ip}\}$$
$$\bar{V}_{3i} = \{x_{iq}\, y_{iq}\, z_{iq}\} - \{x_{ip}\, y_{ip}\, z_{ip}\} \qquad (10.2.2)$$

Substitution of Eqs. 10.2.2 into Eq. 10.2.1 defines element geometry in terms of midsurface nodal coordinates and vectors $\bar{V}_{3i}$:

$$\{x\, y\, z\} = \sum_{i=1}^{8} N_i \{x_i\, y_i\, z_i\} + \sum_{i=1}^{8} N_i \frac{\zeta}{2} \bar{V}_{3i} \qquad (10.2.3)$$

Arguments given in Section 9.2 for reducing the number of d.o.f. from the 60 present in the solid element apply with equal force in the present situation. Accordingly, as was done with the eight-node plate element, strain normal to the midsurface is neglected, and lines originally normal to the midsurface are constrained to remain straight during deformation. But these lines need not remain normal to the midsurface, hence the element is able to represent transverse shear deformation. Element displacements are completely defined by midsurface nodal displacements u_i, v_i, w_i in the x, y, z global coordinate directions, and by rotations α_i and β_i of a line originally normal to the mid-surface about two different axes parallel to the midsurface (Fig. 10.2.2). In Fig. 10.2.2 vectors $\bar{V}_{1i}$, $\bar{V}_{2i}$, $\bar{V}_{3i}$ are mutually perpendicular, and rotation

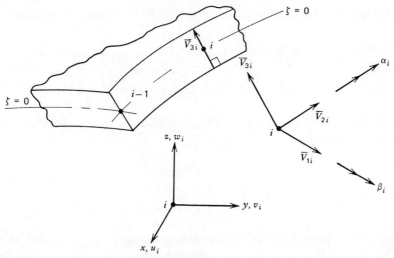

Figure 10.2.2. Nodal degrees of freedom.

vectors β_i, α_i are collinear with $\overline{V}_{1i}$ and $\overline{V}_{2i}$, respectively. It is therefore conceivable that in the assembled structure, no two nodal rotation vectors will have the same direction. Vector $\overline{V}_{3i}$ is defined by input data, and is presumed to span the thickness and be normal to the midsurface. Vector $\overline{V}_{1i}$ is of arbitrary length and might also be defined by input data so that it coincides with a principal direction of an orthotropic material. Or, $\overline{V}_{1i}$ could be defined by the program itself as the cross product $\overline{V}_{1i} = \overline{V}_{3i} \times \hat{\imath}$ or $\overline{V}_{1i} = \overline{V}_{3i} \times \hat{\jmath}$, where $\hat{\imath}$ and $\hat{\jmath}$ are unit vectors in x and y directions. Vector-handling Fortran routines are available if needed (11).

Because of rotations α_i and β_i, in-plane displacements of a point lying on a vector $\overline{V}_{3i}$ are

$$u'_i = \frac{\zeta t_i \alpha_i}{2}, \qquad v'_i = -\frac{\zeta t_i \beta_i}{2} \qquad (10.2.4)$$

where t_i is the shell thickness at a node i. These displacements are in directions $\overline{V}_{1i}$ and $\overline{V}_{2i}$, respectively. Their x, y, z components are

$$\{u \ \ v \ \ w\}_i = \{l_{1i} \ \ m_{1i} \ \ n_{1i}\}u'_i \quad \text{from } u'_i$$

$$\{u \ \ v \ \ w\}_i = \{l_{2i} \ \ m_{2i} \ \ n_{2i}\}v'_i \quad \text{from } v'_i \qquad (10.2.5)$$

where l, m, n represent direction cosines; specifically,

$$\{l_{1i} \ \ m_{1i} \ \ n_{1i}\} = \frac{\overline{V}_{1i}}{|\overline{V}_{1i}|}$$

$$\{l_{2i} \ \ m_{2i} \ \ n_{2i}\} = \frac{\overline{V}_{2i}}{|\overline{V}_{2i}|} \qquad (10.2.6)$$

The complete 40 d.o.f. element displacement field may now be written as

$$\begin{Bmatrix} u \\ v \\ w \end{Bmatrix} = \sum_{i=1}^{8} N_i \begin{Bmatrix} u_i \\ v_i \\ w_i \end{Bmatrix} + \sum_{i=1}^{8} N_i \zeta [\mu_i] \frac{t_i}{2} \begin{Bmatrix} \alpha_i \\ \beta_i \end{Bmatrix} \qquad (10.2.7)$$

where N_i are the shape functions of Eqs. 5.5.3 and $[\mu_i]$ is a matrix of direction cosines at node i:

$$[\mu_i] = \begin{bmatrix} l_{1i} & -l_{2i} \\ m_{1i} & -m_{2i} \\ n_{1i} & -n_{2i} \end{bmatrix} \qquad (10.2.8)$$

In writing Eqs. 10.2.4 we implied that each vector $\bar{V}_{3i}$ is normal to the surface $\zeta = 0$. It is up to the user to ensure that such is the case, at least to a good approximation. The approximation will be poorest where original element shapes are greatly distorted. For example, we might specify nodal quantities that define a cubic shape, such as in the edge view in Fig. 10.2.3a. As the shape functions contain only quadratic terms, the best the eight-node element can do is to yield the linear shape in Fig. 10.2.3b, where *none* of the vectors $\bar{V}_3$ is normal to the midsurface.

The steps in setting up the integrand $[B]^T[E][B]$ of the element stiffness matrix are substantially the same as those given in Section 9.2, so we will not write them out in full detail. The usual strain-displacement relations

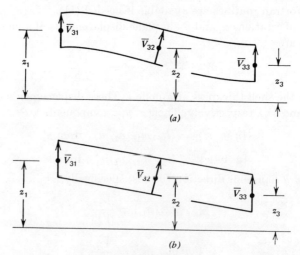

Figure 10.2.3. Desired and actual element shapes.

$\epsilon_x = u_{,x}, \ldots, \gamma_{zx} = u_{,z} + w_{,x}$ are written in matrix form as

$$\{\epsilon_x \quad \epsilon_y \quad \epsilon_z \quad \gamma_{xy} \quad \gamma_{yz} \quad \gamma_{zx}\} = [SD]\{u_{,x} \quad u_{,y} \quad \ldots \quad w_{,z}\} \quad (10.2.9)$$

where $[SD]$ is a 6 by 9 matrix containing 1's and 0's. Displacement derivatives in xyz and $\xi\eta\zeta$ coordinate systems are related by Jacobian matrices:

$$\{u_{,x} \quad u_{,y} \quad u_{,z} \quad \ldots \quad w_{,z}\} = [JA]\{u_{,\xi} \quad u_{,\eta} \quad u_{,\zeta} \quad \ldots \quad w_{,\zeta}\}$$
$$(10.2.10)$$

where $[JA]$ is 9 by 9, composed of matrices $[J]^{-1}$ on-diagonal and zeros elsewhere. Matrix $[J]^{-1}$ is the inverse of the 3 by 3 Jacobian matrix $[J]$. Thus far the manipulations are identical to those outlined for three-dimensional elements in the latter part of Section 5.5. From Eq. 10.2.7 we obtain

$$\begin{Bmatrix} u_{,\xi} \\ u_{,\eta} \\ u_{,\zeta} \\ \cdot \\ \cdot \\ \cdot \\ w_{,\zeta} \end{Bmatrix} = \sum_{i=1}^{8} \begin{bmatrix} N_{i,\xi} & 0 & 0 \\ N_{i,\eta} & 0 & 0 \\ 0 & 0 & 0 \\ \cdot & \cdot & \cdot \\ \cdot & \cdot & \cdot \\ \cdot & \cdot & \cdot \\ 0 & 0 & 0 \end{bmatrix} \begin{Bmatrix} u_i \\ v_i \\ w_i \end{Bmatrix} + \sum_{i=1}^{8} \begin{bmatrix} \zeta N_{i,\xi} & 0 & 0 \\ \zeta N_{i,\eta} & 0 & 0 \\ N_i & 0 & 0 \\ \cdot & \cdot & \cdot \\ \cdot & \cdot & \cdot \\ \cdot & \cdot & \cdot \\ 0 & 0 & N_i \end{bmatrix} [\mu_i] \frac{t_i}{2} \begin{Bmatrix} \alpha_i \\ \beta_i \end{Bmatrix} \quad (10.2.11)$$

By substitution of Eq. 10.2.11 into Eq. 10.2.10 and the result into Eq. 10.2.9 we obtain the strain versus nodal displacement relation

$$\{\epsilon_x \quad \epsilon_y \quad \epsilon_z \quad \gamma_{xy} \quad \gamma_{yz} \quad \gamma_{zx}\} = \sum_{i=1}^{8} [B_i]\{u_i \quad v_i \quad w_i \quad \alpha_i \quad \beta_i\} \quad (10.2.12)$$

Matrix $[B]$ is 6 by 40 and is built of eight 6 by 5 blocks. A typical block $[B_i]$ is of the form

$$[SD]_{6\times9} \begin{bmatrix} (J_{11}^*N_{i,\xi} + J_{12}^*N_{i,\eta}) & 0 & 0 & [(\quad)\zeta + J_{13}^*N_i]\mu_{i11}\dfrac{t_i}{2} & [\quad]\mu_{i12}\dfrac{t_i}{2} \\ (J_{21}^*N_{i,\xi} + J_{22}^*N_{i,\eta}) & 0 & 0 & [(\quad)\zeta + J_{23}^*N_i]\mu_{i11}\dfrac{t_i}{2} & [\quad]\mu_{i12}\dfrac{t_i}{2} \\ (J_{31}^*N_{i,\xi} + J_{32}^*N_{i,\eta}) & 0 & 0 & [(\quad)\zeta + J_{33}^*N_i]\mu_{i11}\dfrac{t_i}{2} & [\quad]\mu_{i12}\dfrac{t_i}{2} \\ \cdot & \cdot & \cdot & \cdot & \cdot \\ \cdot & \cdot & \cdot & \cdot & \cdot \\ \cdot & \cdot & \cdot & \cdot & \cdot \\ 0 & 0 & (\quad) & [(\quad)\zeta + J_{33}^*N_i]\mu_{i31}\dfrac{t_i}{2} & [\quad]\mu_{i32}\dfrac{t_i}{2} \end{bmatrix}_{9\times5}$$
$$(10.2.13)$$

where $[SD]$ defined by Eq. 10.2.9 and $[J^*] = [J]^{-1}$.

It is left to the reader to fill in the omitted terms as an exercise.

10.3   Elastic Properties and Integration

The element stiffness matrix is given by (16)

$$[k] = \int_{-1}^{1} \int_{-1}^{1} \int_{-1}^{1} [B]^T[E][B] \det [J] \, d\xi \, d\eta \, d\zeta \qquad (10.3.1)$$

$$\underset{40 \times 6 \quad 6 \times 6 \quad 6 \times 40}{}$$

Matrix $[E]$ is *not* simply the stress-strain matrix used for three-dimensional solids. We must ensure that $[E]$ provides for zero stress normal to the shell. Let coordinates x', y', z' have the same directions as $\overline{V}_1$, $\overline{V}_2$, $\overline{V}_3$ so that at each point of the shell z' is normal to the midsurface. Taking, for example, an isotropic material, the stress-strain relation $\{\sigma'\} = [E']\{\epsilon'\}$ in $x'y'z'$ coordinates is

$$\begin{Bmatrix} \sigma_{x'} \\ \sigma_{y'} \\ \sigma_{z'} \\ \tau_{x'y'} \\ \tau_{y'z'} \\ \tau_{z'x'} \end{Bmatrix} = \frac{E}{1-\nu^2} \begin{bmatrix} 1 & \nu & 0 & 0 & 0 & 0 \\ \nu & 1 & 0 & 0 & 0 & 0 \\ 0 & 0 & 0 & 0 & 0 & 0 \\ 0 & 0 & 0 & \dfrac{1-\nu}{2} & 0 & 0 \\ 0 & 0 & 0 & 0 & \dfrac{1-\nu}{2} & 0 \\ 0 & 0 & 0 & 0 & 0 & \dfrac{1-\nu}{2} \end{bmatrix} \begin{Bmatrix} \epsilon_{x'} \\ \epsilon_{y'} \\ \epsilon_{z'} \\ \gamma_{x'y'} \\ \gamma_{y'z'} \\ \gamma_{z'x'} \end{Bmatrix} \qquad (10.3.2)$$

where E = elastic modulus and ν = Poisson's ratio. This form of $[E']$ provides for $\sigma_{z'} = 0$ and plane-stress conditions in the $x'y'$ plane. Coordinate transformation is applied to convert $[E']$ to matrix $[E]$ in Eq. 10.3.1, this step being required because matrix $[B]$ yields strains $\{\epsilon\}$ in xyz coordinates rather than strains $\{\epsilon'\}$ in $x'y'z'$ coordinates.[1] Most generally, a separate transformation is required at each Gauss quadrature point, as $x'y'z'$ directions change from one point to another. Accordingly, direction cosines are needed at each Gauss point. These may be established by first finding vectors $\overline{V}_1$, $\overline{V}_2$, $\overline{V}_3$ at the Gauss points by shape-function interpolation from nodal values, then extracting the components of unit vectors $\hat{V}_1$, $\hat{V}_2$, $\hat{V}_3$, as in Eqs. 10.2.6. In other words, the components of vectors at the Gauss points are found by interpolation, then the vectors are normalized, thus yielding their direction cosines.

When the Jacobian is computed by use of Eq. 10.2.3, it is found that ζ

[1] References 13 and 14 of Chapter Nine elect to transform matrix $[B]$ instead, thus replacing $[B]^T[E][B]$ in Eq. 10.3.1 by $[B']^T[E'][B']$. See also Problem 10.10.

to the first power appears in certain terms. Reference 14 of Chapter Nine suggests that these terms may be neglected in comparison with terms to which they are added, provided the thickness to curvature ratio of the shell is small. This approximation implies that derivatives of x, y, and z with respect to ξ and η are substantially the same at either end of a midsurface-normal line. Thus $[J]$ becomes independent of ζ and explicit integration through the thickness is possible, as outlined in the following paragraphs. If ζ terms are retained in $[J]$, two Gauss points in the thickness direction should be adequate; however, the cost of numerical integration is then doubled.

As indicated in Eq. 9.3.3, $[B]$ may be split into a part $[B_o]$ independent of ζ and a part $\zeta[B_1]$ linear in ζ. The products $\zeta[B_o]^T[E][B_1]$ and $\zeta[B_1]^T[E][B_o]$ are linear in ζ and therefore vanish on integration through the thickness. The products $[B_o]^T[E][B_o]$ and $\zeta^2[B_1]^T[E][B_1]$ may be integrated with respect to ζ at once without recourse to numerical integration. Specifically, the integration involves $d\zeta$ and $\zeta^2\,d\zeta$, and introduces factors of 2 and 2/3, respectively. Thus Eq. 10.3.1 is reduced to (16)

$$[k] = \int_{-1}^{1}\int_{-1}^{1} (2[B_0]^T[E][B_0] + \tfrac{2}{3}[B_1]^T[E][B_1])\det[J]\,d\xi\,d\eta \quad (10.3.3)$$

Gauss quadrature with respect to ξ and η completes the integration. Note that although Gauss quadrature is done only two dimensionally, the Jacobian $[J]$ remains 3 by 3. In programming, some advantage may be taken of the many zeros in $[B_o]$ and $[B_1]$ in order to improve efficiency. Because normal-strain terms in $[E]$ represent in-plane action as well as bending action, it is not convenient to regard integration with respect to ζ as producing an in-plane stiffness array $[E]$ and the flexural rigidity array $[D]$ that was used in Section 9.3.

No need for special treatment of $[E]$ arises if the shell is homogeneous. Nonhomogeneous shells, for example, of sandwich construction, may be treated as follows (16). First, use $[E]$ as if the shell were homogeneous (either isotropic or orthotropic) and of thickness $t = 2h$, where $h = $ thickness of each facing. Thus, in-plane stiffness is properly accounted for. However, bending stiffness is underestimated and must be corrected by multiplying the contribution of $[B_1]^T[E][B_1]$ by the bending stiffness ratio (17 of Chapter Nine):

$$\frac{D_{\text{sandwich}}}{D_{\text{homogeneous}}} = \frac{Eh(c+h)^2/2(1-v^2)}{Et^3/12(1-v^2)} = \frac{6h(c+h)^2}{t^3}$$

$$= \frac{3(c+h)^2}{4h^2} \quad (10.3.4)$$

where $c = $ core thickness and $c \gg h$. The rationale for use of this factor is that bending stiffness arises from terms containing ζ^2, that is, from array $[B_1]$ in Eq. 10.3.3. Another necessary correction is that of multiplying the transverse shear stiffnesses $G_{z'x'}$ and $G_{y'z'}$ in $[E']$ by the factor $c/2h$ in order to compensate for the reduction in thickness of the shear-resisting layer from c to $2h$. Finally, nodal moments caused by linear variation of temperature through the thickness are first multiplied by $3(c + h)^2/4h^2$ but must then also be multiplied by $2h/(c + h)$ in order to obtain the correct values.

The foregoing approach of splitting matrices into parts independent of ζ and linear in ζ may also be applied to calculation of nodal loads arising from initial strains $\{\epsilon_o\}$, as $\{\epsilon_o\}$ may usually be split in the same way. Matrix $\{\epsilon_o\}$ must be referred to global coordinate directions. In the generation of nodal loads arising from body force and applied surface pressure, these loads must be also referred to global directions because matrix $[N]$ relates global displacements u, v, w to nodal d.o.f. in global directions.

In the final calculation of element stresses, note that strains computed from $\{\epsilon\} = [B]\{d\}$ are referred to *global* coordinates. The direction cosines needed in $[B]$ are also used in transforming $\{\epsilon\}$ to coordinates $x'y'z'$. Hence the operation $\{\sigma'\} = [E']\{\epsilon'\}$ gives stress caused by displacements, referred to shell-parallel directions $x'y'z'$. If we are dealing with the sandwich construction mentioned above by use of an "equivalent section" of thickness $2h$, we must recognize that strains caused by bending are too low if based on thickness $2h$. These strains vary linearly through the thickness and must therefore be increased by the factor $(c + 2h)/2h$.

10.4   Location of Gauss Quadrature Points

Section 9.4 described why four Gauss quadrature points should be used in the isoparametric plate element for numerical integration with respect to ξ and η. The same argument applies to the shell element, as can be seen qualitatively by noting that a plate is a special case of a shell. A second argument, also presented in Ref. 15 of Chapter Nine and fortunately yielding the same conclusion, is given below. The argument is made with reference to a shallow arch and therefore applies only qualitatively to shell elements of arbitrary shape and greater curvature but is apparently justified by the good results provided by the shell elements.

Let the arch in Fig. 10.4.1 have the equation

$$y = \frac{rx^2}{a^2} \tag{10.4.1}$$

The arch is "shallow," meaning that $r \ll a$. It can be shown that under pure

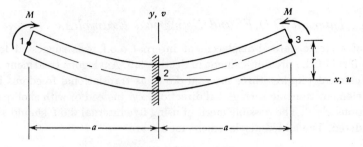

Figure 10.4.1. Shallow arch under moment loading.

bending, a point on the midsurface has displacements

$$u = -\frac{2Mrx^3}{3EIa^2}, \qquad v = \frac{Mx^2}{2EI} \tag{10.4.2}$$

where EI is the bending stiffness. Rotations of cross sections of a thin arch are given by dv/dx. Now let points 1, 2, 3 be regarded as nodes on the edge of our quadratic shell element. If, by use of Eqs. 10.4.2, these nodes are assigned displacements that prevail in the arch, element midsurface displacements are

$$u = -\frac{2Mrx}{3EI}, \qquad v = \frac{Mx^2}{2EI} \tag{10.4.3}$$

It can be shown that strain ϵ_m along the midsurface of the shallow arch is

$$\epsilon_m = \frac{du}{dx} + \frac{dv}{dx}\frac{dy}{dx} \tag{10.4.4}$$

Equations 10.4.1 and 10.4.2 give the correct value $\epsilon_m = 0$. Using Eqs. 10.4.3 and setting $\epsilon_m = 0$, we find

$$0 = -\frac{2Mr}{3EI} + \frac{Mx}{EI}\frac{2rx}{a^2}, \qquad \frac{x}{a} = \pm\frac{1}{\sqrt{3}} \tag{10.4.5}$$

These x/a values are again the Gauss point locations of a two-point quadrature rule and are the only points on the element centerline where membrane strain ϵ_m is correctly evaluated. We conclude that the shell element should employ a 2 by 2 Gauss rule so that displacements corresponding to pure bending will not produce membrane strain and unduly stiffen the element.

The implication for stress computation is that for the sake of accuracy, transverse shear strains and membrane strains parallel to curved sections must be evaluated at the four Gauss points and extrapolated elsewhere if desired.

10.5   Internal D.O.F. and Numerical Examples

Use of a single, lateral-displacement internal d.o.f. was suggested for the plate form of the present element in Section 9.5. Analogous treatment of the shell element requires three internal d.o.f., as translational freedoms in the shell element coincide with global directions xyz instead of with shell-parallel directions $x'y'z'$. The possible merit of using five internal d.o.f. should also be considered. The added displacement modes are

$$\{u \quad v \quad w \quad \alpha \quad \beta\}_{\text{add.}} = (1 - \xi^2)(1 - \eta^2)\{u_c \quad v_c \quad w_c \quad \alpha_c \quad \beta_c\} \quad (10.5.1)$$

where u_c, v_c, etc. are displacements at $\xi = \eta = 0$ *relative* to the displacements at $\xi = \eta = 0$ dictated by the 40 d.o.f. on the element boundary. As with the plate element, Eq. 10.5.1 permits the assembled structure to have zero-energy deformation modes if displacement boundary conditions happen not to suppress such a mode. These may be controlled by the same device as before, namely that of increasing by a small percentage the diagonal coefficients of $[k]$ associated with the internal d.o.f. (See Section 9.5 for a full discussion.)

The problems considered are shown in Figs. 10.5.1 and 10.5.2. These are commonly used test cases for shell elements.

As noted at the end of Section 10.4, certain of the strains must be evaluated at the Gauss points. Computational test must decide whether it is profitable

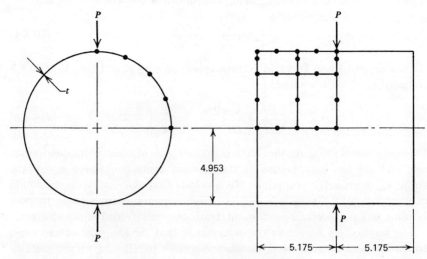

Figure 10.5.1.   Pinched cylinder, showing four identical elements in one octant. Edges of the shell are free.

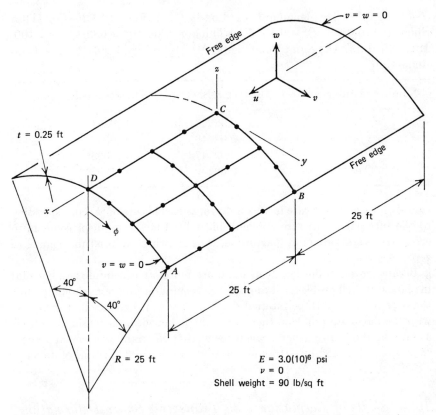

Figure 10.5.2. Cylindrical shell roof, showing four identical elements in one quadrant.

to do likewise for the remaining strains or to calculate them directly at node points. An aspect of this choice is illustrated by the midsurface strain ϵ_x in Fig. 10.5.2. In this case ϵ_x is directly proportional to N_x in Table 10.5.2 because Poisson's ratio $\nu = 0$. It is seen that in the four-element case N_{xB} is more accurate if computed directly, while the opposite trend appears for certain of the one-element values. This may be explained by noting that ϵ_x varies quadratically with coordinate ϕ. Extrapolation imposes a linear variation of ϵ_x with ϕ. Apparently the beneficial effect of calculating more accurate strain values at the Gauss points may or may not be overcome by the detrimental effect of discarding the quadratic variation.

Element stiffness matrices used to obtain the data in Tables 10.5.1 and 10.5.2 (16) were formulated according to Eq. 10.3.3. Table 10.5.1 shows that the pinched cylinder problem of Fig. 10.5.1 is little affected by internal d.o.f. According to Ref. 7, the elements of Refs. 5 and 6 fail when applied to the

Table 10.5.1. Displacement of Load P in Fig. 10.5.1. For Thin Shell, $t = 0.0155$, $P = 0.10$. For Thicker Shell, $t = 0.094$, $P = 100$. Internal d.o.f. Diagonal Coefficients Multiplied by 1.003 Prior to Condensation

Number of Internal d.o.f.	Thin Shell	Thicker Shell
0	0.022271	0.10281
3	0.022465	0.10347
5	0.022472	0.10395
Theory (7)	0.02439	0.1139

thin shell case. A test case in which diagonal coefficients of $[k]$ corresponding to the internal d.o.f. were *not* multiplied by 1.003 gave displacements in error by several orders of magnitude, thus showing the need for control of zero-energy modes.

Because internal displacement d.o.f. are more often helpful than harmful in plate and shell problems, it appears that computer programs should permit the user to employ three internal d.o.f. in the eight-node shell element. As with the flat plate problem, there seems little benefit in also using rotations as internal d.o.f. As a general rule it seems that *all* strains at element corners should be calculated by extrapolation from Gauss points instead of directly.

10.6 Shells of Revolution with Transverse Shear Deformation

The discussion in Chapter Eight presumed that transverse shear deformation could be neglected. In this section it will be retained. We will outline the formulation of an element (12) that in complexity lies between the plate element of Chapter Nine and the general shell element of this chapter. The development is quite similar to that described in Sections 10.2 and 10.3.

Figure 10.6.1a shows a three-node element in cross section. Node 2 is internal, and its d.o.f. are to be condensed before assembly of elements. The rotation d.o.f. at node 2 permits the element to model correctly a linearly varying bending moment. In numerical integration, Gauss points are to be located at $\xi = \pm 1/\sqrt{3}$ for reasons given in Sections 9.4 and 10.4.

Figure 10.6.1b shows that at a typical node i, $\overline{V}_{3i}$ has components

$$\overline{V}_{3i} = (t_i \cos \phi_i, t_i \sin \phi_i) \tag{10.6.1}$$

Hence, by specialization of Eq. 10.2.3, the geometry of the axisymmetric element is given by

$$\begin{Bmatrix} r \\ z \end{Bmatrix} = \sum_{i=1}^{3} N_i \begin{Bmatrix} r_i \\ z_i \end{Bmatrix} + \sum_{i=1}^{3} N_i \frac{\eta t_i}{2} \begin{Bmatrix} \cos \phi_i \\ \sin \phi_i \end{Bmatrix} \tag{10.6.2}$$

Table 10.5.2. Displacements (ft), Membrane Forces N (kips/ft), and Bending Moments M (ft kips/ft) in Shell of Fig. 10.5.2. Italicized Entries Represent Values Calculated By Extrapolation, While Other Values are Calculated Directly At Nodes. Internal d.o.f. Diagonal Coefficients Multiplied By 1.003 Prior to Condensation

Number of Internal d.o.f.	$10w_B$	$100w_C$	$100u_A$	$10v_B$	N_{xB}	$N_{\phi B}$	$M_{\phi B}$	$M_{\phi C}$	M_{xB}
0	−2.901	4.175	1.228	−1.534	75.17		−0.136	2.313	−0.6623
					55.84	*−0.495*	*−0.160*	*2.258*	*−0.7027*
3	−3.027	4.674	1.264	−1.591	77.93		−0.267	2.463	−0.5767
					57.44	*0.142*	*−0.323*	*2.406*	*−0.6208*
5	−3.032	4.663	1.266	−1.594	77.49		−0.236	2.461	−0.5995
					57.33	*0.123*	*−0.292*	*2.427*	*−0.6421*
The following results are produced by a single element in quadrant ABCD									
0	−3.173	6.213	0.781	−1.783	46.82		0.295	2.197	−1.3488
					32.35	*−6.754*	*0.226*	*2.276*	*−1.4356*
3	−3.350	5.900	1.137	−1.883	18.19		0.378	2.830	−0.6276
					33.94	*−0.447*	*0.093*	*2.779*	*−0.7025*
5	−3.351	5.899	1.137	−1.884	18.11		0.360	2.844	−0.6215
					33.94	*−0.447*	*0.094*	*2.779*	*−0.7020*
Theory (4)	−3.086	4.375[a]	1.261	−1.636	76.95	0	0	2.056[b]	−0.9272[c]

[a] A value of 4.7 appears in Ref. 14 of Chapter Nine.
[b] A value of 2.35 appears in Ref. 14 of Chapter Ten.
[c] A value of −0.66 appears in Ref. 14 of Chapter Nine.

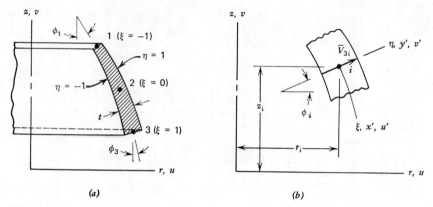

Figure 10.6.1. (*a*) Cross section of axisymmetric element. (*b*) Typical node.

where the shape functions are

$$N_1 = -\frac{\xi(1-\xi)}{2}$$

$$N_2 = (1 - \xi^2) \qquad (10.6.3)$$

$$N_3 = \frac{\xi(1+\xi)}{2}$$

Specialization of Eq. 10.2.7 yields the displacement field for the axisymmetric deformation problem

$$\begin{Bmatrix} u \\ v \end{Bmatrix} = \sum_{i=1}^{3} N_i \begin{Bmatrix} u_i \\ v_i \end{Bmatrix} + \sum_{i=1}^{3} N_i \frac{\eta t_i}{2} \begin{Bmatrix} \sin \phi_i \\ -\cos \phi_i \end{Bmatrix} \alpha_i \qquad (10.6.4)$$

where α_i is the rotation of a midsurface-normal line, positive when clockwise in Fig. 10.6.1*b*. Strains involved in the purely axisymmetric problem are

$$\{\epsilon_{x'} \quad \epsilon_\theta \quad \gamma_{x'y'}\} - \left\{ u'_{;x'} \quad \frac{u}{r} \quad (u'_{;y'} + v'_{;z'}) \right\} \qquad (10.6.5)$$

where θ is the circumferential coordinate.

In the case of nonsymmetric loading we must add to the list of variables the circumferential displacement w and the rotation β of a midsurface-normal line about axis x'. The displacement field for a typical harmonic of loading is (12):

$$
\begin{Bmatrix} u \\ v \\ w \end{Bmatrix} = \begin{bmatrix} \cos n\theta & 0 & 0 \\ 0 & \cos n\theta & 0 \\ 0 & 0 & \sin n\theta \end{bmatrix}
$$

$$
\times \left(\sum_{i=1}^{3} N_i \begin{Bmatrix} u_{in} \\ v_{in} \\ w_{in} \end{Bmatrix} + \sum_{i=1}^{3} N_i \frac{\eta t_i}{2} \begin{bmatrix} \sin \phi_i & 0 \\ -\cos \phi_i & 0 \\ 0 & 1 \end{bmatrix} \begin{Bmatrix} \alpha_{in} \\ \beta_{in} \end{Bmatrix} \right) \qquad (10.6.6)
$$

The strains and strain-displacement relations in $r\theta z$ coordinates are exactly as given in Eq. 7.3.2.

Clearly all the ingredients for a finite-element solution are on hand, and the procedures described in Sections 10.2 and 10.3 may again be followed.

Interelement compatibility prevails when one edge of an axisymmetric quadratic element, Fig. 5.5.1, is joined to the shell element of Fig. 10.6.1 and node 2 of the shell element is not condensed prior to assembly of elements. In contrast, the elements of Chapter 5 are not compatible with the shell element of Section 8.4. These considerations find application in the analysis of rocket motors where the propellant is bonded to the case.

PROBLEMS

10.1. Suppose that a "kink" exists in an actual shell, as, for example, along a ridge line of a folded plate. If nodal d.o.f. consist of displacements and rotations, which freedoms might be eliminated for nodes lying on the ridge? Hence, what conclusions might be drawn regarding the effects of a lack of interelement compatibility for an element with in-plane displacements linear in s and lateral displacement cubic in s, where s is the coordinate along an edge?

10.2. (a) Verify Eq. 10.2.3.
    (b) Write Eq. 10.2.3 in the form

$$
\{x \quad y \quad z\} = \sum [N_i']\{x_i \quad y_i \quad z_i \quad \overline{V}_{3xi} \quad \overline{V}_{3yi} \quad \overline{V}_{3zi}\}
$$

and establish the terms in $[N_i']$.

    (c) Hence, write expressions for each term in the Jacobian matrix $[J]$. What becomes of the terms containing ζ if the element is flat?

    (d) If the element is flat, rectangular, and of constant thickness, how many Gauss points are required to exactly evaluate the integral of $[B]^T[E][B]\,dx\,dy\,dz$?

10.3. (a) Complete Eq. 10.2.13 by deriving the terms that have not been written out.

(b) Show that Eq. 10.2.13 may be specialized to obtain the corresponding result for a flat plate, Eq. 9.2.7.

10.4. A scheme for treatment of nodal loads arising from initial strains $\{\epsilon_o\}$ in non-homogeneous shells is mentioned in Section 10.3. Explain in detail how this may be done.

10.5. (a) Derive Eqs. 10.4.2. *Suggestion.* Find u by summing the appropriate components of v.

(b) Without derivation, give an argument for the correctness of Eqs. 10.4.3 based on the deformation capability of the element.

(c) Derive Eq. 10.4.4. *Suggestion.* Consider the deformation of an infinitesimal length of the midsurface.

10.6. As always, elements should be strain-free under rigid body motion and should exhibit constant strain under appropriate nodal displacements. Does the shell element of Section 10.6 satisfy these criteria? As an example you may wish to consider a cylindrical element ($r_1 = r_2 = r_3$, $\phi_1 = \phi_2 = \phi_3 = 0$ in Fig. 10.6.1), under displacement v and constant strain ϵ_z.

10.7. Show that specialization of the equations for the eight-node general shell element leads to:
(a) Eqs. 10.6.2 and 10.6.3.
(b) Eq. 10.6.4.

10.8. (a) Starting with Eq. 10.6.4, pursue the development of the axisymmetric element to obtain a result analogous to Eq. 10.2.13.

(b, c) Repeat Parts b and c of Problem 10.2 using the analogous equations for the axisymmetric element.

10.9. (To be worked after study of Chapter Eleven.) Irons (13) suggests that in very thin elements the transverse shear strains are effectively made zero at the four Gauss points of the eight-node isoparametric shell element, and that this may promote numerical error. He suggests explicitly imposing the zero shear strain conditions to avoid possible trouble, at the same time reducing the number of boundary d.o.f. from 40 to 32.

(a) Consider the rectangular flat-plate form of the shell element. Let $\{\gamma\}$ be the 8 by 1 array of shear strains γ_{yz} and γ_{zx} at the four Gauss points, and let $\{d_r\}$ be the nodal displacement vector after lateral displacement and rotation about a normal to the edge have been removed at each side node. Explain how to enforce $\{\gamma\} = 0$ and replace $\{d\}$ by $\{d_r\}$. *Suggestion.* Consider a transformation $\{\gamma\ \ d_r\} = [T]\{d\}$.

(b) How would practical computations in Part a be affected if the element is arbitrarily oriented in the xy plane or if it has curved sides? Or if the element is not flat, but part of a general shell?

(c) For an n by n mesh of shell elements, where n is "large," how much storage space and equation-solving time are saved by the reduction in nodal d.o.f.?

See also (15) for further comments on how to avoid numerical error in the presence of transverse shear effects in thin elements.

10.10. (To be worked after study of Chapter Eleven.) The coordinate transformation suggested below Eq. 10.3.2 has the form $[E] = [T_\epsilon]^T [E'][T_\epsilon]$. If in Eq. 10.3.1 $[T_\epsilon]$ is associated with $[B]$ instead of with $[E']$, the integrand matrix becomes $([T_\epsilon][B])^T [E']([T_\epsilon][B])$. Explain why—and how—$[E']$ may be used as a 5 by 5 array and $[T_\epsilon][B]$ reduced to size 5 by 40. Explain whether this computational approach is preferable to that of Eq. 10.3.1.

10.11. The semibandwidth B for an assembly of shell elements having five d.o.f. per node may be calculated as $5(n + 1)$, where n is the maximum node number difference in any element. Alternatively, B may be calculated from array ID after statement 60 of Fig. 2.7.2.

(a) Will the two methods always give the same value of B? *Suggestion.* Consider restraint of all in-plane d.o.f. so as to model plate bending.

(b) Program the calculation of B from array ID.

10.12. A rotation β is introduced in Eq. 10.6.6. Why is the corresponding rotation not needed in the element of Section 8.4?

CHAPTER ELEVEN

Coordinate Transformation

11.1 *Introduction*

Coordinate transformation has been referred to occasionally in preceding chapters, but without explanation or justification. In subsequent chapters it is more frequently needed. A major purpose of coordinate transformation is to permit material or element properties, known with reference to one coordinate system or set of nodal d.o.f., to be used in another coordinate system or with reference to another set of nodal d.o.f. This chapter seeks to explain the techniques involved, largely by means of examples. A "transformation matrix" appears frequently and is denoted by $[T]$, used with or without a subscript. The particular form of $[T]$ depends on the problem at hand.

11.2 *Transformation of Material Properties*

Let two cartesian reference frames xyz and $x'y'z'$ be arbitrarily oriented with respect to one another (Fig. 11.2.1). Also listed in the figure are cosines of the angles between the various axes.

Any state of stress and strain may be expressed in either coordinate system, as $\{\sigma\}$ and $\{\epsilon\}$ in xyz coordinates or as $\{\sigma'\}$ and $\{\epsilon'\}$ in $x'y'z'$ coordinates. We will adhere to the engineering definition of shear strain, for

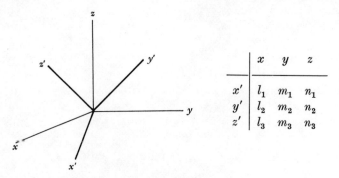

Figure 11.2.1. Cartesian coordinate systems and table of direction cosines.

example, $\gamma_{xy} = u_{,y} + v_{,x}$ and $\gamma_{x'y'} = u'_{,y'} + v'_{,x'}$ where a comma denotes partial differentiation. Stresses $\{\sigma\}$ are arranged in the order

$$\{\sigma_x \quad \sigma_y \quad \sigma_z \quad \tau_{xy} \quad \tau_{yz} \quad \tau_{zx}\}$$

and strains $\{\epsilon\}$ in the order

$$\{\epsilon_x \quad \epsilon_y \quad \epsilon_z \quad \gamma_{xy} \quad \gamma_{yz} \quad \gamma_{zx}\}$$

Stresses and strains are transformed by the operations

$$\{\sigma'\} = [T_\sigma]\{\sigma\}, \qquad \{\sigma\} = [T_\sigma]^{-1}\{\sigma'\} \qquad (11.2.1)$$
$$\{\epsilon'\} = [T_\epsilon]\{\epsilon\}, \qquad \{\epsilon\} = [T_\epsilon]^{-1}\{\epsilon'\}$$

These transformations are derived in texts on the theory of elasticity but are usually not expressed in matrix form. Matrices $[T_\sigma]$ and $[T_\epsilon]$ each have the property that the inverse of one equals the transpose of the other. Matrix $[T_\sigma]$ is

$$[T_\sigma] = \begin{bmatrix} l_1{}^2 & m_1{}^2 & n_1{}^2 & 2l_1m_1 & 2m_1n_1 & 2n_1l_1 \\ l_2{}^2 & m_2{}^2 & n_2{}^2 & 2l_2m_2 & 2m_2n_2 & 2n_2l_2 \\ l_3{}^2 & m_3{}^2 & n_3{}^2 & 2l_3m_3 & 2m_3n_3 & 2n_3l_3 \\ l_1l_2 & m_1m_2 & n_1n_2 & (l_1m_2 + l_2m_1) & (m_1n_2 + m_2n_1) & (n_1l_2 + n_2l_1) \\ l_2l_3 & m_2m_3 & n_2n_3 & (l_2m_3 + l_3m_2) & (m_2n_3 + m_3n_2) & (n_2l_3 + n_3l_2) \\ l_3l_1 & m_3m_1 & n_3n_1 & (l_3m_1 + l_1m_3) & (m_3n_1 + m_1n_3) & (n_3l_1 + n_1l_3) \end{bmatrix}$$

$$(11.2.2)$$

If the above matrix $[T_\sigma]$ is partitioned into 3 by 3 submatrices, matrix $[T_\epsilon]$ may be written in terms of the submatrices as follows:

$$[T_\sigma] = \left[\begin{array}{c|c} T_{11} & T_{12} \\ \hline T_{21} & T_{22} \end{array}\right], \qquad [T_\epsilon] = \left[\begin{array}{c|c} T_{11} & \tfrac{1}{2}T_{12} \\ \hline 2T_{21} & T_{22} \end{array}\right] \qquad (11.2.3)$$

The stress-strain relation may be written in either coordinate system, as $\{\sigma\} = [E]\{\epsilon\}$ or $\{\sigma'\} = [E']\{\epsilon'\}$. Suppose that $[E']$ is known and $[E]$ is desired. Such a situation may arise when dealing with orthotropic materials, where elastic properties are known in principal directions $x'y'z'$ but are needed in global coordinates xyz to generate the element stiffness matrix. Transformation of $[E']$ is also needed for the isoparametric shell element of Chapter Ten. The transformation is derived from the argument that during any virtual displacement, the resulting increment in strain energy density U_o must be the same regardless of the coordinate system in which it is computed. Thus, from Eq. 3.5.4,

$$\delta U_o = \{\delta\epsilon\}^T\{\sigma\} = \{\delta\epsilon'\}^T\{\sigma'\} \qquad (11.2.4)$$

Substitution of Eqs. 11.2.1 and $\{\sigma'\} = [E']\{\epsilon'\}$ into Eq. 11.2.4 yields

$$\{\delta\epsilon\}^T\{\sigma\} = \{\delta\epsilon\}^T[T_\epsilon]^T[E']\{\epsilon'\}$$
$$\{\delta\epsilon\}^T(\{\sigma\} - [T_\epsilon]^T[E'][T_\epsilon]\{\epsilon\}) = 0 \qquad (11.2.5)$$

As the latter relation must be true for *any* $\{\delta\epsilon\}$, the coefficient of $\{\delta\epsilon\}$ must vanish. Thus we obtain $\{\sigma\} = [E]\{\epsilon\}$, where the coefficient of $\{\epsilon\}$ in Eq. 11.2.5 has been identified as the desired matrix $[E]$:

$$[E] = [T_\epsilon]^T[E'][T_\epsilon] \qquad (11.2.6)$$

Transformation matrices for axisymmetric and two-dimensional problems are easily obtained by specialization of the foregoing relations. For example, in the two-dimensional case where $\{\sigma\} = \{\sigma_x \quad \sigma_y \quad \tau_{xy}\}$, we have

$$[T_\epsilon] = \begin{bmatrix} l_1^2 & m_1^2 & l_1 m_1 \\ l_2^2 & m_2^2 & l_2 m_2 \\ 2l_1 l_2 & 2m_1 m_2 & (l_1 m_2 + l_2 m_1) \end{bmatrix} \qquad (11.2.7)$$

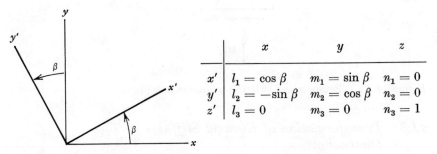

	x	y	z
x'	$l_1 = \cos\beta$	$m_1 = \sin\beta$	$n_1 = 0$
y'	$l_2 = -\sin\beta$	$m_2 = \cos\beta$	$n_2 = 0$
z'	$l_3 = 0$	$m_3 = 0$	$n_3 = 1$

Figure 11.2.2. Coordinates and direction cosines, two-dimensional case.

and similarly for $[T_\sigma]$. Figure 11.2.2 shows the coordinate system and direction cosines.

Moments and curvatures appear in plate and shell problems and may require coordinate transformation. Moments (Fig. 11.2.3) and curvatures transform as do stresses and strains:

$$\{M_{x'} \quad M_{y'} \quad M_{x'y'}\} = \{M'\} = [T_\sigma]\{M\} = [T_\sigma]\{M_x \quad M_y \quad M_{xy}\}$$

$$\{w_{,x'x'} \quad w_{,y'y'} \quad 2w_{,x'y'}\} = \{\kappa'\} = [T_\epsilon]\{\kappa\} = [T_\epsilon]\{w_{,xx} \quad w_{,yy} \quad 2w_{,xy}\}$$

$$(11.2.8)$$

where $[T_\epsilon]$ is given by Eq. 11.2.7 and Fig. 11.2.2. The moment-curvature relations are $\{M\} = -[D]\{\kappa\}$ and $\{M'\} = -[D']\{\kappa'\}$, and the matrix of flexural rigidities $[D']$ transforms as follows:

$$[D] = [T_\epsilon]^T [D'][T_\epsilon] \tag{11.2.9}$$

As with matrix $[E']$, the motivation for transformation of $[D']$ is that $[D']$ may contain properties of an orthotropic material referred to principal material directions $x'y'$, while structural equations are to be generated using stresses and strains referred to global coordinates xy.

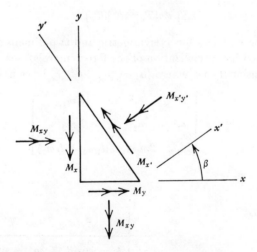

Figure 11.2.3. Bending and twisting moments in a plate.

11.3 Transformation of Element Stiffness in Two Dimensions

Consider a frame element having axial stiffness AE and bending stiffness

EI. The 6 by 6 element stiffness matrix $[k']$ is easily written with reference to a local axis x', which coincides with the member axis (Fig. 11.3.1). The problem we consider is that of transforming $[k']$ to $[k]$ in xy coordinates. In other words, nodal d.o.f. must be changed from u_i', v_i', and θ_i' to u_i, v_i, and θ_i, where $i = 1, 2$. Use of this transformation is as easy a way as any to obtain the stiffness matrix of an arbitrarily oriented frame element.

Nodal d.o.f. in the two coordinate systems are related by the equations

$$u_i' = u_i \cos \phi + v_i \sin \phi$$
$$v_i' = -u_i \sin \phi + v_i \cos \phi \qquad (11.3.1)$$
$$\theta_i' = \theta_i$$

One may easily verify that primed and unprimed nodal forces are related in exactly the same way as are the two sets of nodal displacements. Let nodal d.o.f. and forces be arrayed as

$$\{d\} = \{u_1 \quad v_1 \quad \theta_1 \quad u_2 \quad v_2 \quad \theta_2\}$$
$$\{r\} = \{p_1 \quad q_1 \quad M_1 \quad p_2 \quad q_2 \quad M_2\} \qquad (11.3.2)$$

and similarly for the primed quantities. Forces $\{r\}$ and $\{r'\}$ are regarded as being applied to the nodes by the element. From Eqs. 11.3.1 we may write

$$\{d'\} = [T]\{d\}, \qquad \{r'\} = [T]\{r\} \qquad (11.3.3)$$

in which

$$[T] = \begin{bmatrix} \Lambda & 0 \\ 0 & \Lambda \end{bmatrix}, \qquad [\Lambda] = \begin{bmatrix} \cos\phi & \sin\phi & 0 \\ -\sin\phi & \cos\phi & 0 \\ 0 & 0 & 1 \end{bmatrix} \qquad (11.3.4)$$

Matrix $[T]$ is *orthogonal*; that is, its inverse is the same as its transpose. Therefore,

$$\{d\} = [T]^T\{d'\}, \qquad \{r\} = [T]^T\{r'\} \qquad (11.3.5)$$

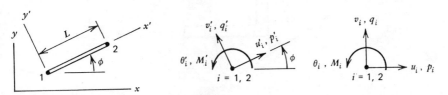

Figure 11.3.1.   Frame element. Nodal d.o.f. and forces.

The latter relation is useful in transforming element nodal forces to global directions prior to assembly of elements. The relation $\{d'\} = [T]\{d\}$ is useful in extracting local displacements from global displacements for the purpose of stress computation. It should be noted that the second of Eqs. 11.3.5 can also be derived by a virtual work argument, as in Eqs. 11.6.2.

Transformation of the stiffness matrix follows by substitution of Eqs. 11.3.3 and 11.3.5 and the stiffness relation $[k']\{d'\} = \{r'\}$ into the stiffness relation $[k]\{d\} = \{r\}$:

$$[k]\{d\} = \{r\} = [T]^T\{r'\} = [T]^T[k']\{d'\} = [T]^T[k'][T]\{d\} \quad (11.3.6)$$

As this relation is presumed valid for *any* $\{d\}$, we conclude that

$$[k] = [T]^T[k'][T] \tag{11.3.7}$$

In this example, as is often the case, matrix $[T]$ is sparse and contains sub-matrices that are identical. Computer programming that takes advantage of these features may provide a worthwhile saving in storage space and execution time. Indeed, the operations of Eq. 11.3.7 may be worked out by hand before programming.

11.4 An Application: Skew Support

Suppose that node 2 of a plane frame is free to displace in only the s direction (Fig. 11.4.1). Thus, displacements at node 2 are not independent, but they satisfy the relation $v_2 = -u_2 \tan \phi$. This awkward boundary condition may be handled by first transforming so that forces and displacements at

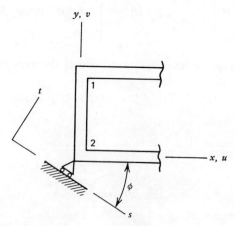

Figure 11.4.1. Skew support.

node 2 are parallel to s and t directions, then imposing the boundary condition of zero t-direction displacement in some standard fashion (see Section 2.7). The necessary transformation is now described.

Let the original and transformed displacement vectors be, respectively,

$$\{D'\} = \{u_1 \quad v_1 \quad \theta_1 \quad u_2 \quad v_2 \quad \theta_2 \quad \ldots\}$$
$$\{D\} = \{u_1 \quad v_1 \quad \theta_1 \quad s_2 \quad t_2 \quad \theta_2 \quad \ldots\}$$

(11.4.1)

where s_2 and t_2 are s and t direction displacements at node 2. Vectors $\{D'\}$ and $\{D\}$ each contain $3n$ terms, where n is the total number of nodes in the structure. Corresponding to $\{D'\}$ and $\{D\}$ are nodal force vectors $\{R'\}$ and $\{R\}$. The transformation may be written as in Eqs. 11.3.3 and 11.3.5:

$$\{D'\} = [T]\{D\}, \qquad \{R'\} = [T]\{R\}$$
$$\{D\} = [T]^T\{D'\}, \qquad \{R\} = [T]^T\{R'\}$$

(11.4.2)

where $[T]$ is the orthogonal matrix

$$[T] = \begin{bmatrix} I & 0 & 0 \\ 0 & \Lambda & 0 \\ 0 & 0 & I \\ & & & \ddots \end{bmatrix}$$

(11.4.3)

in which $[I]$ is a 3 by 3 unit matrix and $[\Lambda]$ is given in Eq. 11.3.4.

Following the argument associated with Eq. 11.3.6 we find the same symbolic result, that the revised matrix $[K]$ associated with skew displacements at node 2 is given by $[K] = [T]^T[K'][T]$, where $[K']$ is the structure stiffness matrix associated with x and y direction displacements at all nodes. The transformed matrix may be transformed again and again if additional skew supports are present. Indeed, all nodes may be so treated, so that perhaps no pair of nodes has the same s and t directions (8). Displacements obtained by solution of structural equations are in the s and t directions of each skew support.

In the particular example of Fig. 11.4.1, only rows 4 and 5 and columns 4 and 5 of the structure stiffness matrix are modified by transformation. This result might be expected on physical grounds (why?). The transformation $[T]^T[K'][T]$ involves row and column operations similar to those in Eqs. 11.8.3. However, in order to efficiently program the transformation one must contend with the sparsity of the matrices and whatever compact storage format may be used to take advantage of the sparsity. These difficulties suggest that it may be preferable to carry out the transformation on each element containing the node in question before assembly of elements into the structure.

Accordingly, taking, for example, element 1–2 of Fig. 11.4.1, the element stiffness matrix is transformed by the operation $[k] = [T]^T [k'][T]$ prior to its assembly into the structure, in which

$$\underset{6\times6}{[T]} = \begin{bmatrix} I & 0 \\ 0 & \Lambda \end{bmatrix}, \qquad \begin{array}{l} [I] = 3 \text{ by } 3 \text{ unit matrix} \\[4pt] [\Lambda] \text{ given by Eq. 11.3.4} \end{array} \qquad (11.4.4)$$

It is, of course, also necessary to transform all element load vectors $\{r'\}$ containing the node in question, as well as any externally applied forces $\{P\}$, so that all forces are referred to st directions at skew supports.

11.5 Transformation of Element Stiffness in Three Dimensions

We take as our example the constant-strain triangle in plane stress (this element was first seen in Fig. 4.4.1). In Fig. 11.5.1 the element 1–2–3 is presumed to lie in the $x'y'$ plane. For the present we presume that the 6 by 6 element stiffness matrix $[k']$ and 6 by 1 element nodal forces $\{r'\}$ are known in coordinates $x'y'$. It is desired to transform these arrays to global coordinates xyz.

In order to transform, direction cosines must be established. This may be done as follows. Using *global* coordinates of nodes 1 and 3, we establish the

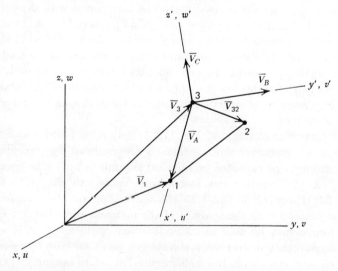

Figure 11.5.1. Arbitrarily oriented plane stress element.

vector $\bar{V}_A$:

$$\bar{V}_A = \{(x_1 - x_3) \quad (y_1 - y_3) \quad (z_1 - z_3)\} \tag{11.5.1}$$

Similarly, vector $\bar{V}_{32}$ may be established. A vector in direction z' is given by the cross product:

$$\bar{V}_C = \bar{V}_A \times \bar{V}_{32} = \begin{Bmatrix} (y_1 - y_3)(z_2 - z_3) - (y_2 - y_3)(z_1 - z_3) \\ (z_1 - z_3)(x_2 - x_3) - (z_2 - z_3)(x_1 - x_3) \\ (x_1 - x_3)(y_2 - y_3) - (x_2 - x_3)(y_1 - y_3) \end{Bmatrix} \tag{11.5.2}$$

Vector $\bar{V}_B$ is given by the cross product $\bar{V}_B = \bar{V}_C \times \bar{V}_A$. When vectors $\bar{V}_A$, $\bar{V}_B$, and $\bar{V}_C$ are normalized by dividing each by its magnitude, the components of the resulting unit vectors are the direction cosines shown in Fig. 11.2.1.

The transformation is most easily explained if element matrices are first expanded by adding rows and columns of zeros. Let the local and global displacement vectors be, respectively,

$$\begin{aligned} \{d'\} &= \{u_1' \quad v_1' \quad w_1' \quad u_2' \quad v_2' \quad w_2' \quad u_3' \quad v_3' \quad w_3'\} \\ \{d\} &= \{u_1 \quad v_1 \quad w_1 \quad u_2 \quad v_2 \quad w_2 \quad u_3 \quad v_3 \quad w_3\} \end{aligned} \tag{11.5.3}$$

In the expanded $[k']$, rows and columns 3, 6, and 9 are zero, as the element has no stiffness in the z' direction. Similarly, terms 3, 6, and 9 in the expanded $\{r'\}$ are zero. Pursuing the same arguments as in Section 11.3, we again obtain the transformations given by Eqs. 11.3.3, 11.3.5, and 11.3.7, in which

$$[T] = \begin{bmatrix} \Lambda & 0 & 0 \\ 0 & \Lambda & 0 \\ 0 & 0 & \Lambda \end{bmatrix}, \qquad [\Lambda] = \begin{bmatrix} l_1 & m_1 & n_1 \\ l_2 & m_2 & n_2 \\ l_3 & m_3 & n_3 \end{bmatrix} \tag{11.5.4}$$

Again, $[T]$ is orthogonal, that is, $[T]^T[T] = [I]$. This may be seen by noting that the columns of $[\Lambda]$ form an orthogonal triple of unit vectors.

Examination of the transformations reveals that direction cosines l_3, m_3, and n_3 are not needed (1). They are multiplied by zero in the transformations $[k] = [T]^T[k'][T]$ and $\{r\} = [T]^T\{r'\}$ and produce only the unnecessary nodal displacements w' in the transformation $\{d'\} = [T]\{d\}$. Accordingly, the expansion in size of $\{d'\}$, $\{r'\}$, and $[k']$ is not necessary; storage space and computation time may be saved by omitting it. Specifically, the transformations may be written

$$\underset{6\times 1}{\{d'\}} = [T]\underset{9\times 1}{\{d\}}, \qquad \underset{9\times 1}{\{r\}} = [T]^T\underset{6\times 1}{\{r'\}}, \qquad \underset{9\times 9}{[k]} = [T]^T\underset{6\times 6}{[k']}[T]$$

$$\underset{6\times 9}{[T]} = \begin{bmatrix} \Lambda & 0 & 0 \\ 0 & \Lambda & 0 \\ 0 & 0 & \Lambda \end{bmatrix}, \qquad \underset{2\times 3}{[\Lambda]} = \begin{bmatrix} l_1 & m_1 & n_1 \\ l_2 & m_2 & n_2 \end{bmatrix} \tag{11.5.5}$$

In order to generate $[k']$ in the first place, nodal coordinates in the local system $x'y'z'$ are required. These must be extracted from nodal coordinates in the global system xyz (2). The transformation is the same as used for displacements: $\{x'y'\} = [T]\{xyz\}$, in which $\{x'y'\}$ is the 6 by 1 vector of local nodal coordinates, $\{xyz\}$ is the 9 by 1 vector of global nodal coordinates, and $[T]$ is given by Eq. 11.5.5. Of course, some entries in $\{x'y'\}$ are known to be zero because of the way local coordinates are defined in Fig. 11.5.1.

It may at times be convenient to establish the transformation by working through a series of local coordinate systems. Each local system is established by a rotation about an axis of the preceding system. Thus if the element is first defined in $x'''y'''z'''$ coordinates we establish

$$[k''] = [T_1]^T[k'''][T_1], \qquad [k'] = [T_2]^T[k''][T_2], \text{ etc.} \qquad (11.5.6)$$

which leads to the final result

$$[k] = [T]^T[k'''][T], \qquad [T] = [T_1][T_2][T_3] \qquad (11.5.7)$$

This approach has been used for curved beam elements (23).

11.6 Joining of Dissimilar Elements

As our first example consider a frame element, which in general may be tapered, and which is to be joined to quadrilateral elements having two translational d.o.f. at each corner (Fig. 11.6.1). D.o.f. of frame and quadrilateral elements differ in both number and type, hence we call the elements "dissimilar." The following scheme for connecting them was suggested in Ref. 4.

The relation between original and transformed displacements is not unique. Various forms may be devised, each of them reasonable. The following transformation should be as palatable as any.

$$
\begin{Bmatrix} d_1' \\ d_2' \\ d_3' \\ d_4' \\ d_5' \\ d_6' \end{Bmatrix} =
\begin{bmatrix}
\frac{1}{2} & 0 & 0 & 0 & 0 & 0 & \frac{1}{2} & 0 \\
0 & \frac{1}{2} & 0 & 0 & 0 & 0 & 0 & \frac{1}{2} \\
\frac{1}{t_1} & 0 & 0 & 0 & 0 & 0 & -\frac{1}{t_1} & 0 \\
0 & 0 & \frac{1}{2} & 0 & \frac{1}{2} & 0 & 0 & 0 \\
0 & 0 & 0 & \frac{1}{2} & 0 & \frac{1}{2} & 0 & 0 \\
0 & 0 & \frac{1}{t_2} & 0 & -\frac{1}{t_2} & 0 & 0 & 0
\end{bmatrix}
\begin{Bmatrix} d_1 \\ d_2 \\ d_3 \\ d_4 \\ d_5 \\ d_6 \\ d_7 \\ d_8 \end{Bmatrix}
\qquad (11.6.1)
$$

or, symbolically $\{d'\} = [T]\{d\}$.

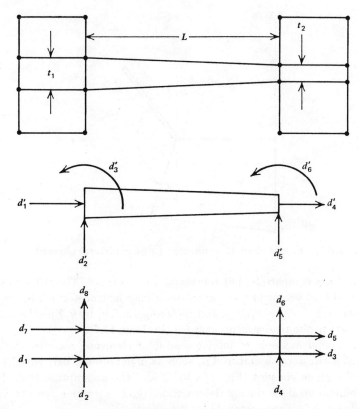

Figure 11.6.1. Frame element joined to quadrilateral elements. Original and transformed d.o.f.

Nodal forces may be present from thermal expansion, distributed load, etc. To transform these forces we argue that the work they do in a virtual displacement must be the same whether referred to d.o.f. $\{d'\}$ or $\{d\}$. Therefore,

$$\{\delta d\}^T\{r\} = \{\delta d'\}^T\{r'\} = \{\delta d\}^T[T]^T\{r'\}$$

$$\{\delta d\}^T(\{r\} - [T]^T\{r'\}) = 0, \qquad \underset{8\times1}{\{r\}} = \underset{8\times6}{[T]^T}\underset{6\times1}{\{r'\}} \qquad (11.6.2)$$

The latter result is extracted from the preceding equation by saying that the work equality must hold for any virtual displacement $\{\delta d\}$.

Transformation of the 6 by 6 stiffness matrix $[k']$ to the 8 by 8 matrix $[k]$ is as stated in Eq. 11.3.7 but with $[T]$ given by Eq. 11.6.1. The argument that produces this result is as stated in Eq. 11.3.6.

An alternate way of connecting frame and plane elements is suggested in Ref. 23 of Chapter Twelve. In the example of Fig. 11.6.2 translational d.o.f.

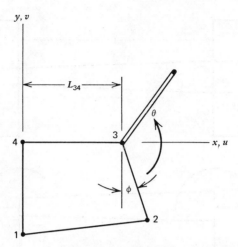

Figure 11.6.2.   Frame element connected to quadrilateral element.

at node 3 are compatible, but rotational d.o.f. θ is not. The stiffness matrix and nodal load vector of the quadrilateral may be transformed by imposing the relation $\theta = (v_3 - v_4)/L_{34}$, and replacing d.o.f. v_4 by θ. Specific formulation is left to the reader as an exercise (Problem 11.7).

As a second example of joining dissimilar elements, consider the placement of a two-force member AB, such as a reinforcing bar, parallel to an edge of a plane element (Fig. 11.6.3). Under the assumption that element sides remain straight during deformation, d.o.f. u_A and u_B of the bar are related to d.o.f. of the rectangle by the equations

$$\frac{u_A - u_1}{s} = \frac{u_4 - u_1}{h}, \qquad \frac{u_B - u_2}{s} = \frac{u_3 - u_2}{h} \tag{11.6.3}$$

Hence, a transformation $\{u_A \quad u_B\} = [T]\{d\}$ may be established, in which $\{d\}$ is the 8 by 1 vector of nodal d.o.f. of the rectangular element. After

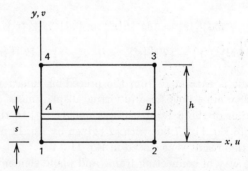

Figure 11.6.3.   Axial force member in plane rectangular element.

transformation, the stiffness matrix of the bar may be directly added to that of the rectangle. Again, details are left to the reader (Problem 11.8).

11.7 Nonstandard End Conditions

Stiffening members are usually placed on one side of a plate or shell instead of having their centroidal axes coincident with the midsurface. Truss members are usually joined by means of gusset plates instead of being pin-connected. In such problems, a useful idealization is that of a rigid connection piece that joins an element node to a nearby structure node. We will take a frame element as our example. Additional discussion appears in Refs. 5 through 7.

The end attachments in Fig. 11.7.1 have lengths e_i and e_j and are assumed to be perfectly rigid. It is desired to transform so that d.o.f. with subscripts i and j replace d.o.f. with subscripts 1 and 2. Nodes i and j are regarded as the structure nodes that lie close to element nodes 1 and 2. For translation of the end pieces we have $u_1 = u_i$, $v_1 = v_i$, etc. For rotation of the end pieces about nodes i and j we have $u_1 = -e_i \sin \phi_i$, $v_1 = e_i \cos \phi_i$, etc. Thus the transformation $\{d'\} = [T]\{d\}$ may be written

$$
\begin{Bmatrix} u_1 \\ v_1 \\ \theta_1 \\ u_2 \\ v_2 \\ \theta_2 \end{Bmatrix}
=
\begin{bmatrix}
1 & 0 & -e_i \sin \phi_i & 0 & 0 & 0 \\
0 & 1 & e_i \cos \phi_i & 0 & 0 & 0 \\
0 & 0 & 1 & 0 & 0 & 0 \\
0 & 0 & 0 & 1 & 0 & e_j \sin \phi_j \\
0 & 0 & 0 & 0 & 1 & -e_j \cos \phi_j \\
0 & 0 & 0 & 0 & 0 & 1
\end{bmatrix}
\begin{Bmatrix} u_i \\ v_i \\ \theta_i \\ u_j \\ v_j \\ \theta_j \end{Bmatrix}
\qquad (11.7.1)
$$

If desired, the substitutions $e_i \sin \phi_i = y_1 - y_i$, $e_i \cos \phi_i = x_1 - x_i$, etc. may be made.

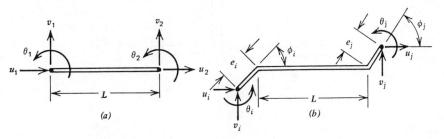

Figure 11.7.1. Frame element with rigid end pieces.

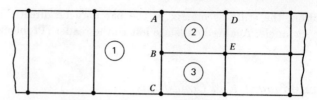

Figure 11.7.2. Grading of a mesh.

A method of grading a mesh from coarse to fine is described as follows (26). The method is similar to the foregoing example in that nodes well outside an element are caused to be associated with the element. An example is shown in Fig. 11.7.2. Here *all* elements have four nodes each, so that node B is not associated with element number 1. What must be done is to modify the matrices of elements 2 and 3 so that they do not depend on d.o.f. at node B. In place of these d.o.f. there appear the d.o.f. of nodes A and C. Thus for element 2 we write

$$[k_{ACDE}] = [T]^{T}[k'_{ABDE}][T] \qquad (11.7.2)$$

where the form of $[T]$ depends on what d.o.f. are used at nodes A and C and on where point B is located along line AC. Reference 26 advises that it is usually easier to obtain $[T]$ numerically instead of by algegraic manipulation.

Another "nonstandard" end condition is shown in Fig. 11.7.3, where rotational d.o.f. of two members are independent of the joint rotation. At joint i there are three independent rotational d.o.f. In general any or all of the members at a joint may be connected in this way, and the rigid end pieces of Fig. 11.7.1 may be used if desired (7). The treatment of such a member end condition is quite simple. Prior to assembly of elements the rotational d.o.f. to be released is regarded as an internal d.o.f. and is condensed, exactly as described in Section 6.1. The element stiffness matrix remains 6 by 6;

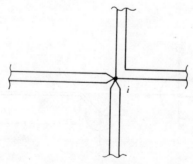

Figure 11.7.3. Pinned connections at a plane frame joint.

the condensed row and column of the transformed $[k]$ are filled with zeros so that on assembly no rotational stiffness is contributed to the structure. After solution for displacements the condensed freedom may be recovered in the usual way.

If all members are free to rotate independently at a joint, or if a rigidly connected joint is free to rotate on a pin support, some further treatment is necessary (Problem 11.9; Ref. 7).

11.8 Enforcing Equality of Displacements

It is sometimes desirable to require that two nodes, usually adjacent, have the same displacement. Two examples are shown in Fig. 11.8.1. The nut and bolt of Fig. 11.8.1a are separate finite element structures; where they are in contact (surfaces ab, cd, ef, etc.) relative displacement may be permitted in the tangential direction but not in the normal direction (9). In other words, sliding is permitted, but interpenetration is not. In Fig. 11.8.1b, axial shortening of the plane frame members may be negligible; therefore vertical displacements at nodes 1 and 2 may be ignored and horizontal displacements u_1 and u_2 made equal.[1] In general, if two d.o.f. are nearly equal

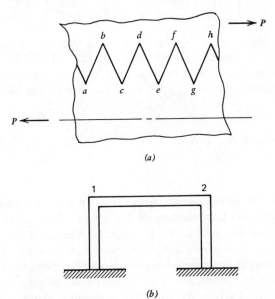

(a)

(b)

Figure 11.8.1.   (a) Meshing threads. (b) Plane frame.

[1] Owing also to the geometric regularity of building frames, use of a special purpose frame program seems indicated. Input data could be minimal, and deformation relations as mentioned above automatically accounted for.

because a very stiff member is present, it is advisable to *enforce* equality of the d.o.f. instead of to hope that the d.o.f. will be correctly computed; a very stiff member, if surrounded by more flexible members, may be the cause of numerical error (see Section 16.5). In an effort to avoid the error one might artificially decrease axial stiffnesses of the frame; axial strains may still be negligible in comparison with bending deformation. A similar observation applies to transverse shear deformations.

A procedure for enforcing equality of displacements is illustrated as follows. The matrices are *arbitrary;* they do not represent any particular structure. Let the vectors in the original system $\lfloor K'\rfloor\{D'\} = \{R'\}$ be

$$\{D'\} = \{D_1' \quad D_2' \quad D_3' \quad D_4'\}$$
$$\{R'\} = \{R_1' \quad R_2' \quad R_3' \quad R_4'\} \tag{11.8.1}$$

Suppose that the equality $D_2' = D_4'$ is to be imposed. We establish the transformation $\{D'\} = [T]\{D\}$, specifically

$$\begin{Bmatrix} D_1' \\ D_2' \\ D_3' \\ D_4' \end{Bmatrix} = \begin{bmatrix} 1 & 0 & 0 & 0 \\ 0 & 1 & 0 & 1 \\ 0 & 0 & 1 & 0 \\ 0 & 0 & 0 & 1 \end{bmatrix} \begin{Bmatrix} D_1' \\ D_2' - D_4' \\ D_3' \\ D_4' \end{Bmatrix} \tag{11.8.2}$$

where, in $\{D\}$, $D_2 = D_2' - D_4'$. By the usual arguments, as in Eqs. 11.3.6 and 11.6.2, we obtain the transformations $\{R\} = [T]^T\{R'\}$ and $[K] = [T]^T[K'][T]$. The transformed relation $[K]\{D\} = \{R\}$ turns out to be the same as the original relation $[K']\{D'\} = \{R'\}$ except for the following terms:

$$K_{4j} = K_{2j}' + K_{4j}' \quad \text{and} \quad K_{i4} = K_{i2}' + K_{i4}' \quad \text{for } i, j = 1, 2, 3$$
$$K_{44} = K_{22}' + K_{24}' + K_{42}' + K_{44}', \qquad D_2 = D_2' - D_4', \qquad R_4 = R_2' + R_4'$$

$$\tag{11.8.3}$$

Finally, equality of displacements is enforced by using a standard method of boundary condition treatment to set the new d.o.f. D_2 equal to zero.

Some comments may now be made. Skew boundaries, if present, must be treated (Section 11.4) before equality of displacements is enforced. Symmetry of the stiffness matrix is preserved. Operations are confined to only a portion of a banded matrix. Bandwidth may be increased, as may be seen in the foregoing example if $K_{14}' = K_{41}' = 0$. The increase is slight if the d.o.f. to be made equal are close to one another in $\{D'\}$. We see from Eqs. 11.8.3 that the transformation is essentially an operation that transfers loads and stiffnesses from one d.o.f. to another. This observation suggests that the operation might be more efficiently programmed by suitable subscripting and data handling instead of by formal transformation (see next to last paragraph of Section 11.10).

11.9 Substructures

Substructuring is the division of a single large structure into component structures. Each component (substructure) is analyzed separately and finally all are reconnected. Reasons for division into substructures are as follows. It offers an approach to analysis of a structure so large that available computers cannot accommodate it without sophisticated programming of one kind or another. The jobs of idealizing the structure for analysis and preparing data are divided into smaller and more manageable parts, each part being largely independent of the others (5.6–5.8 of Chapter Eighteen). Likewise, nearly independent structural design is possible if changes in one substructure have little effect on others. Modification would be particularly easy if, for example, all thicknesses in a substructure built of plane stress elements were multiplied by the same factor F; then, the new substructure stiffness matrix is simply $F[K_{\text{subs}}]$. If a structure can be idealized as an assembly of identical substructures, the effort of formulating the substructure stiffness matrix need be carried out only once.

Treatment of substructures in a static problem proceeds as follows (11, 12, 13). Let lines A–A and B–B be the boundaries of a typical substructure (Fig. 11.9.1). Straight lines are element boundaries. Loading within the substructure and on its boundaries is arbitrary. First, all displacements along A–A and B–B are prohibited, and analysis of the thus isolated substructure is performed to determine the forces it applies to boundaries A–A and B–B. The same is done for all other substructures. Second, equilibrium equations of the substructure boundaries are solved to determine their displacements under the combination of external load (if any) and forces applied by the substructures. Third, boundary displacements found in the second step are applied to the substructure, along with original loads applied within the

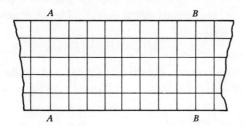

Figure 11.9.1. Substructuring.

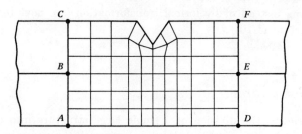

Figure 11.9.2. "Reduced" substructure.

substructure, and internal displacements determined. From this final displacement field, stresses may be calculated.

Clearly, the foregoing analysis is exactly that used in condensation (Sections 6.1 and 6.2), where d.o.f. within an element were removed prior to solving for the remaining d.o.f., then recovered to yield the final element configuration. Accordingly, a substructure may be regarded as a single large element with many internal d.o.f.

Kamel (22) advocates a "reduced substructure" technique. An example appears in Fig. 11.9.2. The substructure, bounded by lines ABC and DEF, has seven nodes on each boundary. The substructure may be joined to the remaining structure, without serious compatibility problems at the two boundaries, if on each boundary the four intermediate nodes have displacements defined in terms of nodal d.o.f. at nodes A, B, C and D, E, F by means of some interpolation function. Thus the substructure may have a fine but regular mesh, and the number of boundary displacements to be processed is reduced. Further details appear in Ref. 22.

A transformation applicable to the reduced substructure should be mentioned. As displacements of some nodes are expressed in terms of others, we may write an equation of the form

$$[H_{AF} \quad H_R]\{D_{AF} \quad D_R\} = 0 \qquad (11.9.1)$$

where $\{D_{AF}\}$ = displacements of nodes A, B, C, D, E, F in Fig. 11.9.2 and $\{D_R\}$ = displacements of the remaining boundary nodes. There are as many constraint equations as displacements $\{D_R\}$. Therefore matrix $[H_R]$ is square and

$$\{D_R\} = -[H_R]^{-1}[H_{AF}]\{D_{AF}\} \qquad (11.9.2)$$

Accordingly the transformation between the total and condensed boundary displacement arrays is

$$\begin{Bmatrix} D_{AF} \\ D_R \end{Bmatrix} = \begin{bmatrix} I \\ -H_R^{-1} \quad H_{AF} \end{bmatrix} \{D_{AF}\} \qquad (11.9.3)$$

where $[I] = $ a unit matrix. This type of transformation will be encountered again (Problem 11.15, Section 12.4, Problem 12.10).

User-oriented computer implementation of substructuring should allow the node numbering of connecting substructures to be independent. Also, division of each substructure into further substructures may be contemplated (12, 15, 27). Substructures may be assembled by use of a transformation $[T]^T[k_d][T]$ (see Problem 11.16); however, formal use of such a transformation is very inefficient. Apparently effective use of substructures requires a large programming effort in organization of data files and internal bookkeeping. Therefore it is not immediately obvious that substructuring is more economical than direct treatment of the complete system (14, 25).

Substructures may be used in dynamic analysis. The method has also been applied to transient heat conduction (16). The procedure is not simple to explain, but procedural details have some similarity to the method used in static analysis and to the "eigenvalue economizer" of Section 12.4. Pertinent references include Refs. 16 through 20 and 24.

11.10   Repeating Substructures

Repeatability and symmetry of geometry and loading permit behavior of an entire structure to be predicted by analysis of only a small part of it. As noted in Ref. 10, "broadly speaking each axis of symmetry halves the size of the system to be analyzed . . . . In numerous cases however *a repetition of structural form and loading* is present while no axes of symmetry exist. For such situations economies of similar type can be made by use of *the principle of repeatability.*"

Consider, for example, Fig. 11.10.1, where the structure is divided into six regions that are identical in geometry and loading. Each region is a substructure. However, the present problem is simpler than that described in Section 11.9, as analysis of a single substructure is all that is required. There is no need to assemble substructures nor to analyze the assembly.

The stiffness equations $[K]\{D\} = \{R\}$ for a typical substructure (Fig. 11.10.1b) may be written (10)

$$\begin{bmatrix} K_{II} & K_{IA} & K_{IB} \\ K_{IA}^T & K_{AA} & K_{AB} \\ K_{IB}^T & K_{AB}^T & K_{BB} \end{bmatrix} \begin{Bmatrix} D_I \\ D_A \\ D_B \end{Bmatrix} = \begin{Bmatrix} R_I \\ R_A \\ R_B \end{Bmatrix} + \begin{Bmatrix} 0 \\ P_A \\ P_B \end{Bmatrix} \qquad (11.10.1)$$

where each term represents a submatrix. Forces $\{P_A\}$ and $\{P_B\}$ are applied along AA and BB, respectively, by the neighboring substructures. The remaining forces $\{R_I\}$, $\{R_A\}$, and $\{R_B\}$ arise from applied loading on the

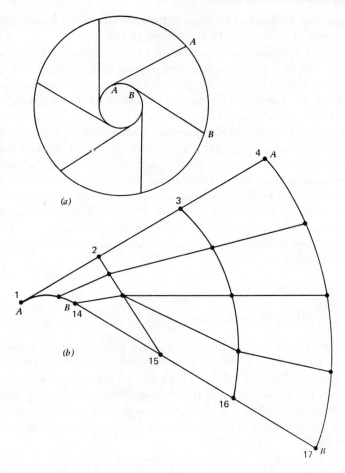

Figure 11.10.1. (a) Plane circular structure. (b) Typical repeating finite element substructure.

structure (concentrated, distributed, thermal, etc.). Forces $\{R_I\}$ act on internal nodes, and $\{R_A\}$ and $\{R_B\}$ act on the boundary nodes along AA and BB. It is convenient to place all external loads along a boundary in either $\{R_A\}$ or $\{R_B\}$; if mistakenly placed in both, the structure receives twice the load intended.

As all substructures are identical, it follows that (10)

$$\{D_A\} = \{D_B\}, \qquad \{P_A\} = -\{P_B\} \tag{11.10.2}$$

The operations described in Section 11.8 enforce the relation $\{D_A\} = \{D_B\}$.

If the d.o.f. $\{D_B - D_A\}$ replaces (say) $\{D_B\}$, the transformed equations are

$$\begin{bmatrix} K_{II} & K_{IA} + K_{IB} & K_{IB} \\ K_{IA}^T + K_{IB}^T & K_{22} & K_{AB} + K_{BB} \\ K_{IB}^T & K_{AB}^T + K_{BB} & K_{BB} \end{bmatrix} \begin{Bmatrix} D_I \\ D_A \\ D_B - D_A \end{Bmatrix} = \begin{Bmatrix} R_I \\ R_A + R_B \\ R_B + P_B \end{Bmatrix}$$

$$(11.10.3)$$

where $[K_{22}] = [K_{AA}] + [K_{AB}] + [K_{AB}]^T + [K_{BB}]$ and $\{P_A + P_B\}$ disappears because of Eqs. 11.10.2. Following a standard method of boundary condition treatment, we enforce $\{D_B\} = \{D_A\}$ by discarding the third row and column in Eq. 11.10.3. This leaves

$$\begin{bmatrix} K_{II} & K_{IA} + K_{IB} \\ K_{IA}^T + K_{IB}^T & K_{AA} + K_{AB} + K_{AB}^T + K_{BB} \end{bmatrix} \begin{Bmatrix} D_I \\ D_A \end{Bmatrix} = \begin{Bmatrix} R_I \\ R_A + R_B \end{Bmatrix}$$

$$(11.10.4)$$

as the set of equations that represents the entire structure of repeating substructures.

Note that Eqs. 11.10.4 can be formed *automatically* by proper node numbering of the substructure. In Fig. 11.10.1b, for example, the desired result follows if nodes 14, 15, 16, and 17 are renumbered 1, 2, 3, and 4 while leaving all other node numbers unchanged.

A warning must be made with reference to Eqs. 11.10.2; these equations presume that similar local coordinate systems prevail at each section. For example, in Fig. 11.10.1 one might adopt tangential and normal axes on AA and separate tangential and normal axes on BB. If a single cartesian axis system is used for the entire substructure, then a coordinate transformation of the form $\{D_A\} = [T]\{D_B\}$ must precede use of Eqs. 11.10.2.

PROBLEMS

11.1. (a) Derive Eqs. 11.2.8 from Eq. 11.2.1.
(b) Verify Eq. 11.2.9.
(c) Write the matrices $[T_\sigma]$ and $[T_\epsilon]$ for the axisymmetric but layered material shown in the sketch, whose principal directions of orthotropy in any $\theta = $ constant plane are $x'y'$.

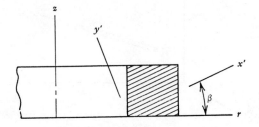

11.2. (a) Verify that matrix $[T]$ of Eq. 11.3.4 is orthogonal.

(b) Write $[k']$ of Eq. 11.3.7, and carry out the transformation to obtain the explicit form of $[k]$, valid for any member orientation.

11.3. Suppose that angle ϕ of Fig. 11.4.1 is redefined, as in the present sketch. What portions of Section 11.4 are affected, and what are the specific changes?

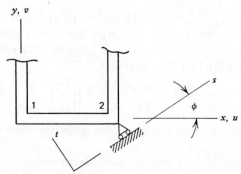

11.4. A triangular plate element of a type we have not discussed (see Refs. 2 and 3) has at the middle of each edge a rotational d.o.f. whose vector is parallel to the edge. Thus, in the assembled structure each such d.o.f. is common to two elements. A *shell* may be modeled by joining these elements to form a faceted surface. Must midside rotations be transformed from local to global coordinates prior to assembly of elements? If so, define the transformation.

11.5. A two-force member is arbitrarily oriented in space, as shown. Global coordinates of the two nodes and axial stiffness AE are known. The element stiffness matrix in local coordinates is 2 by 2.

(a) How may a unique set of local coordinates $x'y'z'$ be established and a complete table of direction cosines computed?

(b) However, only direction cosines l_1, m_1, and n_1 are needed for coordinate transformation. Write the form of Eqs. 11.5.5 appropriate to the present problem.

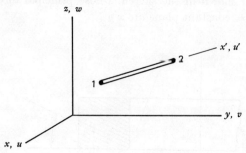

(c)   The stiffness matrix $[k']$ in local coordinates may be decomposed into the form $[k'] = [U]^T[U]$, hence in global coordinates

$$[k] = ([U][T])^T([U][T])$$

Pursue this development to write a compact set (say 13) of Fortran statements to compute $[k]$ from the given information.

11.6.   (a)   Refer to Fig. 11.6.1, and write by inspection a reasonable 8 by 6 transformation matrix $[T_1]$ for the transformation $\{d\} = [T_1]\{d'\}$.

(b)   See if $[T_1]$ agrees with Eq. 11.6.1. That is, see if $[T_1][T]$ is a unit matrix.

11.7.   (a)   Write the transformation matrix $[T]$ in the relation $\{d'\} = [T]\{d\}$ for the problem of Fig. 11.6.2, where $\{d\}$ contains θ in place of v_4.

(b)   If other elements also contain node 4 in Fig. 11.6.2, is further transformation required? Explain.

(c)   Repeat Part a, but consider now that θ relates displacements normal to edge 2–3, which is arbitrarily oriented. Let $\{d\}$ contain θ in place of u_2.

11.8.   (a)   For the problem of Fig. 11.6.3, write $[T]$ for the transformation suggested below Eq. 11.6.3.

(b)   Repeat Part a, but let the bar join arbitrary points on sides 1–2 and 3–4 of a quadrilateral element of arbitrary shape.

11.9.   (a)   Let all members at node i in Fig. 11.7.2 have their d.o.f. θ_i condensed prior to assembly. What deficiency exists in the structure stiffness matrix and how may it be avoided?

(b)   Let (say) three plane frame members be rigidly joined to one another. If the joint is free to rotate on its support, how might this condition be treated?

(c)   Suppose two joints of the type mentioned in Part b are to be pinconnected to one another. How might this condition be treated?

11.10.   Is any alteration of Eq. 11.7.1 required if the member of Fig. 11.7.1 is arbitrarily oriented? If rigid end pieces of Fig. 11.7.1 and the pin connection of Fig. 11.7.2 are *both* used, what specific transformations are required?

11.11.   Spring stiffnesses are k_1 and k_2, nodal displacements are D_1 and D_2, and applied nodal loads are P_1 and P_2. Use the transformation of Section 11.8 to enforce the equality $D_1 = D_2$, then solve for D_1 and D_2.

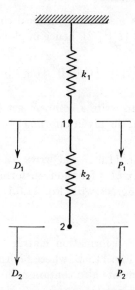

11.12. How are Eqs. 11.10.3 and 11.10.4 changed if $\{D_A - D_B\}$ replaces $\{D_B\}$? And if $\{D_B - D_A\}$ replaces $\{D_A\}$?

11.13. Problem 6.15 asked for modification of shape functions to introduce the "departures from linearity" u_i' and v_i'. Suppose that u_i' and v_i' are to be introduced *after* the element stiffness matrix has been written using the usual d.o.f. Write the required transformation matrix $[T]$.

11.14. Starting with Eqs. 6.3.4, verify that Fig. 6.3.1c results from transformation of Fig. 6.3.1a.

11.15. The elements are unit squares in plane strain and there are two d.o.f. per node. Suppose that no volume change is permitted for any element (21), that is, $\int (\epsilon_x + \epsilon_y)\, dx\, dy = 0$ for each element. How many d.o.f. are thus eliminated? What transformation may be set up at the element level? And at the structure level? Which is preferable?

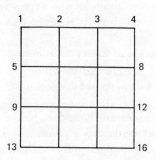

11.16. The sketch below shows two axial force members. Thus with $i = 1, 2$ we have element displacement and force arrays $\{d_i\}$ and $\{r_i\}$, each 4 by 1, and element stiffness matrices $[k_i]$, each 4 by 4. If the two elements are joined by connecting node b to node c, we obtain the six structure equations $[K]\{D\} = \{R\}$.

(a) Write the displacements $\{d_1\}$ and $\{d_2\}$ consecutively in a 8 by 1 vector $\{d_1 \quad d_2\}$. Then write the transformation matrix $[T]$ in the relation $\{d_1 \quad d_2\} = [T]\{D\}$.

(b) Similarly, write an 8 by 1 vector of nodal forces $\{r_1 \quad r_2\}$. Obtain the relation between $\{R\}$ and $\{r_1 \quad r_2\}$. *Suggestion.* In any virtual displacement, the total virtual work of element forces must equal the virtual work of structure forces.

(c) Write the diagonal matrix of element stiffness matrices $[k_d]$ $= [k_1 \quad k_2]$. In the present example $[k_d]$ is 8 by 8 and contains two 4 by 4 null matrices off-diagonal. Derive the transformation relation between $[k_d]$ and $[K]$.

(d) Actually carry out the transformations to verify the correctness of $\{R\}$ and $[K]$.

Note. In early matrix structural analysis, the transformations of this problem were used to assemble a structure from separate elements. The method of direct addition of matrices, described in Chapter Two, was introduced later.

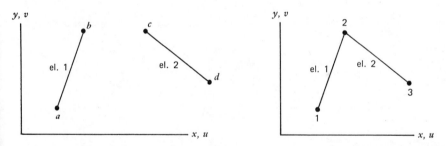

11.17. Verify Eqs. 4.2.16 by means of the virtual work argument (e.g., as used in Eqs. 11.6.2).

11.18. Show that transformation of a mass matrix is given by an equation similar to Eq. 11.3.7, namely $[M] = [T]^T[M'][T]$. *Suggestion.* Consider kinetic energy.

CHAPTER TWELVE

Dynamics and Vibrations

12.1 Introduction

In this chapter attention is directed toward those aspects of dynamics and vibrations most closely associated with finite element work. The subject of dynamics has its own extensive literature. Methods for extraction of eigenvalues and computation of dynamic response have been developing rapidly in recent years, so that many competing schemes are available. The methods of analytical and computational dynamics often presume that mass and stiffness matrices are available as "raw material" to be processed. Accordingly, we will not attempt to be comprehensive, but will formulate the dynamic problem and sample some of the computational methods that can be used for its solution.

12.2 Dynamic Equations. Mass and Damping Matrices

We begin with a derivation of the dynamic equations. In the same notation as used in Chapter Four, the displacement of a typical point in an element and its velocity are

$$\{f\} = \{u \quad v \quad w\} = [N]\{d\}, \qquad \{\dot{f}\} = [N]\{\dot{d}\} \tag{12.2.1}$$

where a dot denotes differentiation with respect to time t. Let $\rho = $ mass density. The kinetic energy T_e of an element is

$$T_e = \int_{\text{vol}} \frac{\rho}{2} \{\dot{f}\}^T \{\dot{f}\} \, dV = \tfrac{1}{2}\{\dot{d}\}^T \int_{\text{vol}} \rho[N]^T[N] \, dV\{\dot{d}\} \tag{12.2.2}$$

The latter integral is called the *mass matrix* of the element, that is,

$$[m] = \int_{\text{vol}} \rho [N]^T [N] \, dV \qquad (12.2.3)$$

Clearly the mass matrix is symmetric. The total kinetic energy of the structure is the sum of element energies:

$$T = \sum T_e = \sum \tfrac{1}{2} \{d\}^T [m]\{d\} = \tfrac{1}{2}\{\dot{D}\}^T [M]\{\dot{D}\} \qquad (12.2.4)$$

where the now familiar expansion of element matrices to "structure size" and summation of overlapping terms is implied.

To obtain a damping matrix we presume the existence of dissipative forces proportional to relative velocities. The dissipation function F (1, 2) is, for one element and for the structure, respectively,

$$F_e = \tfrac{1}{2}\{d\}^T [c]\{d\}, \qquad F = \tfrac{1}{2}\{\dot{D}\}^T [C]\{\dot{D}\} \qquad (12.2.5)$$

where $[c]$ is the element *damping matrix*. The usual assembly process is implied in summing element $[c]$ matrices to form $[C]$. How $[c]$ is to be constructed need not concern us until later. For the moment we simply note that $[c]$ is symmetric, contains quantities independent of the velocities, and that a typical column j of $[c]$ gives the array of generalized forces that arises when the jth generalized coordinate is given a unit velocity while all other coordinates are held fixed.

Lagrange's equations are (1, 2):

$$L = T - \Pi_p$$
$$\frac{d}{dt}\left\{\frac{\partial L}{\partial \dot{D}}\right\} - \left\{\frac{\partial L}{\partial D}\right\} + \left\{\frac{\partial F}{\partial \dot{D}}\right\} = 0 \qquad (12.2.6)$$

where T and F are defined by Eqs. 12.2.4 and 12.2.5, and Π_p is the potential energy, used in Section 4.2 and including strain energy and the potential of applied loads. Substituting Eqs. 4.2.13, 12.2.4, and 12.2.5 into Eq. 12.2.6 we obtain

$$[M]\{\ddot{D}\} + [C]\{\dot{D}\} + [K]\{D\} = \{R\} \qquad (12.2.7)$$

as the governing equation of motion for discretized system. Note that if all time derivatives vanish, then $L = -\Pi_p$, and Eq. 12.2.6 yields again the static equation (4.2.8) in Section 4.2.

As suggested by Eq. 12.2.7, a column j of $[M]$ (or of $[m]$) represents the set of generalized forces applied to the structure (or element) by a node and arising from a unit acceleration of generalized coordinate j while all other d.o.f. have zero acceleration.

Damping matrix $[c]$ is easily evaluated for a Newtonian fluid; its terms have been given by Rayleigh (1). Of more interest in structural mechanics is hysteretic or solid damping and Coulomb or dry friction. In solid damping, forces are opposite in direction to velocity vectors but are dependent on stress, not velocity, for their magnitude (2, 3). Clearly solid damping and dry friction cannot be represented by matrix $[c]$ as we have defined it. Solid or dry friction may be modeled by a fictitious viscous damping of the form

$$[C] = \alpha[M] + \beta[K] \tag{12.2.8}$$

where constants α and β must be chosen to suit the problem at hand. With $[C]$ defined in this way, Eqs. 12.2.7 have normal modes; that is, Eqs. 12.2.7 may be decoupled into independent differential equations, one for each d.o.f. Further information about damping matrices appears in the literature, for example Refs. 1 through 6.

The mass matrix of Eq. 12.2.3 was first derived by Archer (7, 11). When shape functions $[N]$ are identical to those used in formulating the element stiffness matrix, $[m]$ is called the *consistent* mass matrix. It is always positive definite. Alternatively, simpler forms of $[N]$ may be used in computing $[m]$. The simplest mass matrix is that obtained by placing point masses m_i at the displacement d.o.f. The latter form of $[m]$ is called a *lumped* mass matrix. The lumped mass formulation is nearly exact if small but massive objects are placed at nodes of a lightweight structure. The consistent formulation is exact if the actual deformed shape under dynamic conditions is contained in the shape functions $[N]$.

For examples of lumped mass matrices, consider first a plane two-force member of total mass m. Nodal d.o.f. are $\{d\} = \{u_1 \quad v_1 \quad u_2 \quad v_2\}$. The lumped mass matrix is

$$[m] = \begin{bmatrix} \dfrac{m}{2} & 0 & 0 & 0 \\[2ex] 0 & \dfrac{m}{2} & 0 & 0 \\[2ex] 0 & 0 & \dfrac{m}{2} & 0 \\[2ex] 0 & 0 & 0 & \dfrac{m}{2} \end{bmatrix} \tag{12.2.9}$$

For a plane beam element of total mass m and with $\{d\} = \{v_1 \quad \theta_1 \quad v_2 \quad \theta_2\}$, the lumped $[m]$ is 4 by 4 and the only nonzero coefficients are $m_{11} = m_{33} = m/2$. Note that for any valid lumped (or consistent) mass matrix, Newton's second law $\{F\} = [m]\{\ddot{d}\}$ must yield correct forces $\{F\}$ when $\{\ddot{d}\}$ represents any rigid body translational acceleration. For example, with $\{d\} = \{a \quad 0 \quad a \quad 0\}$, Eq. 12.2.9 yields $F_1 = F_3 = ma/2$, so that $F_1 + F_3 = ma$.

The lumped mass matrix is a diagonal matrix. The consistent mass matrix is fully populated and more time consuming to generate.[1] There are computational advantages if mass matrices are diagonal instead of full, as we will see below. However, if the lumped mass approximation is crude, many elements may be needed to obtain sufficiently accurate results. Thus we have the question: can we say which type of mass matrix is generally to be preferred? The answer appears to be "no." Use of lumped masses usually tends to lower natural frequencies, while the excess stiffness present in a mesh of compatible elements tends to raise them. Thus the two opposite tendencies tend to cancel, and for a given mesh size lumped masses may give more accurate results than a consistent formulation (9). If *in*compatible elements are used with lumped masses, computed frequencies may be quite low. In the case of (compatible) beam elements, use of lumped masses gives frequencies that may be about 25% low (7). However, this large error represents a particularly unfavorable test case, and it is more often true that lumped and consistent mass matrices yield errors of about the same magnitude in beam problems. A formulation less crude than that given by lumped masses is preferred for use with higher order elements such as a 21 d.o.f. triangular plate element (10, 35). Clearly lumped masses would not be desirable in the limiting case of a single complicated element without internal d.o.f., which represents an entire structure. Use of the consistent formulation permits bounds to be set on computed frequencies (Section 12.3). However, most workers appear to favor use of lumped masses for the best combination of acceptable accuracy and computational efficiency (4, 5, 8, 9). Felippa (6 of Chapter Six) finds that in plate vibration problems the lumped mass analysis is some 15 times faster than the consistent mass analysis. Reference 5 advises that the lumped mass formulation must be used in any wave propagation problem if reliable answers are to be obtained.

The preceding paragraph suggests that for elements having internal d.o.f. or "extra" incompatible modes, it is not worthwhile to associate mass with the added d.o.f. (the main purpose of the added d.o.f. is to improve the *stiffness* properties of the element). However, if the complete consistent mass matrix is formed, note that the extra d.o.f. may not be simply condensed as if the mass matrix were a stiffness matrix. Condensation of mass requires use of stiffness coefficients, as described in Section 12.4.

In closing we note that mass matrices for shells and beams are occasionally stated explicitly in published papers (7, 11, 12). References for straight and curved beams appear in Section 17.3.

---

[1] This is especially true for nonuniform construction, as for a sandwich plate using the eight-node element of Chapter Nine. Hence we may adopt lumped masses. But to include rotary inertia, which may be important in such problems, each mass m_i may be taken not as a particle but as a prismatic bar that spans the thickness t. Thus, moment of inertia is $m_i t^2/12$ associated with each rotational d.o.f. (8).

12.3 Natural Frequencies. The Eigenvalue Problem

Let us consider natural frequencies of vibration, in the absence of damping and with no externally applied forces (2). All particles execute harmonic motion in phase with one another, therefore

$$\{D\} = \{\bar{D}\} \sin \omega t, \qquad \{\ddot{D}\} = -\omega^2 \{\bar{D}\} \sin \omega t \qquad (12.3.1)$$

where $\{\bar{D}\}$ represents amplitudes of the d.o.f. and ω is the circular frequency. Equations 12.2.7 and 12.3.1 yield

$$([K] - \omega^2[M])\{\bar{D}\} = 0 \qquad (12.3.2)$$

which is called a "linear" eigenvalue problem as neither $[K]$ nor $[M]$ is a function of ω. Equation 12.3.2 is satisfied for nonzero $\{\bar{D}\}$ only if the determinant of $([K] - \omega^2[M])$ vanishes. There are as many eigenvalues (natural frequencies of vibration) as there are d.o.f. if $[M]$ is a consistent mass matrix. If $[M]$ is a lumped mass matrix some of its rows may be zero, and there are as many eigenvalues as there are nonzero rows of $[M]$. To each eigenvalue ω_i there corresponds an eigenvector (natural mode of vibration) $\{\bar{D}_i\}$. Different eigenvectors may have identical or nearly identical eigenvalues. The $\{\bar{D}_i\}$ corresponding to $\omega_i = 0$ is a rigid body motion; however, the structure is usually supported so that eigenvalues $\omega_i = 0$ are not encountered. If $[M]$ is positive definite, all eigenvalues are real; if $[K]$ is also positive definite, all eigenvalues are both real and positive (13). Eigenvectors are orthogonal with respect to the mass and stiffness matrices (2); that is, for $i \neq j$

$$\{\bar{D}_i\}^T[M]\{\bar{D}_j\} = \{\bar{D}_i\}^T[K]\{\bar{D}_j\} = 0 \qquad (12.3.3)$$

The Rayleigh quotient (1)

$$\omega^2 = \frac{\{\bar{D}\}^T[K]\{\bar{D}\}}{\{\bar{D}\}^T[M]\{\bar{D}\}} \qquad (12.3.4)$$

may be obtained by premultiplying Eq. 12.3.2 by $\{\bar{D}\}^T$ and solving for ω^2. It has the useful property that when $\{\bar{D}\}$ approximates a natural mode with error of order ϵ, the corresponding natural frequency is computed with error of only order ϵ^2 (1, 13). The quotient is stationary when $\{\bar{D}\}$ varies in the neighborhood of an eigenvector.

The Rayleigh-Ritz method overestimates the natural frequencies (13), as might be expected because of the constraints that are added by an assumed and approximate displacement field. Compatibility and consistent use of assumed fields are required in the classical Rayleigh-Ritz method. Accordingly we expect the finite element method to also yield upper bounds on the natural frequencies, provided that the formulation is based on assumed displacement fields, the volume of the structure is correctly represented by the mesh layout, elements are compatible and not "softened" by low order integration rules, and consistent mass matrices are used. If any one of these conditions is violated, upper bounds cannot be guaranteed.

The "standard form" for an eigenvalue problem is

$$([A] - \lambda[I])\{Y\} = 0 \tag{12.3.5}$$

where $[A]$ is a *symmetric* matrix and $[I]$ is a unit matrix. Eigenvalues λ_i that satisfy this equation are known as "eigenvalues of $[A]$." Many eigenvalue extraction routines require the standard form, so we will now mention some ways to get it.

Clearly premultiplication of Eq. 12.3.2 by $[M]^{-1}$ is not acceptable as $[M]^{-1}[K]$ is not in general symmetric. But in the case of lumped masses we have a diagonal mass matrix $\lceil M \rfloor$ and may proceed as follows (5). We define

$$\lceil M \rfloor^{1/2} = \lceil \sqrt{M_{11}} \quad \sqrt{M_{22}} \quad \ldots \quad \sqrt{M_{nn}} \rfloor \tag{12.3.6}$$

$$\{D\} = \lceil M \rfloor^{-1/2}\{Y\} \tag{12.3.7}$$

where

$$[M]^{-1/2} = \lceil 1/\sqrt{M_{11}} \quad 1/\sqrt{M_{22}} \quad \ldots \quad 1/\sqrt{M_{nn}} \rfloor$$

Substituting Eq. 12.3.7 into Eq. 12.3.2, then premultiplying by $\lceil M \rfloor^{-1/2}$ and noting that $[M] = \lceil M \rfloor^{1/2}\lceil M \rfloor^{1/2}$, we obtain the standard form of Eq. 12.3.5 with

$$[A] = \lceil M \rfloor^{-1/2}[K]\lceil M \rfloor^{-1/2}, \qquad \lambda = \omega^2 \tag{12.3.8}$$

Natural modes $\{D\}$ are recovered from eigenvectors $\{Y\}$ by use of Eq. 12.3.7.

The preceding reduction fails if any diagonal mass M_{ii} is zero. This happens if rotational d.o.f. are present but not associated with any rotary inertia. In such a case the first step is to remove the rotational d.o.f. by condensation. That is, we follow the procedure described by Eqs. 6.1.2 through 6.1.4 where $\{d_2\}$ represents now the rotational d.o.f. of the structure.

A possible second step involves inversion of the condensed stiffness matrix. Both steps may be combined in a procedure adapted to the sparsity or bandedness of $[K]$ and the need to retain only translational displacement freedoms in the reduced stiffness matrix (5, 14). Let $[F]$ represent the inverse of the *reduced* structure stiffness matrix. We obtain a column $\{F_j\}$ of $[F]$ by solving the equations $[K]\{F_j\} = \{e_j\}$ where $\{e_j\}$ is a unit vector consisting of unity for *translational* d.o.f. j and zero for all other d.o.f. Thus, in physical terms, $[F]$ is built of the displacement vectors resulting from application of a unit force to each translational d.o.f. in turn. Rows of the $\{F_j\}$ corresponding to rotational d.o.f. are discarded, hence assembly of the $\{F_j\}$ yields a reduced, square matrix $[F]$ associated with translational d.o.f.

only. We need compute only the upper or lower triangle of $[F]$ because of its symmetry. Further comments on efficient programming are given in Ref. 14. The reduced form of Eq. 12.3.2 is premultiplied by $[F]/\omega^2$ to yield

$$\left([F][M] - \frac{1}{\omega^2}[I]\right)\{D_r\} = 0 \qquad (12.3.9)$$

where $\{D_r\}$ contains only translational d.o.f., the diagonal of $[M]$ is solidly filled, and in general $[F]$ is (unfortunately) a full matrix. We now proceed much as in Eqs. 12.3.6 through 12.3.8 to obtain the standard form of Eq. 12.3.5:

$$[A] = [M]^{1/2}[F][M]^{1/2}$$

$$\{Y\} = [M]^{1/2}\{D_r\}, \qquad \lambda = \frac{1}{\omega^2}$$

$$(12.3.10)$$

The latter manipulations are more time consuming than forming Eq. 12.3.8 and cannot be applied to unconstrained structures, for which $[K]$ is singular. But Eqs. 12.3.10 have two advantages stemming from the fact that the highest λ's correspond to the lowest ω's. First, the lowest ω's are usually of most interest, but many eigenvalue routines allow the user to find any number of eigenvalues, starting with the *highest*. Second, standard routines often determine higher eigenvalues more accurately than lower ones (5, 14).

If consistent mass matrices are used, the diagonal of $[M]$ is fully populated, and there are off-diagonal terms as well. To obtain the standard form we first premultiply Eq. 12.3.2 by $[M]^{-1}$ to obtain

$$([M]^{-1}[K] - \omega^2[I])\{D\} = 0 \qquad (12.3.11)$$

We define (5, 14)

$$[M] = [L][L]^T, \qquad \{D\} = [L]^{-T}\{Y\} \qquad (12.3.12)$$

where $[L]$ is a lower triangular matrix obtained by Choleski decomposition. Substitution of Eq. 12.3.12 into Eq. 12.3.11 and premultiplication by $[L]^T$ yields the standard form of Eq. 12.3.5, in which

$$[A] = [L]^{-1}[K][L]^{-T}, \qquad \lambda = \omega^2 \qquad (12.3.13)$$

The relation between $[A]$ and $[M]^{-1}[K]$ is

$$[A] = [L]^T[L]^{-T}[A] = [L]^T([M]^{-1}[K])[L]^{-T} \qquad (12.3.14)$$

This operation is known as a "similarity transformation," hence according to the theory of linear algebra $[A]$ and $[M]^{-1}[K]$ have the same eigenvalues (15). Natural modes $\{D\}$ are recovered from eigenvectors $\{Y\}$ by use of Eq. 12.3.12.

If $\lambda_{\max}$ and $\lambda_{\min}$ of $[M]$ are widely separated a procedure different from that of the preceding paragraph may be needed to preserve accuracy (5). Efficient computer operations are discussed in Ref. 14. Instead of decomposing $[M]$ we may decompose $[K]$ into $[L][L]^T$ and again produce the standard form, but with Eq. 12.3.13 replaced by

$$[A] = [L]^{-1}[M][L]^{-T}, \qquad \lambda = \frac{1}{\omega^2} \qquad (12.3.15)$$

Unfortunately matrix $[A]$ is a full matrix.

Two additional standard forms may be produced by inverting the standard form Eq. 12.3.5, using $[A]$ first from Eq. 12.3.13 and then from Eq. 12.3.15.

If a structure is partially or completely unconstrained the formulation of Eqs. 12.3.15 cannot be obtained directly, because $[K]$ is singular. A simple approach in such cases is that of adding and subtracting $c[M]$ from Eq. 12.3.2, where c is a positive constant chosen by the analyst (21, 26). Thus we obtain

$$([K + cM] - (c + \omega^2)[M])\{\bar{D}\} = 0 \qquad (12.3.16)$$

If the diagonal of $[M]$ is full, $[K + cM]$ is nonsingular and Eq. 12.3.16 may be cast in the form of Eq. 12.3.15, where eigenvalues are now $\lambda = 1/(c + \omega^2)$. Disadvantages of Eq. 12.3.16 include some wasted effort because rigid body modes $\omega = 0$ are now extracted by the eigenvalue solution, and slower convergence if the solution is iterative because adjacent eigenvalues are now closer together (21). References 2 and 22 give alternate ways of treating unconstrained structures.

12.4   Condensation (the "Eigenvalue Economizer")

Extraction of eigenvalues and eigenvectors is a much more expensive operation than solution of simultaneous linear equations. It requires roughly twice as long to get a single eigenvalue as to do a single static analysis. A typical structure may have too many d.o.f. for economical treatment. Accordingly we call on the "eigenvalue economizer" (5, 16, 17, 18) to eliminate many or most d.o.f. from the problem. The general method is not limited to vibration analysis or even to dynamics; it is essentially a technique for reducing the size of a system of equations. However, we will describe the method with reference to the vibration problem. In brief, the method intends to model the complete structure with a lesser number of d.o.f., just as the structure itself, with only a finite number of d.o.f., models a reality having infinitely many d.o.f.

Let us suppose that amplitudes $\{\bar{D}\}$ of the complete set of n d.o.f. are related to amplitudes $\{D_r\}$ of m d.o.f. by the equation

$$\underset{n\times 1}{\{D\}} = \underset{n\times m}{[T]}\underset{m\times 1}{\{D_r\}} \tag{12.4.1}$$

where m is usually much smaller than n. Freedoms $\{D_r\}$ may be contained in $\{D\}$. For the moment we defer comment on how to establish matrix $[T]$. By substituting Eq. 12.4.1 into Eq. 12.3.2 and premultiplying by $[T]^T$ we obtain the reduced system[2]

$$([K_r] - \omega^2[M_r])\{D_r\} = 0 \tag{12.4.2}$$

where

$$\underset{m\times m}{[K_r]} = [T]^T\underset{n\times n}{[K]}[T], \qquad \underset{m\times m}{[M_r]} = [T]^T\underset{n\times n}{[M]}[T] \tag{12.4.3}$$

Equation 12.4.2 may be converted to the standard form by one of the techniques described in Section 12.3. *Note that even if $[M]$ is a diagonal matrix, the transformation of Eq. 12.4.3 produces a more fully populated matrix.* When eigenvalues $\{D_r\}$ have been found, the complete vectors $\{D\}$ may be recovered as discussed in the paragraph following Eq. 12.4.9.

An argument that establishes the transformation matrix $[T]$ is as follows (17). Suppose, for the sake of argument, that the structure stiffness matrix $[K]$ has been inverted to yield the flexibility matrix $[F] = [K]^{-1}$, and the static structural equations are then partitioned.

$$\begin{bmatrix} F_{11} & F_{12} \\ F_{21} & F_{22} \end{bmatrix}\begin{Bmatrix} R_1 \\ R_2 \end{Bmatrix} = \begin{Bmatrix} D_1 \\ D_2 \end{Bmatrix} \tag{12.4.4}$$

Now let $\{R_2\} = 0$, hence $\{R_1\} = [F_{11}]^{-1}\{D_1\}$. With $[I] =$ a unit matrix we therefore obtain

$$\begin{Bmatrix} D_1 \\ D_2 \end{Bmatrix} = \begin{bmatrix} F_{11} \\ F_{21} \end{bmatrix}\{R_1\} = \begin{bmatrix} F_{11} \\ F_{21} \end{bmatrix}[F_{11}]^{-1}\{D_1\} = \begin{bmatrix} I \\ F_{21}F_{11}^{-1} \end{bmatrix}\{D_1\} \quad (12.4.5)$$

Or, in the notation of Eq. 12.4.1, $\{D\} = [T]\{D_r\}$, where

$$\{D\} = \begin{Bmatrix} D_1 \\ D_2 \end{Bmatrix}, \qquad \{D_r\} = \{D_1\}, \qquad [T] = \begin{bmatrix} I \\ F_{21}F_{11}^{-1} \end{bmatrix} \tag{12.4.6}$$

Thus matrix $[T]$ is established, at least symbolically. Physically, what has been done is to say that only freedoms $\{D_1\}$ carry applied load, and that freedoms $\{D_2\}$ simply follow along in whatever manner is dictated by the geometry and elastic properties of the structure. Accordingly, freedoms $\{D_2\}$ are designated "slaves" and $\{D_1\}$ "masters" (18). We see from Eqs. 12.4.3 and 12.4.6 that the reduced mass matrix contains a mixture of mass and stiffness coefficients.

---

[2] This operation is a coordinate transformation (see Chapter Eleven).

The meaning of the transformation may be further clarified after explaining the computational process used to get it (17). We write the structural equations for several load vectors, but with $[R_2] = 0$ and $[R_1]$ equal to a unit matrix.

$$\begin{bmatrix} K_{11} & K_{12} \\ K_{21} & K_{22} \end{bmatrix} \begin{bmatrix} D_1 \\ D_2 \end{bmatrix} = \begin{bmatrix} I \\ 0 \end{bmatrix} \tag{12.4.7}$$

Solution of these equations yields, since $[F] = [K]^{-1}$,

$$\begin{bmatrix} D_1 \\ D_2 \end{bmatrix} = \begin{bmatrix} F_{11} & F_{12} \\ F_{21} & F_{22} \end{bmatrix} \begin{bmatrix} I \\ 0 \end{bmatrix} = \begin{bmatrix} F_{11} \\ F_{21} \end{bmatrix} \tag{12.4.8}$$

Matrix $[F_{11}]$ is usually not large and may be inverted at modest cost. Solution of Eq. 12.4.7 for displacements is carried out by (say) Gauss elimination. This also is probably done at modest cost because a static analysis often accompanies a dynamic analysis, and Eq. 12.4.8 then requires only the addition of a few load cases. The partitioning seen in Eqs. 12.4.4 through 12.4.8 will in practice probably be done by suitable subscripting and bookkeeping operations instead of by preliminary rearrangement of the contents of arrays. References 17 and 18 give further comments on computer operations.

The reduced set $\{\bar{D}_r\}$ is capable of representing any displaced shape that may be formed by superposition (with arbitrary amplitudes) of the static modes that result from unit loads applied one at a time to the "master" freedoms. Note that this statement helps motivate the condensation of rotational d.o.f. mentioned in Section 12.3. It is left to the reader as an exercise to relate the present condensation to the condensation of internal d.o.f. described in Section 6.1 (Problem 12.10).

An alternate method for establishing transformation matrix $[T]$ should be mentioned (5). It begins with the assumption of a matrix $[G]$, composed of independent displacement vectors $\{G_j\}$. These vectors may each contain all zeros except for a single 1. In any case a linear combination of these "master" displacements should be able to approximate any vibration mode of interest. We solve for $[P]$ and then $[T]$ in the equations

$$\underset{n \times m}{[P]} = \underset{n \times n}{[M]} \underset{n \times m}{[G]}, \qquad \underset{n \times n}{[K]} \underset{n \times m}{[T]} = \underset{n \times m}{[P]} \tag{12.4.9}$$

where m is considerably less than n. Thus $[T]$ is composed of the static displacement modes induced by forces $[P]$. A column $\{P_j\}$ of $[P]$ contains inertia forces associated with motion at unit frequency in the mode $\{G_j\}$. The coordinate transformation is $\{\bar{D}\} = [T]\{\bar{D}_r\}$, which constrains $\{D\}$ to be a linear combination of amplitudes in $\{\bar{D}_r\}$. Note that the stiffness transformation is simply $[K_r] = [T]^T[P]$. The present formulation differs from that of Eq. 12.4.6 in that it begins with modes $[G]$ instead of unit nodal forces. Also, no matrix such as $[F_{11}]$ is inverted. Modes $\{D_r\}$ are not contained in $\{D\}$ so that $\{D\}$ must be recovered in order to see a mode shape.

Each complete mode $\{D\}$ may be recovered from its corresponding reduced mode $\{D_r\}$ by the transformation $\{D\} = [T]\{D_r\}$ (Eq. 12.4.1). However, if $[T]$ is established according to Eq. 12.4.6, there is a defect in this transformation arising from the fact that the slave nodes carry no load in the static analysis that generates $[T]$. Actually, inertia loads should be applied to slave nodes when $\{D\}$ is recovered (73). We should partition the dynamic equations as

$$\left(\begin{bmatrix} K_{11} & K_{12} \\ K_{21} & K_{22} \end{bmatrix} - \omega^2 \begin{bmatrix} M_{11} & M_{12} \\ M_{21} & M_{22} \end{bmatrix} \right) \begin{Bmatrix} D_1 \\ D_2 \end{Bmatrix} = 0 \qquad (12.4.10)$$

where $\{D_1\} = \{D_r\}$. The second of Eqs. 12.4.10 is then solved for $\{D_2\}$, hence the complete mode $\{D\} = \{D_r \quad D_2\}$ may be established. The transformation $\{D\} = [T]\{D_r\}$ is satisfactory if Eqs. 12.4.9 are used, as $[T]$ then includes the desired inertia forces.

Reference 19 suggests the following procedure to generate a "best possible" transformation $[T]$. Use Eqs. 12.4.9, and use a considerable number of vectors $\{G_j\}$ in matrix $[G]$. In each $\{G_j\}$ have a single nonzero term $G_{ij} = 1$, and select the vectors $\{G_j\}$ so that $[P]$ contains the potentially most significant inertia loads. Next compute $[T]$, but discard columns $\{T_j\}$ which have the smallest norms, so that $[T]$ is reduced to m columns corresponding to m "master" d.o.f.

The success of the eigenvalue economizer relies on the master freedoms being able to adequately represent vibration modes of the actual structure. Fortunately even a mediocre representation may serve well because, as suggested below Eq. 12.3.4, a first-order error in mode shape produces only a second-order error in the associated frequency. Indeed, if inaccuracy is suspected in results from an economized solution, we could recover the full vector of amplitudes $\{D\}$ and then improve the estimate of the associated eigenvalue by use of Eq. 12.3.4 with the complete arrays $[K]$ and $[M]$.

We may conclude that use of the economizer raises the lower frequencies because of the constraints imposed. The highest frequencies are not represented at all because of the freedoms that have been discarded. A theoretical appraisal of error introduced by the economizer has been given (61).

In choosing which freedoms to use as "masters" one should select d.o.f. that, when loaded by a unit force, yield a "basic" displacement mode. Our intent is that superposition of "basic" modes will adequately describe the motion in areas of interest. These d.o.f. are generally associated with a large mass-to-stiffness ratio (17) and distributed over the structure instead of being concentrated in a single area. Rotational d.o.f. are usually among those discarded.

The number of master d.o.f. may be 1/10 or 1/20 of the total (18). Reference 19 suggests that for a reliable determination of NE eigenvalues there should be at least $3NE$ master freedoms if $[T]$ from Eq. 12.4.6 is used, but that the number of masters may be reduced to nearly NE if $[T]$ is chosen in the best possible way. Figure 12.4.1 gives an example of the accuracy provided by the economizer. Triangular elements and consistent mass matrices were used (18).

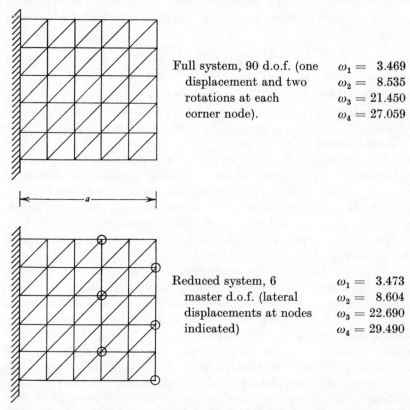

Full system, 90 d.o.f. (one
displacement and two
rotations at each
corner node).

$\omega_1 = 3.469$
$\omega_2 = 8.535$
$\omega_3 = 21.450$
$\omega_4 = 27.059$

Reduced system, 6
master d.o.f. (lateral
displacements at nodes
indicated)

$\omega_1 = 3.473$
$\omega_2 = 8.604$
$\omega_3 = 22.690$
$\omega_4 = 29.490$

Figure 12.4.1. First four natural frequencies of square cantilever plate, $D/\rho t a^4 = 1$.

12.5 Eigenvalue Extraction Methods

Methods of solving for eigenvalues and eigenvectors have developed rapidly in recent years, and at present writing the development continues. For this reason we make no attempt to discuss any method in detail. Instead we survey some schemes currently in use or advocated and refer the reader to the literature for further details. Reference 20 is the most comprehensive single work.

Broadly speaking, there are two classes of eigenvalue extraction methods. First there are transformation methods that operate on a standard form, such as Eq. 12.3.5. Typical methods in this class are Jacobi, Givens, Householder, and Q-R. These methods are described in texts on numerical analysis (20, 25), and one or more are usually found in the standard library package of a computer installation. The second class of extraction methods includes iterative and minimization schemes. The latter methods are often well suited to sparsity or bandedness of the matrices involved. Usually they do not require a preliminary condensation or transformation to standard form; indeed, zeros may appear on-diagonal in a diagonal mass matrix. It is this second class of extraction methods that we will consider in this section. Pertinent references, in addition to those subsequently mentioned, include Refs. 20 and 25 and Ref. 12 of Chapter Two. The techniques discussed are, of course, not limited to the vibrations problem.

Perhaps the best-known iterative method for the eigenvalue problem of Eq. 12.3.5 is the Stodola-Vianello direct iteration method (2, 13, 20). A trial vector $\{Y\}$ is assumed and repeatedly improved. The kth iteration is

$$[A]\{Y\}_{k-1} = \{Y'\}_k, \quad \text{solve for } \{Y'\}_k$$

$$\lambda_k = \frac{\sum_i (Y_i')_k}{\sum_i (Y_i)_{k-1}} \tag{12.5.1}$$

where

$$\{Y\}_k = \{Y'\}_k / (Y_1')_k$$

The summations extend over all elements in the vectors, and $\{Y'\}_k$ is scaled by dividing by its first element. Convergence is to the *highest* eigenvalue λ and its associated eigenvector $\{Y\}$. Unfortunately what is more often sought are the lowest eigenvalues.

A way to extract the lower eigenvalues by direct iteration is suggested by Kamel (65). The highest eigenvalue λ_n of Eq. 12.3.5 is first extracted by direct iteration. Next a new matrix is defined, $[\bar{A}] = \lambda_n[I] - [A]$. Because of this shift, the highest eigenvalue of $[\bar{A}]$ is $(\lambda_n - \lambda_1)$, which may again be found by direct iteration. Thus we obtain λ_1, the lowest eigenvalue of $[A]$. Extraction of intermediate eigenvalues is not so easily explained. The method requires a great number of iterations for each eigenvalue, but each iteration is done very quickly.

The *inverse* iteration method (14, 20, 23, 24, 25) permits extraction of the lowest eigenvalue, or of the eigenvalue closest to a given number. We begin with the eigenvalue problem as stated by Eq. 12.3.2. Let λ represent ω^2, and let λ_0 be a constant known as the "shift point." The kth iteration is (23)

$$([K] - \lambda_0[M])\{D'\}_k = [M]\{D\}_{k-1}$$

$$\{D\}_k = \frac{\{D'\}_k}{C_k}$$

(12.5.2)

where C_k is the element of $\{D'\}_k$ having the largest absolute value. The ratio $1/C_k$ converges to Λ, the "shifted eigenvalue" closest to the shift point λ_0, and $\{D\}_k$ converges to the corresponding eigenvector. The desired eigenvalue λ is computed from the relation $\lambda = \lambda_0 + \Lambda$. Solution for each $\{D'\}_k$ is carried out by elimination instead of by matrix inversion. As an initial guess we may set each element of $\{D\}_1$ to unity. To converge to the lowest eigenvalue we begin with $\lambda_0 = 0$. A new shift point $\lambda_0 + \Lambda$ may replace the old one once every few iterative cycles in order to speed convergence, but each time this is done a new reduction of the coefficient matrix is required. The Rayleigh quotient, Eq. 12.3.4, could be used to compute the new shift point more accurately. Inverse iteration does not require $[K]$ to be nonsingular or the eigenvalue problem to be cast in the standard form of Eq. 12.3.5. However, the matrix $[K] - \lambda_0[M]$ must be nonsingular to the extent that division by zero does not occur during decomposition (25). In other words, provision must be made for the chance that λ_0 is exactly an eigenvalue, but doing so is not difficult (20, 25). The time needed to extract a single eigenvalue is about twice that required for solution of a set of algebraic equations of equal size (23).

Jennings and others have advocated a "simultaneous iteration" method (20, 26, 62, 72, 74) that may be viewed as a form of inverse iteration in which several trial vectors are active simultaneously. The vectors are made to interact in such a way that convergence is speeded and extraction of closely spaced eigenvalues made more reliable. Jennings (75) shows a correspondence between simultaneous iteration and the "eigenvalue economizer," and concludes that while both require about the same computational expense, simultaneous iteration is more versatile and reliable.

Eigenvalue routines may be based on the Sturm sequence property (20, 25, 27, 28, program listings in 29 and 66). In this method one forms the sequence of leading principal minors of $[K] - \omega^2[M]$, then augments the sequence by placing the number $+1$. at the start of it. The number of algebraic sign changes in the augmented sequence is equal to the number of eigenvalues smaller than the ω^2 used in constructing the sequence. By use of this property we may compute to any required accuracy the smallest eigenvalue, or indeed any specified number of eigenvalues in a given range of frequencies. Eigenvectors, if desired, are then computed by inverse iteration. Preliminary transformation to standard form is not required. The method is claimed to be among the more efficient.

The computer routine of Fig. 2.8.1 may be used as an alternative to counting sign changes in the Sturm sequence. After forward reduction of the matrix $[K] - \omega^2[M]$ by Gauss elimination, the number of negative pivot elements equals the number of eigenvalues smaller than the ω^2 used in forming the matrix (77). Thus, in Fig. 2.8.1, we would examine S(1, 1), S(2, 1), etc. after statement 790.

Mathematical programming may be used to minimize the Rayleigh quotient, Eq. 12.3.4 (30, 31, 64). Advantages include the availability of minimization algorithms and the fact that transformation to standard form is not required. Indeed, the structural $[K]$ and $[M]$ matrices need not even be assembled; the numerator of the Rayleigh quotient is twice the maximum potential energy of the structure and the denominator is twice the maximum kinetic energy, and these energies may be found as the sum of potential and kinetic energies of the several elements. However, no strong claims seem to have been made for the efficiency of minimization procedures.

If we were to graph the determinant of $[K] - \lambda[M]$ versus λ we would find several values of λ for which the determinant vanishes. These values of λ are, of course, the eigenvalues. An approximate root can be refined by, say, the regula falsi scheme. This "determinant method" for eigenvalues can obviously be programmed for a computer (23, 58, 59, 72). The rate of convergence can be accelerated (72). A determinant may be evaluated as the product of the pivot terms produced by Gauss elimination reduction of the matrix $[K] - \lambda[M]$. Specifically, in Fig. 2.8.1 the determinant of the matrix in array S is given by the product S(1,1) * S(2,1) * $\cdots$ * S(NSIZE, 1), to be computed after statement 790. Usually the product must be scaled to prevent overflow. The determinant method is best suited to problems where only one eigenvalue is required and matrices $[K]$ and $[M]$ are narrowly banded. Otherwise it tends to be inefficient (23). A major strength of the method is its applicability to nonlinear problems; either $[K]$ or $[M]$ could be a function of λ. For example, we could use the method to find the buckling load in a system having a nonlinear load-displacement relation (see Sections 13.3, 13.4, 14.5). In stability problems $[M]$ represents a "stability matrix," not a mass matrix.

In all methods but the Sturm sequence method, matrices must be "swept" or "deflated" after each eigenvalue is extracted in order to permit extraction of the next one. Sweeping is particularly easy in the determinant method, where the reduced determinant of a matrix $[A] = [K] - \lambda[M]$ used in finding the ith eigenvalue λ_i is (23)

$$\det\left([A]_{\text{reduced}}\right) = \frac{\det[A]}{(\lambda_i^* - \lambda_1)(\lambda_i^* - \lambda_2)\ldots(\lambda_i^* - \lambda_{i-1})} \qquad (12.5.3)$$

where λ_i^* is the current approximation to λ_i. The validity of Eq. 12.5.3 is suggested by the fact that if λ_j is any eigenvalue of $[A]$, then $(\lambda - \lambda_j)$ is a factor of the polynomial det $[A] = 0$. For further information on deflation the reader is referred to the literature (2, 13, 20, 23, 25, 32, 33).

Which method of eigenvalue extraction is best? The transformation methods, mentioned at the beginning of this section, are best suited to extraction of *all* eigenvalues of a small, full matrix. Inverse iteration seems to be preferred for extraction of a few eigenvalues when matrices are large and sparse but tends to loose in accuracy as each successive eigenvalue is extracted and to converge less rapidly if eigenvalues are closely spaced (23, 34). Determinant search is best suited to extraction of the lowest eigenvalues of large, narrowly banded systems that fit in core storage. If the system is large, has large bandwidth, and does not fit in core storage, simultaneous iteration methods are preferred (72). No strong claims for the efficiency of minimization schemes seem to have been made. Effective combinations of methods have been described (77, program listings in 57, 78, 79). A combination of Sturm sequence and inverse iteration methods (78, 79) appears to be three to five times faster than a program based on the Sturm sequence alone (29).

12.6   Calculation of Dynamic Response

The problem we confront is that of solving Eq. 12.2.7, where $\{R\}$ is a set of forces that varies with time, to obtain $\{\ddot{D}\}$, $\{\dot{D}\}$, and $\{D\}$ as functions of time. There are two approaches to the problem. One is the mode-super-position method, which transforms the problem into a set of independent differential equations, one for each degree of freedom. Solution of these equations is followed by superposition of the results. The second approach is to integrate numerically the equations by marching in a series of time steps Δt, evaluating accelerations, displacements, and velocities at each step.

Some advantages and disadvantages of the two approaches are as follows (4, 5, 35, 67). Mode superposition is effective if a few of the low frequency modes are adequate to describe the response, as may happen if loads do not vary rapidly. If all modes are needed extraction of the complete set of eigenvectors becomes a large computational effort. Step-by-step integration is better suited to structures subjected to short duration loads that excite many modes. Also, integration can be applied to nonlinear structures by revising a deformation-dependent stiffness matrix after each time step; mode-superposition, on the other hand, presumes linearity. Step-by-step integration is subject to errors and sometimes numerical instability unless time steps Δt are small, but many small time steps produce long computer run times. In schemes that are not unconditionally stable the maximum permissible Δt is often estimated as a fraction of the shortest period of the structure. It is therefore worth noting that the finer the mesh, the higher the maximum frequencies, and that a structure such as a shell may have very high frequencies in the membrane-stretching modes (35) and (if permitted) the transverse shear deformation modes.

The mode superposition technique will not be described here. It is reasonably well known and is discussed in standard texts (2). In contrast, the step-by-step methods for structural problems are discussed primarily in journal articles.

In recent years step-by-step integration methods have been studied intensively. Dozens of schemes have been proposed. Some are implicit and others explicit; some are unconditionally stable and others not; some introduce artificial damping; frequencies or amplitudes may be distorted. Authors do not agree on the method of choice for a given class of problems. An attempt to clarify matters would have to be extensive and is not appropriate in the present context. Accordingly, in the following we will do little more than refer the reader to some pertinent papers.

In explicit methods, the new vectors to be computed in a given step are functions only of known vectors computed in previous steps. In implicit methods, expressions for the new vectors contain the new vectors, which suggests that an iterative solution is required in each step. However, as noted in connection with Eq. 12.6.3, proper manipulation can sometimes provide explicit expressions for the new vectors, which blurs the distinction between explicit and implicit methods. There seem to be no unconditionally stable explicit methods. To date most structural analysts seem to prefer implicit methods, but there are exceptions (63).

Numerical analysis texts often favor a fourth-order Runge-Kutta method, but structural analysts seem to find it inefficient. Reference 36 finds that extremely small time steps are needed to avoid instability (it is worth noting that stability in linear problems does not guarantee stability in nonlinear problems). Carpenter (37) expresses a preference for Runge-Kutta methods and lists formulas. Reference 43 considers two deVogelaere methods. Nickell et al. (38, 39) evaluate the behavior of certain schemes and conclude that Newmark's method (40) with $\gamma = 0.5$ and $\beta = 1/4$ is a good choice. Further comparisons appear in Refs. 41, 43, 46, 68, and 71. The basic equations of Newmark's method are

$$\{\dot{D}\}_t = \{\dot{D}\}_{t-\Delta t} + (1 - \gamma)(\Delta t)\{\ddot{D}\}_{t-\Delta t} + \gamma(\Delta t)\{\ddot{D}\}_t \qquad (12.6.1)$$

$$\{D\}_t = \{D\}_{t-\Delta t} + (\Delta t)\{\dot{D}\}_{t-\Delta t} + \left(\frac{1}{2} - \beta\right)(\Delta t)^2\{\ddot{D}\}_{t-\Delta t} + \beta(\Delta t)^2\{\ddot{D}\}_t$$

$$(12.6.2)$$

Artificial positive damping is introduced if $\gamma > 0.5$, and artificial negative damping if $\gamma < 0.5$. The method is unconditionally stable if $\beta \geq (2\gamma + 1)^2/16$.

Reference 41 suggests use of a β value that barely exceeds the lower limit for unconditional stability, and damping out the highest (and least important) modes while preserving the lower ones by using γ slightly greater than 0.5. A possible choice is $\gamma = 0.55$ and $\beta = 0.276$. Wilson et al. (4, 35, 42, 67) give clear descriptions of several schemes, including one for nonlinear problems.

If implicit methods are applied to linear problems a large coefficient matrix need be reduced only once, provided the same Δt is used in all steps. Thus successive time steps require only the forward reduction and back substitution of the load vector; the iteration within a time step Δt suggested by Ref. 40 is not necessary (4, 35, 42, 60). This noniterative formulation is obtained by writing Eq. 12.2.7 for time t, then substituting Eqs. 12.6.1 and 12.6.2 into it. It is then solved for $\{\ddot{D}\}_t$ to yield an equation of the form

$$\{\ddot{D}\}_t = [Q]^{-1}\{\bar{R}\}_t \tag{12.6.3}$$

where $[Q]$ is a function of $[M]$, $[C]$, $[K]$, and Δt. Vector $\{\bar{R}\}_t$ is a function of $[C]$, $[K]$, $\{D\}_{t-\Delta t}$, $\{\dot{D}\}_{t-\Delta t}$, $\{\ddot{D}\}_{t-\Delta t}$, and $\{R\}_t$. Matrix $[Q]$ need be "inverted" only once. Dynamic response is calculated by repeated application of Eqs. 12.6.1, 12.6.2, and 12.6.3.

In studies of nonlinear problems, Stricklin (36) concludes that the Houbolt method is best and about four times faster than the method of Chan, Cox, and Benfield. Others (60, 71) opt for a Newmark method, often with $\beta = 1/4$. References 45 and 67 suggest the use of error-correction terms to improve efficiency in nonlinear problems (see below Eq. 15.4.2). Argyris (76) takes $\{\ddot{D}\}$ as cubic in t over step Δt.

Willmert (44) considers what to do if the mass matrix is singular, as may happen when using lumped masses without rotary inertia.

12.7   Concluding Remarks

In addition to the foregoing theoretical work, many applications-oriented papers on vibrations using finite elements have appeared in the *AIAA Journal*, the *Journal of Sound and Vibration*, and others. We will cite only a few. The finite strip element mentioned in Section 9.1 has been used in vibration analysis (47; Ref. 12 of Chapter Nine). Vibration of members stiffened by centrifugal forces has been considered (48, 66). Mei (49, 69, 70) presents an effective formulation for large-amplitude (nonlinear) vibrations. Dynamic analysis of shells of revolution under asymmetric loads is discussed in Ref. 50; suggestions are made for efficient generation of a consistent mass matrix and for treating dynamic response of each harmonic separately. This treatment suggests analogous vibration analysis; that is, find modes and frequencies for a fixed number of circumferential waves, repeat for each of several sets of circumferential waves, and identify the lowest frequency of the lot as the true fundamental frequency.

More advanced developments include addition of the time dimension to finite elements (51, 52). Thus, the time span of interest is divided into finite elements just as are the spacial dimensions. For example, a plane-stress dynamic problem becomes three dimensional, with time t as the third dimension. In this way the initial-value problem of dynamic response is converted to a boundary-value problem. Solutions for all intervals of time are obtained simultaneously, with nodes on each line or surface $t = $ constant defining the configuration of the system at that time. The expansion in problem size that accompanies the added time dimension is a disadvantage of the method.

Nonconservative stability and flutter problems have been analyzed using finite elements (53, 54, 55, 56). In both cases numerical solution requires evaluation of eigenvalues for each of several sets of matrix equations that describe the system. The different sets of equations are generated by use of different values of a parameter that defines the loading.

PROBLEMS

12.1. Using the d'Alembert principle, acceleration may be viewed as producing a "static" body force owing to the mass of an element. Use this viewpoint to derive the expression for $[m]$ given by Eq. 12.2.3.

12.2. Let a uniform two-force bar have mass density ρ, length L, and cross-sectional area A. Nodal freedoms are u_1 and u_2 at the two ends. Find
   (a) The lumped mass matrix.
   (b) The consistent mass matrix.

12.3. Let the bar of Problem 12.2 be fixed at one end. Find the natural frequency and vibration mode using
   (a) The lumped mass matrix.
   (b) The consistent mass matrix.

12.4. Let the bar of Problem 12.2 be free at both ends. Determine the natural frequencies and modes of vibration using
   (a) The lumped mass matrix.
   (b) The consistent mass matrix.

12.5. (a) A uniform beam element has mass density ρ, length L, and cross-sectional area A. Nodal freedoms and the assumed displacement field are given in Section 3.9. Show that the consistent mass matrix is, for $\{d\} = \{v_1 \quad \theta_1 \quad v_2 \quad \theta_2\}$,

$$[m] = \frac{\rho A}{420} \begin{bmatrix} 156L & 22L^2 & 54L & -13L^2 \\ 22L^2 & 4L^3 & 13L^2 & -3L^3 \\ 54L & 13L^2 & 156L & -22L^2 \\ -13L^2 & -3L^3 & -22L^2 & 4L^3 \end{bmatrix}$$

(b) Show that $[m]$ yields the correct nodal forces for a rigid body translational acceleration ($\ddot{v}_1 = \ddot{v}_2 = $ constant, $\ddot{\theta}_1 = \ddot{\theta}_2 = 0$).

12.6. Use the beam element of Problem 12.5, but consider different assumed displacements. Derive the element mass matrix for the following cases.

(a) $v = (1 - x/L)v_1 + (x/L)v_2 + (x/2)(1 - x/L)(\theta_1 - \theta_2)$. (This expression gives the correct curvature and displacement under pure bending and gives a linear displacement if $\theta_1 = \theta_2$.)

(b) $v = (1 - x/L)v_1 + (x/L)v_2$. (The resulting $[m]$ has no coefficients associated with θ_1 and θ_2. Such a formulation is roughly intermediate to lumped and consistent mass formulations in accuracy and computational expense.)

12.7. Find the natural frequencies and modes of vibration for the following problem situations. In each case use the element stiffness matrix based on a cubic displacement (Problem 4.6). Use as the mass matrix the following: (A) consistent, (B) from Problem 12.6a, (C) from Problem 12.6b, (D) lumped. The problems are

(a) One element cantilever beam.

(b) Two-element simply supported beam. Impose symmetry about the midpoint and base calculations on one element.

(c) One-element simply supported beam.

Note. The exact fundamental frequency in Parts b and c is $\omega_1 = \pi^2(EI/\rho AL^4)^{1/2}$.

12.8. (a) What sort of assumed displacement field is implied by a lumped mass approximation?

(b) Should the diagonal coefficients m_{ii} of a lumped mass matrix sum to the total element mass? Explain.

(c) In the lumped mass matrix of a 4 d.o.f. beam element, m_{11} and m_{33} are the only nonzero terms if rotary inertia is neglected. Show that this mass matrix gives correct kinetic energy under translational motion of the beam.

(d) Determine nonzero values of m_{22} and m_{44} that permit a lumped (diagonal) mass matrix of a uniform beam (with m_{11} and m_{33}

unchanged) to yield the correct kinetic energy under pure rotation of the element. Find an example problem for which this mass matrix yields very poor estimates of natural frequencies.

12.9. (a) Derive Eqs. 12.3.15.

(b) Derive the two standard forms mentioned immediately below Eqs. 12.3.15.

(c) Do the standard forms of Part b appear to involve any particular computational advantages or penalties?

12.10. (a) Show that the transformation implied by Eq. 12.4.7 may be written

$$\begin{Bmatrix} D_1 \\ D_2 \end{Bmatrix} = \begin{bmatrix} I \\ -K_{22}^{-1}K_{21} \end{bmatrix} \{D_1\} = [T]\{D_1\}$$

and that the transformation $[T]^T[K][T]$ produces the same result as given by Eq. 6.1.4.

(b) Show that matrix $[T]$ of Part a is in fact identical to $[T]$ as given by Eq. 12.4.6.

12.11. (a, b) Consider the beams of Problem 12.7, Parts a and b. Apply the eigenvalue economizer (as described, say, in Problem 12.10) and in each case determine the lowest natural frequency using the consistent mass matrix.

12.12. What might be the effects of using an eigenvalue economizer with master freedoms all clustered in a small area of the structure?

12.13. Derive the specific form of $[Q]$ and $\{\bar{R}\}_t$ in Eq. 12.6.3 for
(a) $\gamma = 0.5$, $\beta = 1/4$.
(b) General values of γ and β.

CHAPTER THIRTEEN

The Initial Stress Stiffness Matrix and Linear Stability

13.1 Introduction

In this chapter it is tacitly assumed that the structures of interest are thin-walled, admit the possibility of buckling, and may be modeled by beam, plate, or shell elements. Accordingly, the conventional term "membrane force" is used to denote an internal force whose direction is tangent to the element surface. The methods discussed may be applied to more massive structures modeled by other elements, but there is little practical motivation for doing so.

This chapter is restricted to linear problems. By "linear" we mean that (a) membrane forces can be determined by a linear analysis and (b) membrane forces remain constant during the deformation caused by a second set of loads or by buckling. For example, membrane forces in a flat plate subjected to specified in-plane loads and displacements may be found by a standard (linear) plane-stress analysis, provided that bending action is zero or very nearly so. Then, if bending displacements under subsequent lateral loading are small, membrane forces may be taken as known constants, while bending displacements are found by another linear analysis. In a stability analysis, membrane forces are taken as remaining unchanged as the structure displaces an infinitesimal amount into one of its buckling modes. *Nonlinear* action is more often found in doubly curved shells than in flat plates and developable shells. An iterative analysis may be used to relate membrane forces to applied loads in nonlinear problems (Chapter Fourteen).

To investigate the effects of membrane forces on bending action we must introduce a new matrix, designated $[k_\sigma]$ for an element and $[K_\sigma]$ for a

259

structure, and known by such names as "initial stress stiffness matrix," "geometric stiffness matrix," and "stability coefficient matrix." The names derive from its application and from the fact that $[k_\sigma]$ is independent of elastic properties and is instead a function of the element geometry and its internal membrane forces. In brief, $[k_\sigma]$ accounts for the stiffening or weakening effects of membrane forces. For example, $[k_\sigma]$ accounts for the effect of compressive axial force in reducing the resistance of a beam to lateral load. Further comments on the interpretation of $[k_\sigma]$ appear in Section 13.3.

Matrix $[k_\sigma]$ for any given element is derived by the addition of higher order terms to the linear strain-displacement relations we have used previously. In linear problems we elect to retain only the lowest of these additional terms and justify the decision by the practical success of this approach. The particular higher order terms of interest are those that relate membrane strains to rotations that arise from lateral displacement. It is awkward to derive $[k_\sigma]$ for a particular element by specialization of a general formula. In the following we instead consider each class of problem (beam, plate, etc.) as a special case, and select the appropriate high order terms by familiarity with established analytical methods. This approach has been widely and successfully used.

It is perhaps obvious that assembly of element matrices $[k_\sigma]$ to form the corresponding structure matrix $[K_\sigma]$ is done in exactly the same way as assembly of conventional stiffness matrices.

13.2   Initial Stress Stiffness Matrices

As our first example we consider a beam element carrying an axial tensile stress σ_o that is presumed known and constant. We consider deformations caused by bending and proceed, in the usual way, to develop a finite element model by use of the total potential energy expression, Eq. 3.5.10. In the present situation only axial strain ϵ_x and lateral displacement v are needed. Figure 13.2.1 shows a fiber of length dx and cross-sectional area dA in its original and displaced states. The strain-displacement relation is

$$\epsilon_x = -yv_{,xx} + \frac{v_{,x}^2}{2} \tag{13.2.1}$$

where $y =$ distance of the differential element from the centroidal axis. Axial strain $u_{,x}$ of the centroidal axis is omitted as extraneous to our present purpose; if retained it would yield the conventional stiffness matrix of an axial-force member and the corresponding nodal loads due to σ_o. In Eq. 13.2.1 the term $yv_{,xx}$ is familiar from elementary beam theory. The term

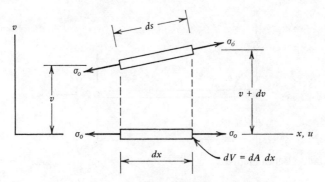

Figure 13.2.1. Differential volume element of a beam.

$v_{,x}^2/2$ expresses the lengthening of the element caused by its rotation and is an approximation based on the assumption that rotation $v_{,x}$ is small:

$$\frac{ds - dx}{dx} = (1 + v_{,x}^2)^{1/2} - 1 \approx \frac{v_{,x}^2}{2} \tag{13.2.2}$$

Equation 13.2.1 is now substituted into the strain-energy integral of Eq. 3.5.10, with $[E] = E$ and $\{\epsilon\} = \epsilon_x$. Then, noting that

$$\int y \, dA = 0, \qquad \int y^2 \, dA = I, \qquad \int \sigma_o \, dA = P \tag{13.2.3}$$

where I = moment of inertia and P = axial tensile force, and discarding the higher order term $v_{,x}^4$, we obtain the strain energy

$$U = \int_0^L \frac{EI}{2} v_{,xx}^2 \, dx + \int_0^L \frac{P}{2} v_{,x}^2 \, dx \tag{13.2.4}$$

The first integral gives the conventional beam element stiffness matrix $[k]$. The second integral leads to the initial stress stiffness matrix $[k_\sigma]$ and accounts for the work of constant force P acting through displacement increments $v_{,x}^2 \, dx/2$. Force P may be large enough for the two integrals to have about the same magnitude.

The finite-element formulation may now be completed in the usual way. For the beam element of Fig. 13.2.2 we may take v as cubic in x, the specific expression being that given in Section 3.9. The second integral of Eq. 13.2.4 becomes

$$\frac{P}{2} \{d\}^T [DA]^{-T} \int_0^L [X']^T [X'] \, dx [DA]^{-1} \{d\} \tag{13.2.5}$$

where

$$\{d\} = \{v_1 \quad \theta_1 \quad v_2 \quad \theta_2\}$$
$$[X'] = [0 \quad 1 \quad 2x \quad 3x^2] \tag{13.2.6}$$

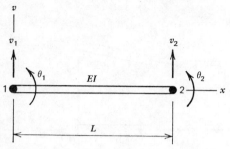

Figure 13.2.2. Beam element.

and $[DA]^{-1}$ is given in Eq. 3.9.5. Minimization of the total potential with respect to nodal displacements yields the familiar terms derived in Chapter Four plus the initial stress stiffness matrix

$$[k_\sigma] = P[DA]^{-T} \int_0^L [X']^T [X']\, dx [DA]^{-1} = \frac{P}{30L} \begin{bmatrix} 36 & 3L & -36 & 3L \\ 3L & 4L^2 & -3L & -L^2 \\ -36 & -3L & 36 & -3L \\ 3L & -L^2 & -3L & 4L^2 \end{bmatrix}$$

$$(13.2.7)$$

The latter matrix appeared in Ref. 1 and subsequently in others (2, 4). As suggested by this development, an initial stress stiffness matrix is always symmetric.

Plate elements yield to the same approach as used for beam elements (1, 3, 4). Let N_x, N_y, and N_{xy} represent normal and shearing membrane forces per unit length in a plate (Fig. 13.2.3). Membrane strains of the mid-surface arising from lateral displacement w are given by plate theory as

$$\epsilon_x = \frac{w_{,x}^2}{2}, \qquad \epsilon_y = \frac{w_{,y}^2}{2}, \qquad \epsilon_{xy} = w_{,x} w_{,y} \qquad (13.2.8)$$

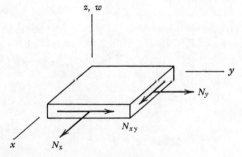

Figure 13.2.3. Plate element.

where displacement w is normal to the plate and the small-rotation approximation of Eq. 13.2.2 is used again. Strain energy associated with Eq. 13.2.8 is

$$\int \int \frac{1}{2} [w,_x \ w,_y] \begin{bmatrix} N_x & N_{xy} \\ N_{xy} & N_y \end{bmatrix} \begin{Bmatrix} w,_x \\ w,_y \end{Bmatrix} dx \, dy \qquad (13.2.9)$$

the latter expression being analogous to Eq. 13.2.5. When w is given as a function of element nodal displacements $\{d\}$, we may form matrix $[G]$ in the expression $\{w,_x \ w,_y\} = [G]\{d\}$; hence we have

$$[k_\sigma] = \int \int [G]^T \begin{bmatrix} N_x & N_{xy} \\ N_{xy} & N_y \end{bmatrix} [G] \, dx \, dy \qquad (13.2.10)$$

In some elements displacement w may be expressed in terms of isoparametric coordinates $\xi\eta$; then numerical integration may be used in Eq. 13.2.10.

Initial stress stiffness effects in shells may be accounted for by use of expressions quite similar to Eqs. 13.2.8 and 13.2.9. For example, in the case of axisymmetric shells the midsurface strains $(\epsilon)_o$ of Eq. 8.3.1 are augmented by the higher order terms (5, 6)

$$(\epsilon_s)_o = \frac{1}{2} e_{13}^2, \qquad (\epsilon_\theta)_o = \frac{1}{2} e_{23}^2, \qquad (\gamma_{s\theta})_o = e_{13}e_{23}$$

$$e_{13} = w,_s + u\phi,_s, \qquad e_{23} = \frac{1}{r} w,_\theta - \frac{v \cos \phi}{r} \qquad (13.2.11)$$

When an initial stress stiffness matrix is constructed using the same displacement functions used in constructing the conventional stiffness matrix, the resulting $[k_\sigma]$ may be termed "consistent." However, as suggested by the discussion of mass matrices in Chapter Twelve, other displacement functions may be used. For example, if the beam element of Fig. 13.2.2 is assumed to have the linear displacement $v = (1 - x/L)v_1 + (x/L)v_2$, we obtain (2, 4)

$$[k_\sigma] = \frac{P}{L} \begin{bmatrix} 1 & 0 & -1 & 0 \\ 0 & 0 & 0 & 0 \\ -1 & 0 & 1 & 0 \\ 0 & 0 & 0 & 0 \end{bmatrix} \begin{matrix} v_1 \\ \theta_1 \\ v_2 \\ \theta_2 \end{matrix} \qquad (13.2.12)$$

(Note that this is the *consistent* $[k_\sigma]$ for a pin-connected truss member.) Use of lower order displacements seems to be a generally recommended practice that gives a good balance between accuracy and computational effort (6; 35 and 9 of Chapter Twelve). For elements where numerical integration is required, it is sufficient to use a lower order quadrature rule than required to form an exact $[k_\sigma]$. Numerical examples in the last-mentioned reference suggest that the "best" $[k_\sigma]$ is intermediate to the consistent $[k_\sigma]$ and the simplest possible $[k_\sigma]$. The same numerical examples show in *increase* in computed buckling loads when the consistent $[k_\sigma]$ is replaced by the simplest possible $[k_\sigma]$ (based on a linear displacement field). This behavior is contrary to the trend found in vibration analysis, where use of lumped masses tends to lower natural frequencies.

Initial stress stiffness matrices are explicitly stated in the literature for prismatic members of arbitrary cross section (7, 8; program statements in 53 of Chapter Twelve). The expressions are sufficiently general that problems of lateral and torsional buckling may be treated.

The product $[k_\sigma]\{d\}$ represents zero forces only when nodal displacement vector $\{d\}$ represents a motion associated with zero membrane strain. If $\{d\}$ represents rigid body rotation, forces $[k_\sigma]\{d\}$ are not zero. Nodal rotations are associated with zero force if elementary forms such as Eq. 13.2.12 are used, because a linear displacement field is independent of rotation. Rigid body translations always produce zero nodal forces. Had we used exact expressions for strain and carried all higher order terms into $[k_\sigma]$, forces $[k_\sigma]\{d\}$ would also vanish when $\{d\}$ represents rigid body rotation.

Finally we note that the consistent $[k_\sigma]$ is generally indefinite. This is in contrast to the consistent mass matrix, which is always positive definite.

13.3   The Eigenvalue Problem

We assume that during an infinitesimal buckling displacement from an initially stressed state, membrane forces remain substantially unchanged. Suppose that matrix $[K_\sigma]$ of the structure has been computed based on a known distribution but arbitrary level of membrane stresses. Membrane stresses are assumed to maintain constant relative magnitudes, so that changing one by a factor λ changes all others by the same factor and produces a geometric stiffness matrix $\lambda[K_\sigma]$. The question posed in the stability problem is: if structure displacements $\{D\}$ of the initially stressed state are augmented by virtual displacements $\{\delta D\}$ while external loads remain constant, what must λ be so that states $\{D\}$ and $\{D + \delta D\}$ are both equilibrium configurations? Thus, with $[K]$ the conventional structure stiffness matrix,

$$([K] + \lambda[K_\sigma])\{D\} = ([K] + \lambda[K_\sigma])\{D + \delta D\} = \{R\} \qquad (13.3.1)$$

By subtracting the first equation from the second we obtain the linear eigenvalue problem

$$[K_{\text{total}}]\{\delta D\} = ([K] + \lambda[K_\sigma])\{\delta D\} = 0 \qquad (13.3.2)$$

The $+$ sign may be changed to $-$ if it is understood that compressive membrane stresses are used in forming $[K_\sigma]$.

Equation 13.3.2 may be interpreted in various ways, each of which helps clarify the physical meaning of $[K_\sigma]$. Three viewpoints are:

1. Some λ and corresponding $\{\delta D\}$ exist such that $[K_{\text{total}}]\{\delta D\} = \{\delta R\} = 0$. Forces $\{\delta R\}$ produced by displacement being zero, we may say that the value of λ is sought that reduces the total stiffness to zero with respect to the buckling mode $\{\delta D\}$.

2. Premultiplication by $\{\delta D\}^T/2$ yields

$$\frac{1}{2}\{\delta D\}^T[K]\{\delta D\} + \frac{\lambda}{2}\{\delta D\}^T[K_\sigma]\{\delta D\} = 0 \qquad (13.3.3)$$

   This equation states that the increment in strain energy produced by $\{\delta D\}$ is equal in magnitude to the work done by membrane stresses acting through displacements caused by $\{\delta D\}$. The total energy change is zero, as external forces have done no work on the structure. Note also that by solving for λ we obtain the Rayleigh quotient of Eq. 12.3.4.

3. Writing Eq. 13.3.2 as

$$[K]\{\delta D\} = -\lambda[K_\sigma]\{\delta D\} \qquad (13.3.4)$$

   and viewing the right side as a load vector, we can say that we ask for a λ and $\{\delta D\}$ such that the elastic resistance of the structure exactly balances the pseudoloads. If a column is built of beam elements and Eq. 13.2.12 is used, the pseudoloads are nodal forces directed perpendicular to the column axis.

In-plane displacements of flat plates and straight beams may be discarded from the displacement array $\{\delta D\}$, since they play no role during buckling and are needed only to compute the initial membrane forces. If retained, a larger eigenvalue problem results. If discarded, the cost of rearranging arrays is present.

It is perhaps obvious that buckling of a structure does not imply buckling of each element. That is, at the structure's critical load λ_{cr},

$$\{\delta d\}^T([k] + \lambda_{cr}[k_\sigma])\{\delta d\} \geq 0$$

Equality prevails only in the highest buckling modes or for one-element structures.

The statement made in Chapter Twelve regarding bounds on eigenvalues applies to the stability problem as well. That is, if we use compatible displacement fields, an element mesh that correctly represents the volume of the structure, consistent initial stress stiffness matrices, and avoid low order integration rules, the computed buckling load is guaranteed to be an upper bound on the true buckling load.

To treat the eigenvalue problem of Eq. 13.3.2 we may apply the methods discussed in Sections 12.3, 12.4, and 12.5. The "determinant method" of eigenvalue extraction is attractive (3, 7, 12), since the lowest critical load is usually the only one of interest. As a numerical example of this method, consider a uniform one-element column, fixed at one end and free at the other. For compressive loads $P = 0.5$ and 1.0 on a certain column the values of det $[K_{\text{total}}]$ are, respectively, 9.40 and 6.95 if the consistent $[k_\sigma]$ is used. By linear extrapolation, det $[K_{\text{total}}] = 0$ at $P = 2.42$. The exact critical load is $P_{\sigma r} = \pi^2/4 = 2.47$. Note, however, that det $[K_{\text{total}}]$ is not a linear function of λ. The determinant method may also be used to establish collapse loads of nonlinear structures (see Chapter Fourteen).

Shells of revolution generally do not have axisymmetric buckling modes even if the loading is axisymmetric. Upon buckling, many waves are likely to appear in each hoop circle. Accordingly, the Fourier series treatment must be used. If the loading is axisymmetric or has symmetry about a plane $\theta = 0$, only θ-symmetric terms such as appear in Eq. 8.4.8 are needed. Antisymmetric loading such as torsion requires the complete Fourier series of Eq. 7.3.8, because the higher order displacement terms destroy the axisymmetric shape and cause symmetric and antisymmetric Fourier terms to interact.

Chapter Seven describes how nonsymmetric loads on solids of revolution could be treated by analysis of one harmonic at a time. The same general approach has been used to determine the critical load of thin shells of revolution (5, 9; 12 of Chapter Eight). The first step is to establish the in-plane forces and matrices $[K]$ and $[K_\sigma]$ for n circumferential waves. Next, choose a numerical value of n and solve for the corresponding eigenvalue λ_{cr}. Repeat for $n + 1$ circumferential waves, then for $n + 2$ waves, etc. In this way a series of eigenvalues λ_{cr} is generated, the lowest of which corresponds to the desired buckling load. Unless a good estimate of the actual n is somehow known, many solutions are needed; Ref. 5 mentions a case where the critical load corresponds to $n = 39$.

The computational process just described can be applied to structures built of the finite-strip elements mentioned in Section 9.1. An application to lightweight stiffened panels has been described (10, 15). Alternatively, such panels could be modeled as axisymmetric structures of small width but very large radius (12, 14 of Chapter Eight).

13.4 Concluding Remarks

Numerical examples of buckling analysis appear in many papers. Data given in Table 13.4.1 appear in Ref. 3 and are based on the element of Eq.

Table 13.4.1. Error in Computed Buckling Load for Simply Supported Square Plate Under Uniform Compression in One Direction (3)

Mesh	3 by 3	4 by 4	6 by 6	8 by 8	10 by 10	12 by 12
% Error	−8.88	−5.75	−2.80	−1.68	−1.00	−0.58

9.1.1 and a consistent matrix $[k_\sigma]$. The element is incompatible, which accounts for the fact that convergence is "from below." Results obtained by use of an eigenvalue economizer are given in Ref. 18 of Chapter Twelve. The plate element of Ref. 5 of Chapter Nine has been used in buckling analysis (11) and gives excellent results even in a coarse mesh.

Natural frequencies ω of structures are modified by the presence of membrane forces. The eigenvalue problem remains essentially as stated by Eq. 12.3.2, except that the structure stiffness is modified by the presence of the initial stress stiffness matrix:

$$([K + K_\sigma] - \omega^2[M])\{\bar{D}\} = 0 \tag{13.4.1}$$

where $[K]$ is the conventional stiffness matrix, $[M]$ is the mass matrix, and $\{\bar{D}\}$ is a vibration mode.

Let us close by reviewing the limitations of this chapter. We have assumed that prior to buckling, loads and displacements are linearly related. Thus, as loading begins, a plot of loading parameter λ versus a typical in-plane (membrane) displacement quantity D is a straight line, namely line OAE in Fig. 13.4.1. (If D represents instead the *lateral* displacement of a straight column or a flat plate, path OA coincides with the vertical axis.) The buckling

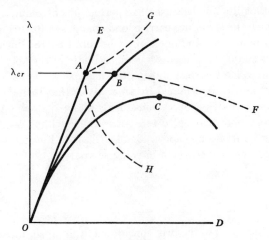

Figure 13.4.1. Load versus displacement plots.

load has been defined in the classical way; it is that load for which two infinitesimally close configurations $\{D\}$ and $\{D + \delta D\}$ are both possible. The point corresponding to this load, point A of the figure, is a "bifurcation point." Path AE is only a theoretical possibility. One of the postbuckling paths AF, AG, or AH is always generated by any attempt to exceed λ_{cr}. The shape of the postbuckling path is determined by the nature of the problem at hand. Postbuckling path AG represents a structure that has postbuckling strength. In contrast, a structure that displays path AH may fail by "snap-through" at $\lambda < \lambda_{cr}$, because the actual structure has imperfections while the bifurcation load λ_{cr} pertains to an ideal structure.

Nonlinear effects lead to the nonlinear path OB prior to bifurcation at B (Ref. 9 considers this possibility) and to path OC, for which point C represents a collapse load but without an adjacent equilibrium state $D + \delta D$. Loads corresponding to points B and C may, of course, be found; the determinant method of finding the load for which stiffness becomes zero remains effective.

The total stiffness $[K + K_\sigma]$ of this chapter applies to the structure in its original configuration, that is, to the configuration existing before *any* deformations take place. Actually, deformation under load may alter structural geometry to the extent that $[K + K_\sigma]$ should be written with respect to the deformed configuration. It is this deformation-dependent stiffness that gives rise to paths such as OB and OC in Fig. 13.4.1. That OB and OC are curves can be seen by imagining that loading is increased in equal steps $\{\Delta R\}$. Since stiffness depends on accumulated displacements, the resulting displacement steps $\{\Delta D\}$ will be unequal.

The nonlinearity described in the preceding two paragraphs is to be expected in practical structures. Because of the shape of the structure and the manner of loading, or because of imperfections of geometry and loading, the way in which the structure carries load changes as load is increased. Thus the problem is nonlinear, and one cannot be sure that the linearized buckling analysis of this chapter represents a good estimate of the actual collapse load (12). The collapse load corresponds to the first peak on the load versus displacement plot and is determined by an analysis that includes the pertinent nonlinear terms. Analyses of various shells show that collapse may occur at loads of from one quarter to ten times the buckling load predicted by classical linear bifurcation theory (13, 14). Chapter Fourteen introduces the nonlinear theory needed to perform collapse analysis.

PROBLEMS

13.1. In Chapter Twelve we considered a lumped mass matrix $[m]$, having only diagonal terms. Construct a diagonal initial stress stiffness matrix for a beam element, such that zero force is associated with rigid body translation and exact results are obtained for buckling of a one-element pin-ended column. Hence, use this matrix to compute the buckling load of a two-element pin-ended column.

13.2. (a) Derive Eq. 13.2.4.
      (b) Derive Eq. 13.2.7.
      (c) Retain the axial strain term $u_{,x}$ in Eq. 13.2.1 and derive the equation analogous to Eq. 13.2.4.

13.3. Derive Eq. 13.2.9 by the same procedure used to obtain Eq. 13.2.5. That is, specialize the strain energy portion of Eq. 3.5.10 for $\sigma_z = \tau_{yz} = \tau_{zx} = 0$, use initial membrane stresses σ_{xo}, σ_{yo}, and τ_{xyo}, and add the strains of Eq. 13.2.8 to the strains produced by bending.

13.4. Let the lateral displacement of a plate be written in terms of nodal displacements $\{d\}$ as $w = [N_1 \quad N_2 \quad N_3 \quad \ldots \quad N_n]\{d\}$, where each shape function N_i is a function of isoparametric coordinates $\xi\eta$ in the plane of the plate. Express the integrand of Eq. 13.2.10 in the form appropriate to this situation.

13.5. Derive Eq. 13.2.12.

13.6. Construct an initial stress stiffness matrix for a beam element by using the displacement function given in Part a of Problem 12.6.

13.7. Use one element to model the cantilever beam under axial tensile load. Determine displacement v at the loaded end using
      (a) The consistent $[k_\sigma]$ (Eq. 13.2.7).
      (b) The matrix $[k_\sigma]$ determined in Problem 13.6.
      (c) The two-force member $[k_\sigma]$ (Eq. 13.2.12).

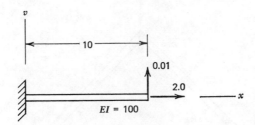

13.8. (a, b, c) Repeat Problem 13.7 but with a *compressive* axial force of 2.0.

13.9. (a, b, c) Consider a pin-ended column of length L and bending stiffness EI under compressive axial load (actual buckling load $= \pi^2 EI/L^2$). Model the column by one beam element and determine the buckling load given by use of the initial stress stiffness matrices listed in Parts a, b, and c of Problem 13.7.

13.10. (a, b, c) Repeat Problem 13.9a, b, c using two elements to model the column. Impose symmetry about the midpoint and base calculations on one element (that is, use two nonzero d.o.f.).

13.11. Make use of the eigenvalue economizer (as described, say, in Problem 12.10) to eliminate rotational d.o.f., and recompute the solutions to
   (a)   Problem 13.7a,b,c.
   (b)   Problem 13.8a,b,c.
   (c)   Problem 13.10a,b,c.

13.12. In Chapter Four we discussed equivalent nodal loads arising from distributed loading on an element (Eq. 4.2.10 and Problem 4.8). If the beam element shown carries an axial compressive force P, are equivalent nodal loads altered by P? If so, how might they be more correctly evaluated?

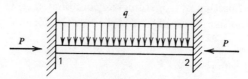

13.13. Two identical beam elements are pin-connected at points A, B, C. Give qualitative values of angle θ and element stiffnesses so that load P and its displacement D produce
   (a)   Path OA of Fig. 13.4.1.
   (b)   Path OB of Fig. 13.4.1.
   (c)   Path OC of Fig. 13.4.1.

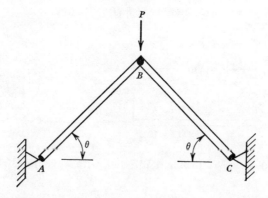

CHAPTER FOURTEEN

Geometric Nonlinearity

14.1 Introduction

Many kinds and combinations of behavior fall under the heading "nonlinear," since the term serves only to say what a problem is not. Nonlinear behavior may arise because of time-dependent or time-independent material nonlinearity or because of large displacements that alter the shape of the structure so that applied loads alter their distribution or magnitude. Nonlinearity may also be associated with smaller displacements, as in the problems of contact stress and flat plates whose displacements exceed their thickness. Nonlinearity may be mild or severe; the problem may be static or dynamic; and buckling may or may not be involved. It is therefore not surprising that many methods of solution have been devised, with no single method being superior for all types of nonlinear problems.

A comprehensive treatment of nonlinear finite element problems requires more than one chapter; indeed, an entire book is devoted to the subject (1). In this chapter we present general comments and consider specific methods only if they are clearly useful and easily understood. Papers referenced often contain extensive literature surveys that direct the reader to other approaches and further detail. In the following, unless otherwise noted, we consider that strains are small, material stress-strain relations are linear, and problems are static.

The basic feature of a problem having geometric nonlinearity is that equilibrium equations must be written with respect to the *deformed* geometry, which is not known in advance. Strictly, equilibrium equations of the deformed configuration should be used in *all* problems. But in so-called linear problems the original configuration may be used because the essential features of the problem are practically unchanged by the deformation.

We begin by describing a solution method that has considerable physical appeal and in which the problem formulation and its numerical solution are intertwined. In subsequent sections, formulation and solution are examined further. It should be mentioned that problems of geometric and material nonlinearity demand similar solution techniques, and in several papers the two nonlinear effects are combined. We note also that nonlinear effects do not demand wholesale revision of preceding chapters where linearity was presumed, because nonlinear problems are commonly solved as a series of linear steps.

14.2   Iterative Solution with Moving Coordinates

The method is described in Refs. 2 through 5. Plate problems are treated in Refs. 2 and 3, where any element may rotate up to 90 degrees out of its original plane. A particular advantage of using moving coordinates is that large rotations may be accommodated.

A general description of the method is as follows. We first recall the obvious fact that the stiffness of a structure depends on its geometry. Thus the column of Fig. 14.2.1 has a different stiffness matrix in position AC than in position AB. Consequently, we might write structural equations as $[K]\{\Delta D\} = \{\Delta R\}$ to emphasize that displacements must be small; that is, $\{\Delta D\}$ cannot represent displacement of the column from position AB to position AC, because $[K]$ is not constant for so large a step. Next we elect to regard $\{\Delta R\}$ as a resultant loading applied to the nodal d.o.f., $\{\Delta R\}$ being composed of the specific external loading plus loads applied to the nodes by the distorted elements. Vector $\{\Delta R\}$ is therefore an array of unbalanced forces, and for equilibrium of the nodal d.o.f. we must have $\{\Delta R\} = 0$. The iterative solution seeks a configuration such that $\{\Delta R\} = 0$.

Many procedural details are conveniently explained with reference to a beam element. Several such elements would be used to form the column of Fig. 14.2.1. However, the method is general and may be applied to structures

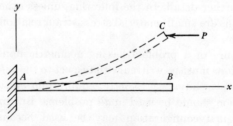

Figure 14.2.1.   Large deflection of column.

other than beams, columns, and frames (2, 3, 39). We first describe how distortions of an element and the forces it applies to the nodes may be computed (5). Figure 14.2.2a shows an undeformed element in global coordinates xy. From the given original global coordinates of nodes 1 and 2 we may easily compute x_o, y_o, L_o, and ϕ_o. After global total displacements $D_1, D_2, \ldots, D_6$ have taken place, the element appears as in Fig. 14.2.2b. The displacement is a combination of rigid body motion and distortion. To subtract out the rigid body motion, a local axis x' is established through nodes 1 and 2 of the displaced element by the calculation

$$x_L = x_o + D_4 - D_1, \qquad y_L = y_o + D_5 - D_2$$

$$\phi = \text{arc tan}\left(\frac{y_L}{x_L}\right) \tag{14.2.1}$$

Element distortions, expressed in the local system $x'y'$, are

$$u_2 = L - L_o = (x_L{}^2 + y_L{}^2)^{1/2} - L_o$$

$$\theta_1 = D_3 - (\phi - \phi_o), \qquad \theta_2 = D_6 - (\phi - \phi_o) \tag{14.2.2}$$

Forces $\{r\}$ applied to nodes 1 and 2 by the distorted element are $\{r\} = -[k]\{d\}$, where $\{d\} = \{0 \quad 0 \quad \theta_1 \quad u_2 \quad 0 \quad \theta_2\}$, and $\{r\}$, $[k]$, and $\{d\}$ are all referred to local coordinates $x'y'$. Matrix $[k]$ may or may not include initial stress stiffness effects, a matter we will comment on below. Arrays $\{r\}$ and $[k]$ may be transformed through angle ϕ to global coordinates xy by the usual operations (Section 11.3).

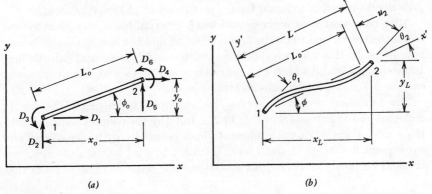

Figure 14.2.2. Beam element. (a) Undistorted. (b) Deformed and displaced.

A typical iterative cycle involves the following steps (2, 3). Let the structure be deformed, but not in equilibrium, under a specified level of external load.

1. Establish local coordinates by use of global displacements $\{D\}$.
2. Compute element distortions, that is, complete element nodal d.o.f. $\{d\}$ in local coordinates.
3. Establish element stiffness $[k]$ and forces $\{r\} = -[k]\{d\}$, both in local coordinates.[1]
4. Transform $\lfloor k \rfloor$ and $\{r\}$ to global coordinates.
5. Repeat steps 1 through 4 for all other elements and assemble structural arrays $[K] = \Sigma\,[k]$ and $\{R_r\} = \Sigma\,\{r\}$. Matrix $[K]$ is the structure stiffness in its current configuration.
6. Compute unbalanced loads $\{\Delta R\}$ as the vector of applied loads plus forces $\{R_r\}$.
7. Solve structural equations $[K]\{\Delta D\} = \{\Delta R\}$ for displacement increments $\{\Delta D\}$.
8. Add increments $\{\Delta D\}$ to global displacements $\{D\}$ accumulated in preceding iterations; this gives the updated estimate of the equilibrium configuration.
9. Test for convergence. If not satisfied, return to step 1.

The foregoing steps are summarized by the equations

$$[K]_i\{\Delta D\}_{i+1} = \{R\} - \Sigma\,[k]_i\{d\}_i$$
$$\{D\}_{i+1} = \{D\}_i + \{\Delta D\}_{i+1}$$

$$(14.2.3)$$

where $\{R\}$ represents external loads $\{P\}$ plus any body forces, initial strain forces, etc. Also, it is understood that $[k]_i$, $\{d\}_i$, and $[K]_i$ are based on current displacements $\{D\}_i$ and are updated every cycle. If displacements are prescribed instead of external forces, part or all of $\{R\}$ may be zero.

After convergence at a given load level, external loads may be increased to another level, and iteration begun again to find the new equilibrium configuration. In this way we combine iterative and incremental calculations. In reaching a final load level we may take many steps with few iterations in each or fewer steps with more iterations in each.

The foregoing procedure serves, for example, to calculate the highly deflected elastic curve AB of Fig. 14.2.3. Initially the column coincides with the x axis. Force P exerts a moment about point A only in the second and subsequent iterations. A small force Q is used to start lateral deflection, but it could be removed after a few iterations. An *essential requirement* is that

---

[1] Alternatively, forces $\{r\}$ may be computed by integration of $[B]^T[E]\{\epsilon\}$ over the element volume. Here $[B]$ and $\{\epsilon\}$ are referred to local coordinates, and $\{\epsilon\}$ is computed from local distortions $\{d\}$. This form of force calculation is also seen in Eqs. 15.4.1 and 15.4.4.

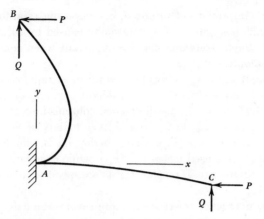

Figure 14.2.3.   Deflected columns.

elements be small enough that rotations and displacements in each local coordinate system are small. Thus the fineness of the mesh required depends on the expected severity of deformation. In the present solution method, all essential nonlinearities are accounted for by the coordinate transformations. Accordingly, matrices such as $[K_{NL}]$ of Section 14.4 do not appear explicitly.

Convergence of the iterative process could be said to occur when the ratio of unbalanced forces to externally applied loads is reduced below a given limit. However, Ref. 8 advises that a criterion based on displacements is preferable. A recommended criterion is to compute, for each structural d.o.f. i, the ratio $e_i = |\Delta D_i / D_{i\text{ref.}}|$ and terminate iteration when the largest e_i is less than 10^{-2} or 10^{-6}, depending on the accuracy required. In this expression ΔD_i is the most recently computed increment and $D_{i\text{ref.}}$ is the largest total displacement of the same type. For example, if ΔD_i is a rotation increment, $D_{i\text{ref.}}$ is the largest nodal rotation in the structure.

It was previously mentioned that the element stiffness matrix $[k]$ may or may not include initial stress stiffness effects. Such effects may be neglected if membrane forces are sufficiently small that their tendency to increase or decrease the bending stiffness of the structure may be neglected. Configuration AB of Fig. 14.2.3 is such a case; excellent results are obtained by taking $[k]$ as the conventional stiffness matrix. If the column is nearly straight under $P \approx P_{\text{buckling}}$, an initial stress stiffness matrix $[k_\sigma]$ must be included for the sake of accuracy and a good convergence rate. The local stiffness matrix $[k]$ is merely the sum of $[k_\sigma]$ and the conventional stiffness matrix used for linear small-displacement analysis. Membrane forces needed to evaluate $[k_\sigma]$ are calculated from elastic properties and strains computed in the local coordinate system. Accordingly $[k_\sigma]$ is null in the first iteration, roughly approximated in the second, etc. Computations with columns (4, 41) indicate that no improvement results if, in computing axial force, elongation u_2 of Fig. 14.2.2b is augmented by stretching caused by rotations θ_1 and θ_2 (see Eq. 8.4.6). Such stretching may be important when computing stresses in plate elements, however (2).

The question of what $[k_\sigma]$ (if any) to use for most rapid convergence is not easily answered. Some theoretical considerations appear in Ref. 5. The writer's numerical work with axially loaded columns indicates the following. Let $[k_{\sigma b}]$ = matrix of Eq. 13.2.7 and $[k_{\sigma a}]$ = matrix of Eq. 13.2.12. Quite near the buckling load, use $[k_{\sigma b}]$. Under higher loads, convergence is about twice as fast if $[k_{\sigma a}]$ is used instead of $[k_{\sigma b}]$, while complete omission of $[k_\sigma]$ gives an intermediate rate. If displacements are specified instead of loads, it is best to omit $[k_\sigma]$.

Unless the convergence tolerance is quite severe, the deformed configuration may not be known exactly enough to produce accurate values of stresses in the structure. However, continued iteration to a severe tolerance is expensive, since a complete solution of the structural equations is required at each step. We may therefore adopt the following scheme. Use a coarse convergence tolerance, and iterate according to Eqs. 14.2.3 until convergence. Then commence a second sequence of iterations, also according to Eq. 14.2.3, but without updating $[K]$. Matrices $[k]_i$ and $\{d\}_i$ are updated every cycle, but the repeated assembly and reduction of $[K]$ is avoided.

When using either Eq. 14.2.3 or the modified iteration, convergence difficulties may be encountered near a critical load. For example, if in Fig. 14.2.3 the *initial* loading is $P > P_{\text{critical}}$, convergence is to configuration AC instead of to AB. Both are equilibrium configurations, but AC is unstable. The remedy is to use more load levels or a larger force Q in the neighborhood of an instability. Numerical tests on columns indicate that if initial stress stiffness effects are included in matrix $[K]_i$ of Eq. 14.2.3, and if convergence is toward an unstable configuration such as AC, then negative pivots appear when $[K]_i$ is reduced in a Gauss elimination solution. Recalling the Sturm sequence property (Section 12.5), we see that a negative pivot implies that a critical load has been exceeded. This in turn implies an error, since we are seeking a stable configuration.

14.3   Further Comments on Iterative Solutions

Solution by use of Eqs. 14.2.3 is an application of the Newton-Raphson method for solving algebraic and transcendental equations. To show the resemblance in a simple way, consider a single d.o.f. system, Fig. 14.3.1. Suppose that after convergence under load R_A, the load is increased to R_B. The corresponding displacement D_B is sought. A first-order Taylor series

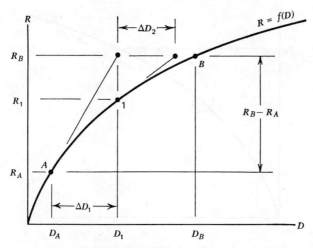

Figure 14.3.1. Newton-Raphson method.

expansion about point A yields

$$f(D_A + \Delta D_1) = f(D_A) + (\Delta D_1)\left(\frac{dR}{dD}\right)_A \qquad (14.3.1)$$

The derivative in this expression is the slope of the curve at A and is recognized as K_A, the structure tangent stiffness at point A. We seek ΔD_1 such that $f(D_A + \Delta D_1) = R_B$. Thus, since $f(D_A) = R_A$, Eq. 14.3.1 may be written as

$$K_A(\Delta D_1) = R_B - R_A \qquad (14.3.2)$$

which is the single d.o.f. form of Eq. 14.2.3. The new estimate of D_B is $D_A + \Delta D_1$, which is not exact because the Taylor series (Eq. 14.3.1) has been truncated. The next iteration uses the tangent stiffness K_1 at point 1 and the force unbalance $R_B - R_1$ to obtain the next correction ΔD_2. Here R_1 represents the resisting force provided by the structure when displacement D_1 prevails.

The Newton-Raphson method requires that the tangent stiffness matrix be formed and then triangularized in each step. This is expensive if the problem has many d.o.f. Accordingly, various modifications have been proposed.

A modified iterative scheme is illustrated in Fig. 14.3.2. The tangent stiffness K_A is used in *all* iterations. As compared with Fig. 14.3.1, Fig. 14.3.2 requires more steps for convergence, but each step is done much more quickly. This method is known as a "modified Newton-Raphson method," for obvious reasons. In another modification the stiffness is held constant

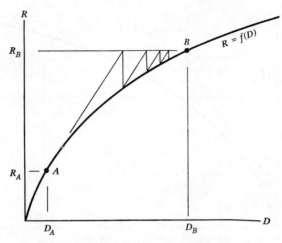

Figure 14.3.2. Modified Newton-Raphson iteration.

for several iterations or load levels and is updated only when the rate of convergence has begun to deteriorate (9).

Yet another variation employs a single Newton-Raphson iteration at each load level (10). Thus there are as many load levels at iterations. The solution proceeds as shown by the solid tangents in Fig. 14.3.3, where points on the computed curve are indicated by dots. This procedure might also be described as incremental, as successive displacement increments ΔD_1, ΔD_2, etc., are produced by successive load increments $\Delta R_1 = R_1$,

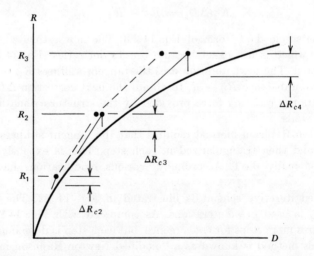

Figure 14.3.3. Incremental solution with one-step Newton-Raphson correction.

$\Delta R_2 = R_2 - R_1$, $\Delta R_3 = R_3 - R_2$, etc. But by adding ΔR_{c2} to ΔR_2, ΔR_{c3} to ΔR_3, etc., the computed displacement increments are made larger and more nearly correct. In other words, addition of the force unbalances ΔR_{ci} to load increments ΔR_i prevents large drift from the correct curve. The method is usually called "incremental with a one-step Newton-Raphson correction." A *purely* incremental scheme omits the corrective terms ΔR_{ci} and is represented by the equations

$$[K]_{i-1}\{\Delta D\}_i = \{\Delta R\}_i, \qquad \{D\}_i = \{D\}_{i-1} + \{\Delta D\}_i \qquad (14.3.3)$$

which yield the dashed line in Fig. 14.3.3.

14.4 A General Formulation

We now consider a more mathematical statement of the nonlinear problem than given in Section 14.2. The present formulation permits further insight into the problem and has the advantage of eliminating the need for coordinate transformations between the global system and local systems that travel with each element. However, element equations are more complicated. The following development is extracted from that of Refs. 6 and 7.

We begin by writing the total potential energy of a structure, Eq. 4.2.13, in the form

$$\Pi_p = U - \{D\}^T\{R\} = U_L + U_{NL} - \{D\}^T\{R\} \qquad (14.4.1)$$

where U = strain energy of the structure, and U_L and U_{NL} are, respectively, the parts of U arising from linear and nonlinear strain-displacement expressions. Despite nonlinearity, static equilibrium still prevails when displacements $\{D\}$ satisfy the equations

$$\left\{\frac{\partial \Pi_p}{\partial D}\right\} = 0 \qquad (14.4.2)$$

provided that the system is conservative. Thus the material must remain elastic (although its stress-strain relation need not be linear) and loads $\{R\}$ must be independent of $\{D\}$ (2 of Chapter Three). More specifically, at the equilibrium configuration $\{R\}$ must be independent of $\{D\}$ to the extent that *changes* in loads $\{R\}$ caused by a virtual displacement $\{\delta D\}$ are negligible in comparison with the loads themselves. Thus, changes in $\{R\}$ are regarded in the same way as changes in stress in Eq. 3.5.4.

From the strain energy U_L and Eq. 4.2.13 we obtain

$$\left\{\frac{\partial U_L}{\partial D}\right\} = [K_L]\{D\}, \qquad \frac{\partial^2 U_L}{\partial D_i\, \partial D_j} = K_{Lij} = K_{Lji} \qquad (14.4.3)$$

where $[K_L]$ is the conventional linear stiffness matrix that was written without any subscript in preceding chapters and is not a function of $\{D\}$. Symbolically we may write

$$[K_L] = \left[\frac{\partial^2 U_L}{\partial D_i \, \partial D_j}\right], \quad \text{also} \quad [K_{NL}] = \left[\frac{\partial^2 U_{NL}}{\partial D_i \, \partial D_j}\right] \quad (14.4.4)$$

Matrix $[K_{NL}]$ arises from nonlinear terms in the strain-displacement relations and is a function of $\{D\}$. It is defined here for later use.

The nonlinear equilibrium equations of the structure are, from Eqs. 14.4.1, 14.4.2, and 14.4.3,

$$[K_L]\{D\} + \left\{\frac{\partial U_{NL}}{\partial D}\right\} = \{R\} \quad (14.4.5)$$

A Newton-Raphson solution of these equations is derived from Eq. 14.4.5 with $\{R\}$ transferred to the left side so that the equation reads $f\{D\} = 0$. Based on an estimate $\{D\}_0$ of displacements under loads $\{R\}$, we write a first-order Taylor series expansion about $\{D\}_0$ to find $\{\Delta D\}_1$ such that $f(\{D\}_0 + \{\Delta D\}_1) = 0$. Thus,

$$([K_L] + [K_{NL0}])\{\Delta D\}_1 = \{R\} - [K_L]\{D\}_0 - \left\{\frac{\partial U_{NL0}}{\partial D}\right\} \quad (14.4.6)$$

The next estimate of displacements is $\{D\}_1 = \{D\}_0 + \{\Delta D\}_1$, and the next iteration for $\{\Delta D\}_2$ is based on $[K_{NL1}]$, U_{NL1}, and $\{D\}_1$. If $[K_{NL0}]$ is not updated, we have the modified Newton-Raphson solution. The resemblance between Eqs. 14.3.2 and 14.4.6 is seen if the single d.o.f. example is written as $f(K_L D + \partial U_{NL}/\partial D - R_B) = 0$, with ΔD_1 sought for which $f(D_A + \Delta D_1) = 0$. If $\{R\}$ in Eq. 14.4.6 is replaced by $\{R\}_0 + \{\Delta R\}$ we obtain the incremental procedure discussed in connection with Fig. 14.3.3. Here $\{\Delta R\}$ is an increment of applied loading and $\{R\}_0$ is the current load level. The three terms that augment $\{\Delta R\}$ on the right side of this modified equation correspond to a force ΔR_c in Fig. 14.3.3 and prevent large drift from the correct curve. (If these three terms are discarded, the purely incremental scheme represented by the dashed line is obtained.) Similar corrective terms may be used in nonlinear dynamic analysis (45, 67 of Chapter Twelve).

Further discussion regarding Eqs. 14.4.5 and 14.4.6 and their solution is given in connection with Eqs. 14.4.14 through 14.4.19. Prior to this discussion we attempt to clarify the nature of the nonlinear terms by means of an example (7). Consider a pin-ended bar element of cross-sectional area A and elastic modulus E (Fig. 14.4.1).

The axial strain is written as

$$\epsilon_x = e + \frac{\theta^2}{2}, \quad \text{where} \quad e = \frac{d_3 - d_1}{L}, \quad \theta = \frac{d_4 - d_2}{L} \quad (14.4.7)$$

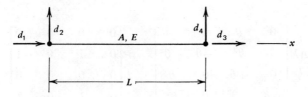

Figure 14.4.1. Bar element.

This approximation is the same as that given by Eq. 13.2.2 and restricts the development to small rotations. The total strain energy is

$$U_L + U_{NL} = \frac{AE}{2} \int_0^L \epsilon_x{}^2 \, dx = \frac{AE}{2} \int_0^L e^2 \, dx + \frac{AE}{2} \int_0^L \left(e\theta^2 + \frac{\theta^4}{4} \right) dx \quad (14.4.8)$$

From U_L we obtain the conventional linear element stiffness matrix $[k_L]$. On the element level, the nonlinear contribution to the ith nodal force is, from Eq. 14.4.8,

$$\frac{\partial U_{NL}}{\partial d_i} = \frac{AE}{2} \int_0^L \left(\theta^2 \frac{\partial e}{\partial d_i} + 2e \frac{\partial \theta}{\partial d_i} + \theta^3 \frac{\partial \theta}{\partial d_i} \right) dx \quad (14.4.9)$$

where $i = 1, 2, 3, 4$. Substitution of e and θ from Eq. 14.4.7 into the partial derivatives yields

$$\left\{ \frac{\partial U_{NL}}{\partial d} \right\} = \frac{AE}{2} \begin{Bmatrix} -\theta^2 \\ -2e\theta - \theta^3 \\ \theta^2 \\ 2e\theta + \theta^3 \end{Bmatrix} \quad (14.4.10)$$

Another differentiation and substitution of partial derivatives yields (7)

$$[k_{NL}] = \left[\frac{\partial^2 U_{NL}}{\partial d_i \, \partial d_j} \right] = \frac{AE}{L} \begin{bmatrix} 0 & \theta & 0 & -\theta \\ \theta & e + \dfrac{3}{2}\theta^2 & -\theta & -e - \dfrac{3}{2}\theta^2 \\ 0 & -\theta & 0 & \theta \\ -\theta & -e - \dfrac{3}{2}\theta^2 & \theta & e + \dfrac{3}{2}\theta^2 \end{bmatrix} \quad (14.4.11)$$

It is tempting to discard the θ^2 terms as being negligibly small, but there is evidence that they are important in shell problems (12).

The expression for $[k_{NL}]$ in Eq. 14.4.11 may be made to look somewhat more familiar if we introduce the axial force $P = AE\epsilon_x = AE(e + \theta^2/2)$.

Thus (7),

$$[k_{NL}] = \frac{AE}{L}\begin{bmatrix} 0 & \theta & 0 & -\theta \\ \theta & \theta^2 & -\theta & -\theta^2 \\ 0 & -\theta & 0 & \theta \\ -\theta & -\theta^2 & \theta & \theta^2 \end{bmatrix} + \frac{P}{L}\begin{bmatrix} 0 & 0 & 0 & 0 \\ 0 & 1 & 0 & -1 \\ 0 & 0 & 0 & 0 \\ 0 & -1 & 0 & 1 \end{bmatrix} \qquad (14.4.12)$$

The second matrix in $[k_{NL}]$ is the initial stress stiffness matrix $[k_\sigma]$ of Eq. 13.2.12, and as usual it accounts for the effect that membrane force has on lateral displacements. For $\theta = 0$, corresponding to an initial unrotated configuration, $[k_{NL}]$ reduces to $[k_\sigma]$. The first matrix in Eq. 14.4.12 must now be explained. To do so we write the conventional linear stiffness matrix of a deformed member, $[k'_L]$, and transform this matrix to the undeformed local coordinates xy by the standard transformation $[k] = [T]^T[k'_L][T]$ (Fig. 14.4.2). For small θ, so that $\sin\theta \approx \theta$ and $\cos\theta \approx 1$, the matrices are

$$[T] = \begin{bmatrix} 1 & \theta & 0 & 0 \\ -\theta & 1 & 0 & 0 \\ 0 & 0 & 1 & \theta \\ 0 & 0 & -\theta & 1 \end{bmatrix}, \qquad [k'_L] = \frac{AE}{L}\begin{bmatrix} 1 & 0 & -1 & 0 \\ 0 & 0 & 0 & 0 \\ -1 & 0 & 1 & 0 \\ 0 & 0 & 0 & 0 \end{bmatrix}$$

$$(14.4.13)$$

The matrix $[k]$ produced by the transformation is found to be $[k'_L]$ plus the first matrix on the right side of Eq. 14.4.12. Therefore, generalizing from Eq. 14.4.12, we argue that the formulation of this section includes the essential features described in Section 14.2: referencing of stiffness matrices to deformed coordinates and inclusion of initial stress stiffness effects. It should be emphasized that assembly of elements into the structure is accomplished

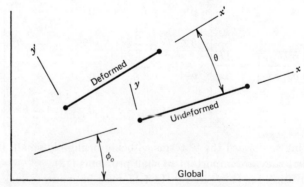

Figure 14.4.2. Coordinate systems.

from the undeformed element position (angle ϕ_o in Fig. 14.4.2); local co-ordinates $x'y'$ are not established, and in every iteration the element-to-structure transformation is based on the same angle ϕ_o in Fig. 14.4.2.

If the latter matrix in Eq. 14.4.12 is viewed as an initial stress stiffness matrix $[k'_\sigma]$ in deformed coordinates $x'y'$, then transformed to coordinates xy by the transformation $[k_\sigma] = [T]^T[k'_\sigma][T]$, terms θ and θ^2 appear in $[k_\sigma]$. This result seems to disagree with our argument that the formulation based on U_{NL} includes implicitly the necessary coordinate transformation. However, the θ and θ^2 terms may be neglected as being of higher order (7). Indeed, as suggested in Section 14.2, it is common practice to base U_{NL} on simpler displacement fields than those used for U_L, so that $[k_\sigma]$ is already approximate or "inconsistent." Adopting this point of view, we may use the nonlinear terms of Eqs. 14.4.9 through 14.4.11 for beam elements as well as for pin-jointed bars. Use of simplified displacement fields for U_{NL} makes derivation easier and saves considerable computer time.

What we have termed "initial stress stiffness effects" appear as matrix $[k_\sigma]$ in Chapter Thirteen and are incorporated in matrix $[k_{NL}]$ of this section. The two developments are not in conflict. In Chapter Thirteen, membrane stresses $\{\sigma_o\}$ were presumed given at the outset and constant during a subsequent linear analysis. In the initial configuration, nodal forces produced by these stresses merely equilibrate the loading that produced them, and these loads play no further role in the subsequent linear analysis or linear buckling problem. In $[k_{NL}]$ of this section, initial stress stiffness is expressed in terms of strain, not stress. These strains change and accumulate as deformation proceeds. Stresses $\{\sigma_o\}$, if present at the outset, contribute only to nodal forces; in the first step of a numerical solution, deformations and $[k_{NL}]$ are both zero. In subsequent steps $\{\sigma_o\}$ contributes to $[k_{NL}]$ through the accumulated deformations.

The following symbolism for nonlinear terms is preferred by many workers. Taking Eq. 14.4.10 as an example, we note that it may be written in the form

$$\left\{ \frac{\partial U_{NL}}{\partial d} \right\} = \left(\frac{1}{2} [N1] + \frac{1}{3} [N2] \right) \{d_1 \quad d_2 \quad d_3 \quad d_4\} \qquad (14.4.14)$$

where $[N1] + [N2] = [k_{NL}]$ of Eq. 14.4.11. Matrix $[N1]$ contains terms linear in e and θ and $[N2]$ contains the θ^2 terms. Equation 14.4.5 becomes (6)

$$\left([K_L] + \frac{1}{2} [N1] + \frac{1}{3} [N2] \right) \{D\} = \{R\} \qquad (14.4.15)$$

and the purely incremental form of Eq. 14.4.6 becomes

$$([K_L] + [N1] + [N2])\{\Delta D\} = \{\Delta R\} \qquad (14.4.16)$$

Apparently Eq. 14.4.16 may be obtained by differentiation of Eq. 14.4.15, noting that $[N1]$ and $[N2]$ are implicitly linear and quadratic functions of the displacements (6, 42).

Equation 14.4.15 may be used to evaluate the resisting forces developed by the deformed structure (37). For example, the resisting force R_1 in Fig. 14.3.1 is found by evaluating Eq. 14.4.15 with $\{D\} = D_1$ and $[N1]$ and $[N2]$ evaluated at D_1.

Equations 14.4.15 and 14.4.16 suggest a further interpretation. For a hypothetical one degree of freedom system, these equations become

$$\frac{R}{D} = c_1 + \frac{c_2}{2}D + \frac{c_3}{3}D^2$$

$$\frac{\Delta R}{\Delta D} = c_1 + c_2 D + c_3 D^2$$

(14.4.17)

where c_1, c_2, c_3 are constants. Quantities R/D and $\Delta R/\Delta D$ are, respectively, the secant and tangent to the curve at point A in Fig. 14.4.3. The true curve, whatever it may be, is approximated in the neighborhood of point A as a cubic in D. The secant R/D suggests a solution of Eq. 14.4.15 known as "direct iteration" (see Section 14.5).

The example of Eqs. 14.4.15 and 14.4.16 may be further explained as follows. After extracting $\{D\}$ from $\{\partial U_{NL}/\partial D\}$, Eq. 14.4.5 may be written as

$$\left([K_L] + \left[\frac{\partial U_{NL}}{\partial D}\right]\right)\{D\} = \{R\}$$

(14.4.18)

We identify the coefficient of $\{D\}$ as a *secant stiffness matrix*, since it multiplies the total displacements $\{D\}$ instead of their increments $\{\Delta D\}$. The *incremental* form of this equation, corresponding to Eq. 14.4.16, is obtained

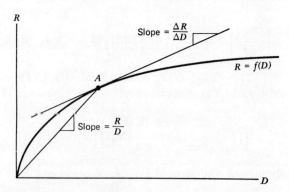

Figure 14.4.3   Single d.o.f. system.

by subtracting Eq. 14.4.18 from the equation

$$\left([K_L] + \left[\frac{\partial U_{NL}}{\partial D}\right]\right)\{D + \Delta D\} = \{R + \Delta R\} \qquad (14.4.19)$$

in which $[\partial U_{NL}/\partial D]$ is a function of $\{D + \Delta D\}$ and it is assumed that $\{\Delta D\} \ll \{D\}$. In practice it is not necessary to obtain Eq. 14.4.18 by "extracting" $\{D\}$ from $\{\partial U_{NL}/\partial D\}$, since the separation occurs naturally in the process of deriving element stiffness matrices. The incremental (tangent) stiffness matrix is symmetric, but the secant stiffness matrix from which it is derived may be unsymmetric. Accordingly, there would seem to be less chance for error if Eqs. 14.4.18 and 14.4.19 are used to derive the incremental form instead of the implicit differentiation that led to Eq. 14.4.16. The concepts of the present paragraph are applied, for example, in Ref. 37.

Finally, it is worth repeating that commonly used strain-displacement expressions such as Eqs. 14.4.7 restrict the development to small rotations. This restriction is implied in the way matrices $[N1]$ and $[N2]$ of Eqs. 14.4.14 through 14.4.16 are defined. The restriction is overcome by use of moving coordinates, which introduce additional nonlinearities (6). Specifically, in Fig. 14.4.2 the element-to-structure transformation is then based on the angle $\phi_o + \theta$, as in Section 14.2.

14.5 Further Solution Techniques

Solution of the governing equations, Eq. 14.4.5 or 14.4.15, may be obtained by so-called *direct iteration*. The governing equation may be written in the form

$$[K_L]\{D\}_{i+1} = \{R\} - \left(\frac{1}{2}[N1] + \frac{1}{3}[N2]\right)_i \{D\}_i \qquad (14.5.1)$$

This equation implies the following iterative solution. Prescribe a loading $\{R\} = \{R\}_A$ and assume $\{D\}_0 = 0$; hence $[N1]_0 = [N2]_0 = 0$. Solve for $\{D\}_1$ by, say, Gauss elimination. In the next step, maintain $\{R\} = \{R\}_A$, update $[N1]$ and $[N2]$ by use of $\{D\}_1$, and solve for $\{D\}_2$. The solution proceeds horizontally through points C, D, E, etc., in Fig. 14.5.1. After convergence under loading $\{R\}_A$, loading may be increased to a level $\{R\}_B$ and iteration begun again. In the first iteration of this second set, $[N1]$ and $[N2]$ are based on displacements $\{D\}_A$.

In the example of Fig. 14.5.1, the right side of Eq. 14.5.1 exceeds R_A in successive iterations. Thus, while always basing computations on the slope of line OC, the *effective* loads are successively R_A, R_H, R_I, etc. The "added loads," which are functions of the total displacements, form the sequence

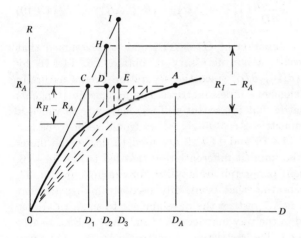

Figure 14.5.1. Direct iteration, single d.o.f. example.

$R_H - R_A = f(D_1)$, $R_I - R_A = f(D_2)$, etc. The sequence must converge if the solution is to converge to $D = D_A$.

Direct iteration may also be applied to Eq. 14.4.15 without transferring the nonlinear terms to the right side. Iteration then generates line OB and the successive secants shown by dashed lines in Fig. 14.5.1. This form of direct iteration is expensive, as the coefficient matrix must be computed and reduced anew in each iteration. However, the rate of convergence of direct iteration is likely to be slowed by shifting one or more nonlinear terms to the right side.

When using direct iteration it is possible to "sneak up" on the desired load level by iterating at each of several intermediate load levels. Also, under-relaxation may be applied to the successive iterates (11, 13), for example, $\{D\}_i = (1 - \beta)\{D\}_{i-1} + \beta\{D\}_i$, where $0 < \beta < 1$. Either or both techniques may be needed to assure convergence. Direct iteration is said to be useful if the nonlinear solution differs from the linear solution by a factor of less than 2.5 (9). For more severe nonlinearities, the method does not converge.

It is possible to attack the nonlinear equations as an initial value problem, thus permitting use of the step-by-step numerical integration methods mentioned in Section 12.6. This is clearly the case if the problem really is one of dynamic response; the governing equations to be solved are still Eqs. 12.2.7 except for the addition of nonlinear terms $\{\partial U_{NL}/\partial D\}$ from Eq. 14.4.5 to the right side (12). The entire right side is then a forcing function that is dependent on both deformation and time. Formally, these dynamic equations may be derived as in Eqs. 12.2.6 if U_{NL} is incorporated into the Lagrangian L. If rotations are sufficiently small that moving coordinates need not be attached to each element, nonlinear terms are confined to expressions for strain; hence mass and damping matrices are unchanged from the linear problem.

The initial-value approach may also be applied to the nonlinear static problem (9, 14, 15). The method is very briefly outlined as follows. Letting $\{R_{NL}\} = \{\partial U_{NL}/\partial D\}$, we write Eq. 14.4.5 in the form

$$[K_L]\{D\} = \bar{P}\{\bar{R}\} - \{R_{NL}\} \qquad (14.5.2)$$

where $\{\bar{R}\}$ represents loading evaluated at some convenient reference level and $\bar{P}$ is a scalar multiplier. It is assumed that $\{\bar{R}\}$ is independent of displacements $\{D\}$. With a dot representing differentiation with respect to $\bar{P}$ we write, by chain rule differentiation,

$$\{\dot{R}_{NL}\} = \left[\frac{\partial R_{NLi}}{\partial D_j}\right]\left\{\frac{\partial D_j}{\partial \bar{P}}\right\} = \left[\frac{\partial^2 U_{NL}}{\partial D_i\,\partial D_j}\right]\{\dot{D}\} = [K_{NL}]\{\dot{D}\} \qquad (14.5.3)$$

Hence Eq. 14.5.2 becomes $([K] + [K_{NL}])\{\dot{D}\} = \{\bar{R}\}$. Thus a step-by-step integration may be used, with load increments $\Delta\bar{P}$ instead of time increments Δt, to establish $\{D\}$ as a function of load. Second-order forms, involving $\{\ddot{D}\}$, have also been devised. The corrective terms on the right side of Eq. 14.4.6 may be introduced to prevent drift of the computed solution from the correct curve; the procedure may then be termed "self corrective." The simplest form of first-order self-corrective solution is that of Fig. 14.3.3—incremental with one-step Newton-Raphson correction (9, 10, 14, 15).

Opinions have been ventured as to the method of choice for the static nonlinear problem (9, 14, 15), but opinions change as knowledge accumulates. Some current viewpoints are as follows. If only geometric nonlinearity is present, a modified Newton-Raphson method may be chosen. The first- and second-order self-correcting procedures are also good, as is direct iteration if nonlinearity is not severe. If only material nonlinearity is present, the first-order self-correcting procedure is preferred. If both material and geometric nonlinearity are present, either the first- or second-order self-correcting procedure may be chosen.

The nonlinear stability problem has been introduced in the last three paragraphs of Chapter Thirteen. If a curve such as OC in Fig. 13.4.1 is obtained for any d.o.f., the occurrence of a peak C indicates an instability. But whether it is point A, B, or C that marks the limit of stability, we have

$$([K_L] + [K_{NL}])\{\Delta D\} = \{\Delta R\} = 0 \qquad (14.5.4)$$

That is, as the load increment $\{\Delta R\}$ approaches zero, displacements $\{\Delta D\}$ still exist, because at buckling the tangent stiffness $[K_L] + [K_{NL}]$ becomes zero for some buckling mode. Thus, at buckling, the determinant of $[K_L] +$ $[K_{NL}]$ vanishes. The determinant-plotting method is useful for establishing the critical load (6; 23, 58, 59 of Chapter Twelve; 3, 7, 12 of Chapter Thirteen).

Upon buckling, deflections increase under negative load increments, as seen in the descending path of curve 3 in the single d.o.f. example of Fig. 14.5.2. Entry into this region is signaled by the determinant of the tangent stiffness matrix becoming negative. In this postbuckling range solution methods fail unless special precautions are taken. A procedure suggested in Ref. 16 keeps the tangent stiffness positive at all d.o.f. and may be used with various numerical solution methods. Figure 14.5.2 shows the use of the procedure for a single d.o.f. (16). Addition of the linear spring eliminates the snap-through behavior and converts curve 3 to curve 1. If κ is large, we have a stiff linear structure augmented by a soft nonlinear one, the nonlinear part being essentially dragged along as the linear part is loaded. For any computed

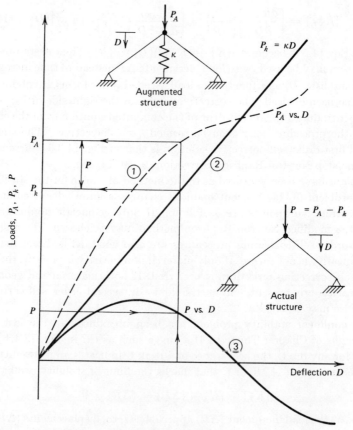

Figure 14.5.2.   Structure augmentation scheme.

value of D, only a portion of the total load P_A is needed to deform the nonlinear part of the augmented structure. This portion is $P_A - P_k$; in the figure, it is curve 1 minus curve 2.

Generalization of the method to several d.o.f. is as follows (16). Define a matrix

$$[K_A] = \frac{\kappa}{\sigma^2} \{R\}\{R\}^T \tag{14.5.5}$$

where κ = a constant, $\{R\}$ = load vector for an arbitrarily chosen load level, and σ = norm of vector $\{R\}$. Thus $[K_A]$ is defined at the outset and is not a function of displacements. Stiffness $[K_L]$ in Eq. 14.4.5 is augmented by $[K_A]$, so that this equation becomes

$$([K_L] + [K_A])\{D\} + \left\{\frac{\partial U_{NL}}{\partial D}\right\} = \{R\} \tag{14.5.6}$$

Substitution of Eq. 14.5.5 into Eq. 14.5.6 yields

$$[K_L]\{D\} + \left\{\frac{\partial U_{NL}}{\partial D}\right\} = \Lambda\{R\}$$

where $\tag{14.5.7}$

$$\Lambda = 1 - \frac{\kappa}{\sigma^2} \{R\}^T\{D\}$$

After solving Eq. 14.5.6 for $\{D\}$ under some loading $\{R\}$, we compute from Eq. 14.5.7 the loads $\Lambda\{R\}$ that produce the same displacements $\{D\}$ in the original unaugmented structure. Advice on the choice of κ and further comments appear in Ref. 16. Note that if $\{R\}$ contains many nonzero entries, matrix $[K_A]$ will not be sparse.

Other methods of tracing the postbuckling path are mentioned in Refs. 16, 29, and 33, and in Ref. 59 of Chapter Twelve. These methods use displacement increments or negative load increments in unstable regions.

A postbuckled configuration may not be unique. If several equilibrium configurations are possible, the one selected by the computation process may be a function of how the system was initially perturbed. For example, a slight asymmetry of loading may prejudice results in favor of an asymmetric mode.

14.6   Miscellaneous Problems and Techniques

Membranes and cables commonly deform under load to such an extent that their behavior must be considered nonlinear (17–19). Indeed, a membrane may be blown up like a balloon from an initially flat shape. The simplest elements to use for analysis of membranes are the constant-strain triangles of Eq. 4.4.1. Two aspects of such problems deserve comment. First, if the structure is initially flat its initial resistance to lateral loading is zero. To start a numerical solution some stiffness is required and may be provided by adding linear springs to all d.o.f. (17). The diagonal $[K_L]$ thus produced may be discarded after the first iteration, as the structure acquires stiffness of its own owing to displacement. Second, as a membrane element stretches and rotates, pressure loading changes in both magnitude and direction (19, 20). Forces were required to be independent of displacement in the derivation of Eq. 14.4.5. However, we may argue that during a small displacement increment the loading remains essentially constant, so that incremental methods may be used. The argument of Section 14.2 that equilibrium is a state of forces in balance indicates that the method of Section 14.2 is also applicable.

Cable networks tend to become stiffer with increasing displacement. Underrelaxation techniques seem to enhance the convergence rate of Newton-Raphson solutions (40).

An interesting application of membrane analysis to shell design has been described (21). Noting that a shell acts most efficiently when loads are supported by membrane action rather than bending, the authors propose to define the shell geometry as that of a membrane loaded by the same forces that must be supported by the actual shell. As the object is only to find a suitable shape that carries load by membrane action, and not to analyze an actual membrane, the hypothetical membrane is assumed to have a linear stress-strain relation.

Roughly similar to the membrane problem is that of kinematic analysis of folded plate structures (4.7 of Chapter Eighteen).

Contact-stress problems usually involve small displacements and rotations. However, the problem is nonlinear because the size of contact areas depends on load. An iterative analysis is required to determine the proper contact region between adjacent finite-element structures and to prevent nodes of one structure from penetrating into the other (22–25). Coordinate transformations may be employed to eliminate excess d.o.f. as two nodes become coincident or to enforce equality of displacements normal to a frictionless surface (see Section 11.8).

Imperfection sensitivity has been considered in a finite element context (26, 27, 34, 36). Apparently departures from an ideal geometry may be introduced at will in the mesh layout or concentrated loads applied to simulate them; however, this appears to be a "brute-force" approach to the imperfection problem. A special problem of departure from an ideal geometry is that of plates with slight initial curvature (28, 29, 35, 37).

Schemes for accelerating the convergence of iterative procedures have been advocated (30, 31, 32).

PROBLEMS

14.1. Explain the similarities or differences between Eqs. 13.3.4 and 14.2.3.

14.2. Figures 14.3.1 and 14.3.2 illustrate the Newton-Raphson methods for the case where the R versus D curve is concave down. Make similar sketches to explain the two methods for the case where the curve is concave up.

14.3. A single d.o.f. system has the load-deflection curve shown. At peak A, $R_A = 241$ and $D_A = 0.211$. Also, $R_B = 250$ and $D_B = 1.080$. After convergence at $R = 200$, R is increased to 250 and the following sequence of displacements D is generated by a Newton-Raphson solution: 0.173, 0.219, 0.071, 0.143, 0.190, 0.249, 0.199, 0.294, 0.235, 0.175, 0.222, 0.108, 0.166, 0.210, 1.178, 1.096, 1.080, 1.080.

(a) Explain how this path to convergence arises.

(b) How could the given curve between A and B be generated by the Newton-Raphson method?

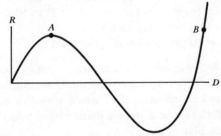

14.4. The single d.o.f. system has a nonlinear spring whose tangent stiffness is $k = f(D) = 10 - 2D$. Find D for $P = 24$ by the following solution methods.

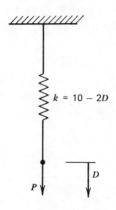

(a)   Newton-Raphson (Fig. 14.3.1).

(b)   Modified Newton-Raphson (Fig. 14.3.2).

(c)   Incremental with one-step Newton-Raphson correction (Fig. 14.3.3), with $\Delta P = 6$.

Note. The spring simulates a geometric nonlinearity whose nature has not been described. Therefore, in this contrived exercise, one must integrate $dP/dD = 10 - 2D$ to find the spring force for a given D.

14.5.   (a)   Derive Eq. 14.4.6.

(b)   Derive Eq. 14.4.11.

(c)   Verify Eq. 14.4.14.

14.6   (a)   For the spring system of Problem 14.4, derive the appropriate form of Eq. 14.4.5.

(b)   Find D for $P = 24$ by direct iteration, if possible (Eq. 14.5.1).

(c)   Cast the governing equation of Part a into the form of Eq. 14.4.15. Solve for $P = 24$, if possible, by iteration (successive secant solution of Fig. 14.5.1).

14.7   Apply the modification scheme of Fig. 14.5.2 to the single d.o.f. system of Problem 14.4. Use $\kappa = 20$. Find the P and D coordinates of a point on the P versus D curve in the range $5 < D < 10$ by means of

(a)   Direct iteration (Eq. 14.5.1).

(b)   Newton-Raphson method (Fig. 14.3.1).

14.8.   (a)   Consider use of a plane-stress *quadrilateral* element for analysis of membranes. How might one formulate the transformation between global coordinates and coordinates that move with the element? *Caution.* Actually, it is not advisable to distort a plane element so that its nodes do not all lie in the same plane (38).

(b)   When using membrane and cable elements, is there any need for use of the initial stress stiffness matrix $[k_\sigma]$ defined in Chapter Thirteen? Explain.

14.9.   Show that if $R_A = 0$ in Fig. 14.3.2, modified Newton-Raphson iteration and the direct iteration of Eq. 14.5.1 are essentially the same method.

CHAPTER FIFTEEN

Material Nonlinearity

15.1 Introduction. Plasticity Relations

This chapter is devoted largely to elastic-plastic problems. Unless otherwise stated, problems are static and without geometric nonlinearity. We will present the essentials of accepted analysis methods but forgo mathematical rigor and consideration of much published material. These restrictions are made in order to provide an introductory treatment of a large topic.

We begin with a summary of elastic-plastic stress-strain relations used for isotropic materials such as common metals (1–6). According to the von Mises criterion, yielding begins under any state of stress when the effective stress $\bar{\sigma}$ exceeds a certain limit, where

$$\bar{\sigma} = \frac{1}{\sqrt{2}} [(\sigma_x - \sigma_y)^2 + (\sigma_y - \sigma_z)^2 + (\sigma_z - \sigma_x)^2 + 6(\tau_{xy}^2 + \tau_{yz}^2 + \tau_{zx}^2)]^{1/2}$$

(15.1.1)

An effective plastic strain increment $d\bar{\epsilon}^p$ is defined as a combination of the separate plastic strain increments:

$$d\bar{\epsilon}^p = \frac{\sqrt{2}}{3} \Big[(d\epsilon_x{}^p - d\epsilon_y{}^p)^2 + (d\epsilon_y{}^p - d\epsilon_z{}^p)^2 + (d\epsilon_z{}^p - d\epsilon_x{}^p)^2 + \frac{3}{2}(d\gamma_{xy}^p)^2$$

$$+ \frac{3}{2}(d\gamma_{yz}^p)^2 + \frac{3}{2}(d\gamma_{zx}^p)^2 \Big]^{1/2} \quad (15.1.2)$$

The engineering definition of shear strain is used, for example, $\gamma_{xy} = u_{,y} + v_{,x}$. For uniaxial stress σ_x, $\bar{\sigma} = \sigma_x$, and in the plastic range where Poisson's ratio is 0.5, $d\bar{\epsilon}^p = d\epsilon_x{}^p$. The relation between effective stress and effective strain may be established by a uniaxial tension test and is shown in Fig. 15.1.1. Notation in the figure defines the elastic component $\bar{\epsilon}^e$ of effective strain, and the relation between slope H' and elastic modulus E and tangent modulus E_T. The latter relation is found by substitution of $E_T = d\bar{\sigma}/d\bar{\epsilon}$ into the relation $d\bar{\epsilon} = d\bar{\sigma}/E + d\bar{\sigma}/H'$. Compact expressions for $\bar{\sigma}$ as a function of $\bar{\epsilon}$ may be written for use in digital computer applications by means of Ramberg-Osgood or similar formulas (7).

Yielding first begins when $\bar{\sigma}$ exceeds $\bar{\sigma}_o$ in Fig. 15.1.1. As $\bar{\epsilon}^p$ grows, the value of $\bar{\sigma}$ which must be exceeded to produce further yield also grows. This is, of course, the strain-hardening phenomenon. If unloading occurs we may assume that no matter what the subsequent state of stress, yielding resumes only when $\bar{\sigma}$ exceeds its previous maximum value. This is the assumption of isotropic hardening and ignores the Bauschinger effect. Other assumptions are, of course, possible but not as easily handled.

If we differentiate both sides of Eq. 15.1.1 and substitute the deviatoric stresses

$$\sigma'_x = \sigma_x - \frac{\sigma_x + \sigma_y + \sigma_z}{3} = \frac{1}{3}(2\sigma_x - \sigma_y - \sigma_z), \text{ etc.} \qquad (15.1.3)$$

we obtain

$$d\bar{\sigma} = \left\{\frac{\partial \bar{\sigma}}{\partial \sigma}\right\}^T \{d\sigma_x \, d\sigma_y \, d\sigma_z \, d\tau_{xy} \, d\tau_{yz} \, d\tau_{zx}\} = \left\{\frac{\partial \bar{\sigma}}{\partial \sigma}\right\}^T \{d\sigma\} \qquad (15.1.4)$$

where

$$\left\{\frac{\partial \bar{\sigma}}{\partial \sigma}\right\} = \left\{\frac{3\sigma'_x}{2\bar{\sigma}} \, \frac{3\sigma'_y}{2\bar{\sigma}} \, \frac{3\sigma'_z}{2\bar{\sigma}} \, \frac{3\tau_{xy}}{\bar{\sigma}} \, \frac{3\tau_{yz}}{\bar{\sigma}} \, \frac{3\tau_{zx}}{\bar{\sigma}}\right\} \qquad (15.1.5)$$

The Prandtl-Reuss relations state that

$$\{d\epsilon^p\} = \{d\epsilon_x{}^p \, d\epsilon_y{}^p \, d\epsilon_z{}^p \, d\gamma_{xy}^p \, d\gamma_{yz}^p \, d\gamma_{zx}^p\} = \left\{\frac{\partial \bar{\sigma}}{\partial \sigma}\right\} d\bar{\epsilon}^p \qquad (15.1.6)$$

The latter equation defines the six plastic strain increments that result when the effective plastic strain increment $d\bar{\epsilon}^p$ occurs under a known state of stress.

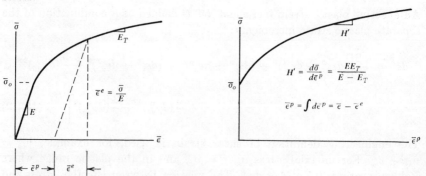

Figure 15.1.1. Effective stress-effective strain relations.

A strain increment $\{d\epsilon\}$ is the sum of its elastic component $\{d\epsilon^e\}$ and its plastic component $\{d\epsilon^p\}$. Hence, from Hooke's law,

$$\{d\sigma\} = [E]\{d\epsilon^e\} = [E](\{d\epsilon\} - \{d\epsilon^p\}) \qquad (15.1.7)$$

where $[E]$ is the conventional matrix of elastic constants that first appeared in Eq. 1.3.4 but is here restricted to isotropy.

A relation that yields $d\bar{\epsilon}^p$ from the total strain increment $\{d\epsilon\}$ is obtained as follows. Substitute Eq. 15.1.6 into Eq. 15.1.7, then premultiply both sides by $\{\partial\bar{\sigma}/\partial\sigma\}^T$. Figure 15.1.1 and Eq. 15.1.4 yield $\{\partial\bar{\sigma}/\partial\sigma\}^T\{d\sigma\} = H'\,d\bar{\epsilon}^p$, and upon substituting this there results

$$d\bar{\epsilon}^p = \frac{\left\{\dfrac{\partial\bar{\sigma}}{\partial\sigma}\right\}^T [E]}{H' + \left\{\dfrac{\partial\bar{\sigma}}{\partial\sigma}\right\}^T [E]\left\{\dfrac{\partial\bar{\sigma}}{\partial\sigma}\right\}}\{d\epsilon\} = [W]\{d\epsilon\} \qquad (15.1.8)$$

An incremental stress-strain relation, analogous to Hooke's law but valid beyond the proportional limit, is obtained by substitution of Eq. 15.1.8 into Eq. 15.1.6 and the result into Eq. 15.1.7. The relation is

$$\{d\sigma\} = \left([E] - [E]\left\{\frac{\partial\bar{\sigma}}{\partial\sigma}\right\}[W]\right)\{d\epsilon\} = [E_{ep}]\{d\epsilon\} \qquad (15.1.9)$$

Examination of the matrices involved shows that $[E_{ep}]$ is symmetric. Note that Eqs. 15.1.8 and 15.1.9 are both valid for elastic-perfectly plastic materials, as nothing in the derivation "blows up" if $H' = 0$.

As defined by Eq. 15.1.9, matrix $[E_{ep}]$ is 6 by 6. A form of $[E_{ep}]$ applicable to axisymmetric problems is obtained by deleting the two rows and two columns corresponding to $d\tau_{z\theta} = d\tau_{\theta r} = d\gamma_{z\theta} = d\gamma_{\theta r} = 0$. A similar modification produces a 3 by 3 matrix $[E_{ep}]$ for plane strain problems (8). For plane stress problems, $\{d\sigma\} = \{d\sigma_x\,d\sigma_y\,d\tau_{xy}\}$, and by rederivation instead of specialization $[E_{ep}]$ is (4, 8):

$$\frac{E}{Q}\begin{bmatrix} \sigma'_y\sigma'_y + 2P & -\sigma'_x\sigma'_y + 2\nu P & -\dfrac{\sigma'_x + \nu\sigma'_y}{1+\nu}\tau_{xy} \\[2ex] & \sigma'_x\sigma'_x + 2P & -\dfrac{\sigma'_y + \nu\sigma'_x}{1+\nu}\tau_{xy} \\[2ex] \text{symmetric} & & \dfrac{R}{2(1+\nu)} + \dfrac{2H'}{9E}(1-\nu)\bar{\sigma}^2 \end{bmatrix} \qquad (15.1.10)$$

where

$$P = \frac{2H'}{9E}\,\bar{\sigma}^2 + \frac{\tau_{xy}^2}{1 + \nu}$$

$$Q = R + 2(1 - \nu^2)P \qquad\qquad (15.1.11)$$

$$R = \sigma_x'\sigma_x' + 2\nu\sigma_x'\sigma_y' + \sigma_y'\sigma_y'$$

and $\nu = $ Poisson's ratio, $E = $ elastic modulus.

An approximate form of $[E_{ep}]$ is simply $[E_{ep}] = [E]E_T/E$, where $[E]$ is the matrix of *elastic* properties for the problem at hand and E_{T} and E are tangent and elastic moduli (Fig. 15.1.1). Modulus E_T is determined from ϵ^*, defined by Eq. 15.2.1, and a plot of $\bar{\sigma}$ versus ϵ^* (Fig. 15.2.1). This approximation (15) resembles that of the secant stiffness matrix of Section 15.2 and has many of its limitations. However, available test cases (15) show good accuracy.

Formulations similar to those of this section but applicable to anisotropic materials have been developed (9, 10). In soil and rock mechanics, where nonlinear material behavior is common, failure and flow rules other than Eqs. 15.1.1 and 15.1.6 may be used (11, 35). Proper treatment of the Bauschinger effect is discussed in Refs. 11 and 36. Several papers directed toward the practitioner appear in Ref. 35.

15.2 Direct Iteration Solution

Direct iteration provides a very simple but approximate solution of an elastic-plastic problem. The equations of Section 15.1 are largely ignored. In brief, the procedure is as follows. Assume that all material is elastic, compute the conventional structure stiffness matrix, apply the full load to the structure, and solve for displacements. Compute strains in each element. For those elements strained beyond their proportional limit, appropriately reduce their elastic stiffness, and re-form the structure stiffness matrix based on the modified elastic properties. Again apply the full load and compute displacements. Stop iteration when some convergence test is satisfied.

It should be noted that direct iteration yields a *correct* equilibrium configuration if material behavior is elastic but nonlinear. "Elastic" means that the stress-strain curve is the same for both loading and unloading, so that no permanent set exists on release of load.

The modified elastic stiffness needed for a direct iteration solution may be computed as follows. Define an effective strain ϵ^* as (15)

$$\epsilon^* = \frac{\sqrt{2}}{2(1 + \nu)}\left[(\epsilon_x - \epsilon_y)^2 + (\epsilon_y - \epsilon_z)^2 + (\epsilon_z - \epsilon_x)^2 + \frac{3}{2}(\gamma_{xy}^2 + \gamma_{yz}^2 + \gamma_{zx}^2)\right]^{1/2}$$

$$(15.2.1)$$

With this definition[1] the elastic modulus E is, in the elastic range, $E = \bar{\sigma}/\epsilon^*$, where $\bar{\sigma}$ is defined by Eq. 15.1.1. Experiment establishes the $\bar{\sigma}$ versus ϵ^* relation (Fig. 15.2.1). In a typical element, the first (elastic) solution establishes line OA. From strains $\{\epsilon\}_A = \{\epsilon\}_B$ we compute ϵ^*, and from the $\bar{\sigma}$ versus ϵ^* relation we find the values of $\bar{\sigma}$ and secant modulus E_s corresponding to point B. In forming element stiffnesses for the next iteration, the elastic property matrix $[E]$ is multiplied by the ratio E_s/E. The successive secants in Fig. 15.2.1 correspond to those in Fig. 14.5.1.

The foregoing method requires complete reassembly and reduction of the structure stiffness matrix in each iteration. To avoid this inefficiency, all nonlinearities may be transferred to the right side, as in Eq. 14.5.1. If $[k]$ is the elastic stiffness matrix of an element, the reduced element stiffness matrix is

$$[k_r] = \frac{E_s}{E}[k] = \left(1 + \frac{E_s - E}{E}\right)[k] = [k] + [k_m] \qquad (15.2.2)$$

where $[k_m] = [k](E_s - E)/E$. Thus the structural equations $[K_r]\{D\} = \{R\}$ may be rewritten as

$$[K]\{D\} = \{R\} - \sum [k_m]\{d\} \qquad (15.2.3)$$

where the summation extends over all modified elements. As noted in Section 14.5, underrelaxation may have to be used to achieve convergence.

The principal advantage of direct iteration is its simplicity; it is quite easy to make the necessary changes in a computer program for elastic

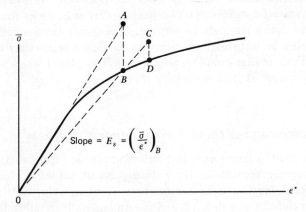

Figure 15.2.1. Stress-strain plot and direct iteration.

---

[1] Quantities $\bar{\sigma}$ and ϵ^* are proportional to the second invariants of deviatoric stress and deviatoric strain, respectively. Note from Eqs. 15.1.2 and 15.2.1 that $d\bar{\epsilon}^p = d\epsilon^*$ if $\nu = 0.5$.

analysis. However, there are several limitations. The method may not converge if plastic action is extensive. Possible elastic unloading of plastically deformed material is not accounted for. The flow rules of plasticity are ignored by the presumption that the solution is uniquely defined by whatever final load level is applied. It is known that the final state of stress and strain depends on how the stresses and strains develop during loading as well as on how much load is applied (1).

Direct iteration is an application of deformation theory, which is exact only if all stresses increase proportionally, that is, if in reality, at any point and throughout loading, $\{\sigma\} = (\text{constant})\{\sigma_o\}$, where $\{\sigma_o\}$ is an arbitrary reference state (1). Deformation theory may be expected to provide the poorest approximation when the final load consists of quite different components that are applied one after the other.

The elementary direct iteration solution described in this section adopts a convenient but essentially ad hoc stress-strain relation. More rational stress-strain relations for deformation theory are possible; the flow rules may be manipulated to produce relations valid in both elastic and elastic-plastic ranges, for both isotropic materials (1, 30) and anisotropic materials (16). These relations account for the change of Poisson's ratio to $\nu = 0.5$ in the plastic range. (Note that division by zero occurs if we simply set $\nu = 0.5$ in matrix $[E]$. However, an ad hoc scheme would serve to change ν to near 0.5 for large plastic strain. See Problem 15.2.)

In all but the simplest elements, the amount of plastic action is likely to vary from one point in the element to another. Hence the elastic-plastic stiffness coefficients also vary. At what locations will we sample plastic strains and material stiffnesses? This question arises in direct iteration and also in other solution methods. In very simple elements, the centroid is a good sampling point. In isoparametric elements the Gauss points may be chosen. The general rule is that sampling points should be placed where strains are likely to be computed most accurately (5, 6).

15.3   Incremental Method Using Tangent Stiffness

By applying load in increments and following plastic action as it develops, one may properly account for the path-dependent nature of plasticity. A procedure for doing this is known as the tangent modulus method and is described in Refs 3 through 6, 12, and 26. An especially detailed description appears in Ref. 6. An alternative plasticity formulation and computational procedure, which appears to be more efficient in the presence of kinematic hardening, is described in (36).

A typical step of the solution is described with the aid of Fig. 15.3.1, which is a single d.o.f. representation of an actual multi-d.o.f. problem. Let

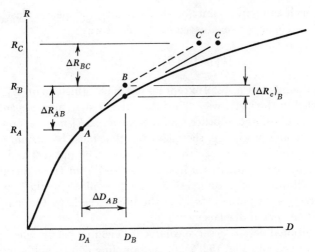

Figure 15.3.1. Load versus displacement plot of typical point.

us suppose that under loads $\{R\}_A$, the correct displacements $\{D\}_A$ and structure tangent stiffness $[K]_A$ are known. Displacements produced by the next load increment are computed:

$$[K]_A\{\Delta D\}_{AB} = \{\Delta R\}_{AB} = \{R\}_B - \{R\}_A$$
$$\{D\}_B = \{D\}_A + \{\Delta D\}_{AB}$$

$$(15.3.1)$$

Stresses, strains and tangent stiffnesses must now be updated. Displacements $\{\Delta d\}$ of an element are extracted from $\{\Delta D\}_{AB}$. Then, for each sampling point in the element, we proceed as follows. Evaluate strains $\{\Delta \epsilon\} = [B]\{\Delta d\}$, where $\{\Delta \epsilon\}$ in general contains both elastic and plastic parts. Next, by use of equations and a figure in Section 15.1,

Eq. 15.1.8: $\qquad \Delta \bar{\epsilon}^p = \displaystyle\int_A^B [W]\{d\epsilon\} \approx [W]_A\{\Delta \epsilon\}$

Eq. 15.1.6: $\qquad \{\Delta \epsilon^p\} = \displaystyle\int_A^B \left\{\frac{\partial \bar{\sigma}}{\partial \sigma}\right\} d\bar{\epsilon}^p \approx \left\{\frac{\partial \bar{\sigma}}{\partial \sigma}\right\}_A \Delta \bar{\epsilon}^p$

$$(15.3.2)$$

Fig. 15.1.1: $\qquad \Delta \bar{\sigma} = \displaystyle\int_A^B H'\, d\bar{\epsilon}^p \approx H'_A\, \Delta \bar{\epsilon}^p$

Eq. 15.1.7: $\qquad \{\Delta \sigma\} = [E](\{\Delta \epsilon\} - \{\Delta \epsilon^p\})$

(Some drift from the true response is developing. More will be said about this later.) The updated quantities are $(\bar{\epsilon}^p)_B = (\bar{\epsilon}^p)_A + \Delta \bar{\epsilon}^p$, $\{\epsilon^p\}_B = \{\epsilon^p\}_A + \{\Delta \epsilon^p\}$, etc. Next read H'_B corresponding to $(\bar{\epsilon}^p)_B$ from Fig. 15.1.1. Evaluate $\{\partial \bar{\sigma}/\partial \sigma\}_B$ and $[E_{ep}]_B$ from Eqs. 15.1.5 and 15.1.9. Using the appropriate

$[E_{ep}]_B$ for each sampling point (or Gauss point), the elastic-plastic tangent stiffness of an element is evaluated in the usual way:

$$[k] = \int_{\text{vol}} [B]^T [E_{ep}][B] \, dV \qquad (15.3.3)$$

which corresponds to Eq. 4.2.9 except for use of $[E_{ep}]$ instead of $[E]$. Elements are assembled in the usual way. Thus the current structure tangent stiffness matrix is produced, and another load increment $\{\Delta R\}$ may be applied.

In the foregoing procedure, the quantities that must be current and available at each sampling point are the plastic strains $\bar{\epsilon}^p$ and $\{\epsilon^p\}$ and the maximum stress $\bar{\sigma}$. Other quantities may be derived from these quantities and from the following: the uniaxial $\bar{\sigma}$ versus $\bar{\epsilon}$ plot, Hooke's law $\{\sigma\} = [E](\{\epsilon\} - \{\epsilon^p\}) + \{\sigma_o\}$ where $\{\sigma_o\}$ are the initial (perhaps thermal) stresses, and the current nodal displacements (6).

Some additional details deserve comment. Yielding begins when $\bar{\sigma}$ exceeds a value we will call $\bar{\sigma}_o$ (Fig. 15.3.2). Some sampling points will be elastic prior to application of the load increment and elastic-plastic afterwards. One way to account for this effect is as follows (3, 12). Estimate the next strain increment from results of the previous load increment and compute the expected change in $\bar{\sigma}$ using *elastic* coefficients. Thus we move, say, from A to B in Fig. 15.3.2. As $\bar{\sigma}_B > \bar{\sigma}_o$, the appropriate material stiffness matrix for use in the current load increment is approximately, from Eq. 15.1.9,

$$[E_{ep}] = [E] - (1 - m)[E]\left\{\frac{\partial \bar{\sigma}}{\partial \sigma}\right\}[W] \qquad (15.3.4)$$

where m is the fraction of $\Delta \bar{\epsilon}$ required to cause yield,

$$m = (\bar{\epsilon}_o - \bar{\epsilon}_A)/(\bar{\epsilon}_B - \bar{\epsilon}_A) \qquad (15.3.5)$$

and $\left\{\dfrac{\partial \bar{\sigma}}{\partial \sigma}\right\}[W]$ is based on strains corresponding to point B. Actual analysis

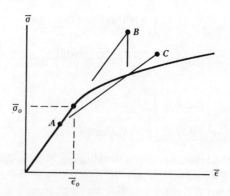

Figure 15.3.2. Elastic-plastic transition.

under the current load increment is now carried out, and stresses $\{\Delta\sigma\} = [E_{ep}]\{\Delta\epsilon\}$ are computed using the most recent estimate of $[E_{ep}]$. The current state may now correspond to point C in Fig. 15.3.2. Again we see if the updated stresses exceed yield, and if so update m and $[E_{ep}]$, and again apply the *same* load increment. At this time we may be "close enough," but another iteration may be desirable. Convergence should be rapid because the transition zone is probably only a small portion of the structure. After convergence, calculation of plastic strains in Eqs. 15.3.2 must be based on $(1 - m)\{\Delta\epsilon\}$ instead of on the full increment $\{\Delta\epsilon\}$.

If $\Delta\bar{\epsilon}^p < 0$ in a load step, elastic unloading is indicated. When this happens, $\Delta\bar{\epsilon}^p$ is set to zero in Eqs. 15.3.2, and in the next step stiffness is based on elastic constants $[E]$ instead of on $[E_{ep}]$. If isotropic hardening is assumed, plastic action is resumed only if $\bar{\sigma}$ exceeds its previous maximum magnitude. If during loading $\Delta\bar{\epsilon}^p < 0$ at many sampling points, plastic collapse may be indicated (4, 5, 6). Computations during *unloading* are detailed in Ref. 26.

Drift from the proper path is evident in Fig. 15.3.1. A corrective device based on unbalanced loads is explained in the next section. A portion of the drift arises in consequence of the approximations in Eqs. 15.3.2. The latter source of drift is largely avoided by simply evaluating the integrals more accurately (5, 6, 11). To do so one might proceed as follows (6). Drop the superscript p from terms in Eq. 15.1.2 and compute a "dummy" reference value $\Delta\bar{\epsilon}$ from the just-computed total strain increments $\{\Delta\epsilon\}$. Divide the integrals into M subincrements, where, for example, $M = \Delta\bar{\epsilon}/0.0002$, but M is an integer greater than zero. Integrands are updated for each subincrement. These refinements apply to the plastic part of the total strain increment, which may be separated from the elastic part prior to yield by use of a device such as factor m of Eq. 15.3.5.

15.4 Modifications of the Incremental Method

Drift that appears as the dashed line BC' of Fig. 15.3.1 may be corrected for. A rigorous argument is possible (10 of Chapter Fourteen), but the simpler argument that follows leads to the same result (11). Upon reaching R_B in Fig. 15.3.1, the computed displacement D_B is too small, hence the resulting stresses are also too small and are not adequate to resist the applied load R_B. If a "corrective" load ΔR_c were added to ΔR, the computed D_B would be larger and more nearly correct. Load ΔR_c plays the same role as a typical unbalanced force ΔR_c in Fig. 14.3.3. As in Section 14.2, the corrective load is the difference between applied loads and resistance produced by distorted elements. This resistance is evaluated from the $\{\sigma_o\}$ term of Eq. 4.2.10, with $\{\sigma_o\}$ now regarded as the *total* accumulated stress

$\{\sigma\}$ in the element. Therefore, in general,

$$\{\Delta R_c\} = \{R\} - \sum \int_{\text{vol}} [B]^T \{\sigma\} \, dV \qquad (15.4.1)$$

where the summation extends over all elements of the structure. Forces $\{R\}$ and stresses $\{\sigma\}$ are those at the end of a step. Correction $\{\Delta R_c\}$ is applied in the subsequent step. For example, equations for the step from point B in Fig. 15.3.1 are, for several d.o.f.,

$$[K]_B \{\Delta D\}_{BC} = \{\Delta R\}_{BC} + \{\Delta R_c\}_B \qquad (15.4.2)$$

In this way we arrive at point C instead of the less correct point C'. If d'Alembert forces $-[M]\{\ddot{D}\}$ are added to the right hand side of Eq. 15.4.1 we have forces $\{\Delta R_c\}$ applicable to a nonlinear dynamic problem (45, 67 of Chapter Twelve). If applied to Eq. 15.4.2, accelerations $\{\ddot{D}\}$ should be evaluated at point B.

The procedure of Eq. 15.4.2 is the same as the "incremental with one-step Newton-Raphson correction" method of Fig. 14.3.3. This observation suggests that for still greater accuracy (at some computational expense), Newton-Raphson iteration may be applied *within* a load step $\{\Delta R\}$ so that essentially zero $\{\Delta R_c\}$ exists at the beginning of the following step (10 of Chapter Fourteen). Thus, within a step we analyze for the successive loads $\{\Delta R\}_{AB}$, $\{\Delta R_c\}_1$, $\{\Delta R_c\}_2$, etc., updating stiffnesses, strains, and stresses each time (Fig. 15.4.1).

Still further modification has been suggested. As correction loads $\{\Delta R_c\}$ serve to return drifting values to the proper path, it is not essential that the tangent stiffness matrix be used. We may use the modified Newton-Raphson method within a step $\{\Delta R\}$, thus replacing Fig. 15.4.1 by Fig. 15.4.2. Indeed, the *original, elastic* structure stiffness matrix may be in *all* load steps.

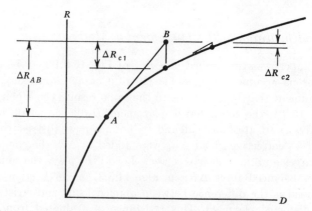

Figure 15.4.1.   Newton-Raphson correction.

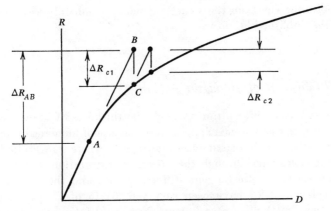

Figure 15.4.2.   Modified Newton-Raphson method.

The latter form of solution has been called the "initial stress" method (2, 11). The significance of the name may be seen as follows. For an increment such as ΔR in Fig. 15.4.2, Eqs. 15.1.7 and 15.4.1 yield

$$\{\Delta R_c\}_1 = \{\Delta R\}_{AB} - \sum \int_{\text{vol}} \int_A^B [B]^T[E](\{d\epsilon\} - \{d\epsilon^p\})\, dV \qquad (15.4.3)$$

Since the step was made using elastic stiffness, $\{\Delta R\}_{AB}$ is exactly balanced by the integral over $\{d\epsilon\}\, dV$, hence

$$\{\Delta R_c\}_1 = \sum \int_{\text{vol}} \int_A^B [B]^T[E]\{d\epsilon^p\}\, dV \qquad (15.4.4)$$

Thus each analysis is an elastic analysis with plastic strains $\{d\epsilon^p\}$ playing the same role as initial strains $\{\epsilon_o\}$ (see Eq. 4.2.10, where $-\{\sigma_o\}$ replaces $[E]\{\epsilon_o\}$). Iteration within each step is mandatory, as computed points such as B in Fig. 15.4.2 are far from the correct path.

It is now apparent that solution techniques of Chapter Fourteen apply to elastic-plastic problems as well as to geometrically nonlinear problems. Current opinion seems to favor some form of self-correcting tangent modulus method for elastic-plastic problems. References 9, 10, 14, and 15 of Chapter Fourteen approve of a first-order, self-correcting, initial-value method, such as the incremental method with one-step Newton-Raphson correction of Figs. 14.3.3 and 15.3.1. Others (11) suggest the modified Newton-Raphson method of Fig. 15.4.2, with the structure tangent stiffness matrix computed once per load step, in the first or second iteration. The "initial stress" method converges rapidly only when nonlinearity is mild, so that a tangent modulus method is preferred when large areas have yielded (14).

In conclusion we remark that incremental solution methods for problems with material and/or geometric nonlinearity have an essential similarity that is not apparent from the variety of names commonly associated with them in the published literature.

15.5　Treatment of Bending Action

In elastic problems, integration through the thickness of beam, plate, and shell elements is easily done explicitly without use of numerical integration. Such is not the case in elastic-plastic problems, where stresses *and* elastic-plastic properties vary through the thickness. Integration methods for obtaining an elastic-plastic tangent stiffness matrix and the contribution to force unbalance $\{\Delta R_e\}$ have been surveyed (5; 15 of Chapter Fourteen). Some of these methods are as follows. Numerical integration through the thickness has been carried out by Gauss quadrature, trapezoidal rule, and Simpson's rule, with 3 to 21 stations through the thickness, depending on the nature of the problem. Three to five stations might serve well for most problems. Simpson's rule permits the smallest number of stations and seems preferable to Gauss quadrature because sampling points lie on the outer surfaces, where yielding begins. *Interior* points must yield before plastic action is detected by Gaussian quadrature. For the sake of economy, material properties may be assumed constant over the lateral dimensions of an element, that is, $[E_{ep}]$ is taken as a function of ζ only, where ζ is the surface-normal coordinate. Standard numerical integration may be replaced by a conceptual model in which the thickness is divided into several layers, with material properties assumed constant through the thickness of each layer (10; 9 of Chapter Eight). The latter reference also considers unsymmetric plastic action in shells of revolution.

With ζ measured from the neutral surface of bending in linearly elastic problems, integrals of $[E]\zeta$ through the thickness are zero. In elastic-plastic problems $[E_{ep}]\zeta$ is not linear in ζ, and in general the surface of zero normal stress will shift toward one side or the other. Therefore, $[E_{ep}]\zeta$ does not vanish upon integration but contributes additional terms to the element tangent stiffness matrix. In Ref. 15, where the approximation $[E_{ep}] = [E]E_T/E$ is used, these terms are made to disappear anyway by introducing a constant b. The substitution is

$$\zeta = \zeta' + b, \quad \text{where} \quad b = \frac{\displaystyle\int_h E_T \zeta \, d\zeta}{\displaystyle\int_h E_T \, d\zeta} \tag{15.5.1}$$

and h = element thickness. Thus, integration over h is done with respect to ζ', and, according to Eqs. 15.5.1, integrals of $E_T \zeta' \, d\zeta'$ over h are zero. It is assumed that element shape functions always apply to the current reference surface $\zeta = b$.

As outlined above, the gradual development of plastic action in bending may be properly accounted for. But in some problems the assumption of an elastic-perfectly plastic hinge is useful. In this idealization the moment-curvature is linear until the extreme fiber stress reaches the yield point, after which the moment remains at a value M_p as curvature increases indefinitely. Plastic hinges in a frame may be accounted for as follows (8).

Suppose that fully plastic moment M_p is developed at node B of a typical plane frame element AB (Fig. 15.5.1). Moment M_p is applied to both the element and the remaining structure at B, as shown. Next, rotational d.o.f. θ_B is removed from element AB by condensation, exactly as described in Section 11.7. If beam AB is uniform, condensation automatically produces nodal moment $M_p/2$ at A and nodal loads $\pm 3M_p/2L$ at A and B. The nodal loads and reduced element stiffness matrix, now effectively having a row and column of zeros corresponding to θ_B, are assembled into the structure. Rotations at B in element and structure are now independent. Structural analysis in the presence of plastic hinges proceeds in a series of linear steps, the beginning of each step being marked by introduction of another plastic hinge and the consequent reduction of structure stiffness and modification of nodal loads. Collapse is indicated by very large displacements or by the structure stiffness matrix not being positive definite.

The need to modify the stiffness matrix of the frame by changing the stiffness of only one element in each step suggests the advantage of some special algorithm for this purpose. Such algorithms are used in automated structural design. Inclusion of geometric nonlinearities in framed structures, which can be significant, and other economies, refinements and extensions, are considered by Majid (33).

The effect of membrane forces on bending stiffness is included by addition of the initial stress stiffness matrix, as was done for elastic problems in Chapter Thirteen. Buckling in the presence of plastic action may occur in

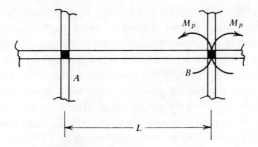

Figure 15.5.1. Plastic hinge at B.

the same way as collapse in the presence of geometric nonlinearities, that is, by vanishing stiffness but without bifurcation buckling.

For such structures as flat plates, bifurcation analysis may be useful. The problem statement is the usual one, as in Eq. 13.3.2:

$$([K] + \lambda[K_\sigma])\{\Delta D\} = 0 \qquad (15.5.2)$$

But now, because of plastic action and the consequent redistribution of stress upon loading, both $[K]$ and $[K_\sigma]$ are functions of the loading parameter λ. Accordingly, if plastic collapse is viewed as a (nonlinear) eigenvalue problem, collapse is defined by $\lambda = 1$. Analyses of this type, based on deformation theory, appear in Ref. 13.

If a postbuckling analysis is required, a procedure for traversing the unstable region may be invoked (Section 14.5). Widespread yielding in strain-softening materials may be treated similarly. Reference 11 uses strain increments for such a problem.

15.6   Other Problems with Material Nonlinearity

Creep makes the material stress-strain relation dependent on time. An isochronous stress-strain curve may be extracted from creep curves by plotting stress versus strain at a given time (Fig. 15.6.1). The relation $\sigma = \sigma_0$ arc sinh (ϵ/ϵ_0) has been suggested (17) where σ_0 and ϵ_0 are functions of time. Additional relations are given in an extensive survey (20). A simple but approximate solution method for creep problems now suggests itself. We may use the direct iteration scheme of Section 15.2, with the isochronous $\bar\sigma$ versus ϵ^* relation for the particular time of interest. The solution for a different time is found by shifting to another $\bar\sigma$ versus ϵ^* relation and beginning iteration with the configuration established in the preceding solution. This approach, while simple, is subject to the limitations mentioned in Section 15.2.

However, if detailed information regarding material yield and flow behavior is not available, an accurate solution using incremental theory may not be possible (16). Also, under a constant temperature field and steady loads of long duration, several materials tend to "forget" their elastic history and assume a steady-state deformation process. In this process of "stationary creep," stresses are constant and the material flows as a viscous fluid whose coefficient of viscosity is a function of stress. The problem may be solved as if it were an elastic problem, but with nonlinear stress-strain relation and strains replaced by strain *rates*. Strains at any time are therefore the product of strain rates and elapsed time. In such problems, direct iteration with the appropriate material stiffnesses and Poisson's ratio should provide reliable answers.

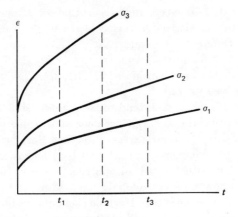

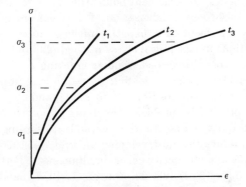

Figure 15.6.1.   Creep and isochronous stress-strain curves.

Incremental solution of a creep problem proceeds in basically the same way as incremental solution of an elastic-plastic problem. The flow rules of Section 15.1 may be used again, with the essential difference that plastic strains are now regarded as creep strains. The increment in effective creep strain in $d\bar{\epsilon}^c = (F)\, dt$ where F is most generally a function of effective stress, effective strain, time, and temperature. One may assume that stresses remain constant over a time step Δt, compute $\Delta\bar{\epsilon}^c = (F)\, \Delta t$, and hence compute creep strain increments $\{\Delta\epsilon^c\} = \{\partial\bar{\sigma}/\partial\sigma\}\,\Delta\bar{\epsilon}^c$ according to Eq. 15.1.6. Stress increments follow from Eq. 15.1.7. Nodal forces produced by current stresses may be compared with applied loads to obtain the unbalanced or corrective loads $\{\Delta R_c\}$ of Eq. 15.4.1. As before, corrective loads serve to prevent large drift in a tangent modulus solution and comprise essential loads in an "initial stress" solution. For further information on creep analysis the reader is referred to the literature (1, 17–20).

A problem related to creep analysis is that of temperature-dependent material properties in elastic-plastic problems where temperature changes during loading. Different temperatures T_1, T_2, T_3 produce different stress-strain curves, just as do different times t_1, t_2, t_3 in Fig. 15.6.1. The calculation of corrective loads has been considered (21, 31, 38). This work finds application in creep analysis as well.

Brittle materials such as rock tend to crack under sufficiently high tensile stress. Upon cracking the material becomes anisotropic (if not already so) with zero stiffness in the tensile-stress direction. Shear stiffness parallel to the crack does not completely vanish. Such problems may be solved by direct iteration but with the risk of appreciable error. In general the final state depends on how cracks develop during loading, so that an incremental solution should be more accurate. If whole elements are idealized as cracked, the crack is modeled as a wide band. Alternatively, cracks might be modeled by disconnecting nodes along element boundaries to be separated by a crack, this being accomplished by condensation of d.o.f. on one side of the crack prior to assembly of elements. But here the cracks are forced to follow element boundaries. Solutions by the initial stress method (22) and by use of special elements (32) have been described. Similar cracking action occurs in concrete beams and slabs, where yielding of reinforcing steel may also be involved. Sample references are Refs. 23 to 25, 34, and 40.

As mentioned previously, problems involving geometric or material nonlinearity demand quite similar solution methods. Accordingly, there are no major additional difficulties in developing an analysis scheme for problems in which both types of nonlinearity occur simultaneously. Pertinent references include Refs. 3, 5, 6, 15, 27, 28, 29, and 37; 1, 10, 14, and 15 of Chapter Fourteen. Some references that discuss the dynamic response problem in the presence of nonlinearity are Refs. 27; 35, 36, 37, 45, and 60 of Chapter Twelve; 39 of Chapter Fourteen. Convergence-acceleration schemes may be applied to problems of geometric or material nonlinearity (39; 30, 31, 32 of Chapter Fourteen).

PROBLEMS

15.1. (a) Derive the forms of Eqs. 15.1.1 and 15.1.2 appropriate to plane stress conditions. *Suggestion.* Plastic flow takes place at constant volume.

   (b) Derive Eq. 15.1.4.

   (c) Derive Eq. 15.1.8.

   (d) Derive $[E_{ep}]$ for the case of uniaxial stress.

   (e) Derive Eq. 15.1.10.

15.2. With reference to the direct iteration method of Section 15.2:

(a) Can strain-softening behavior be treated? Explain.

(b) Devise an approximate scheme to account for the change of Poisson's ratio to nearly 0.5 in the plastic range. *Suggestion.* Assume that the bulk modulus $B = E/(3 - 6\nu)$ remains constant.

(c) Someone proposes to incorporate flow theory (rather than deformation theory) by iterating until convergence at each of several monotonically increasing load levels. Comment on this proposal.

15.3. The two bars carry uniaxial stress only. They are fixed to rigid walls and joined where load P is applied. The material has yield points of 10 units in tension, 30 units in compression, and elastic modulus $E = 10^4$ units.

(a) Use direct iteration (Fig. 15.2.1) and solve for displacement D and stresses under $P = 30$ kips.

(b) Repeat Part a using modified direct iteration (Eq. 15.2.3).

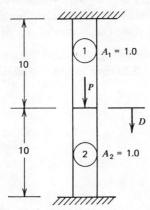

15.4. Solve Problem 15.3 by the incremental method of Section 15.3, using

(a) Two steps $\Delta P = 8$ units, starting from $P = 12$ units.

(b) Two steps $\Delta P = 8$ units, starting from $P = 16$ units.

15.5. Solve Problem 15.3 by the "initial stress" method of Section 15.4 (elastic stiffness used throughout), using one step $\Delta P = 8$ units starting from $P = 16$ units.

15.6. (a–f) Solve the problems posed in Problems 15.3, 15.4, and 15.5, but use the stress-strain relation given in the sketch at the top of p. 310.

15.7. Suppose that correction forces $\{\Delta R_c\}$ are to be computed for a mesh of elements having internal d.o.f. Should forces resulting from $\{\sigma\}$ be associated with the internal d.o.f.? If so, should they be distributed to remaining d.o.f. (that is, condensed) by means of the *elastic* element stiffness equations? What bearing does your answer have on computational efficiency?

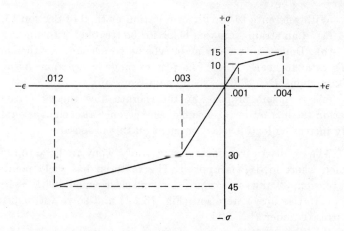

15.8.   Describe in detail a procedure for analysis of elastic-plastic pin-jointed plane trusses. Include the appropriate form of constitutive law, the computational steps, and state the nature of any approximations involved.

15.9.   Treatment of plastic hinges in plane frames without strain hardening is summarized in Section 15.5. How might the effects of strain hardening be included?

CHAPTER SIXTEEN

Detecting and Avoiding Errors

16.1 Introduction

By "error" we mean a difference between computed results and exact results, or whatever leads to this difference. Errors are of several kinds. We will use the following terminology. A "mistake" is a slip or oversight, such as an input data error, or misunderstanding the purpose of an available program. A "discretization error" may arise from an error of judgment, as in selecting a poor mesh of elements. It may also be the inherent approximate nature of the finite element method. A "computation error" is produced by the digital computer as it manipulates data. Some computation error arises even in a logically correct and properly used program but is aggravated by poor discretization on the part of the user.

Perhaps inevitably, authors do not agree on the merit of certain error checks. Notably, the usefulness of residual error measures and the condition number have been questioned. Procedures discussed in the following seem to be favored by most authors. Our discussion is an attempt to clarify and categorize major sources of error and to summarize error tests and good computing practices that are readily understood and easily used. For proofs and derivations the reader is referred to the literature.

Unless stated otherwise, the following discussion pertains to errors in solving linear equations and not to eigenvalue problems. Of course, mistakes may be made in setting up eigenvalue problems, and the discretization may lead to high or closely spaced eigenvalues, the latter being hard to separate. However, computation error should be a minor problem, since a real symmetric matrix is well conditioned with respect to eigenvalue extraction (1, 2). By "well conditioned" we mean that small changes in the matrix produce only small changes in the results. It is intuitively reasonable that there is

311

less numerical difficulty in eigenvalue extraction than in equation solving; in stress analysis accurate displacements are essential, as strains are computed from the displacement *differences* of neighboring nodes; in eigenvalue analysis the lowest eigenvalues are of greatest interest, and these are accurately computed from approximate mode shapes (see the Rayleigh quotient, Section 12.3). References for eigenvalue problems include Refs. 2 and 25, which discuss computational errors; Refs. 3 and 4, which discuss error produced by use of an "eigenvalue economizer"; and Ref. 5, which considers discretization error. Discretization error is of the same order for both equation solving and eigenvalue extraction (5, 11). The error in eigenvalues (5) is the same as the strain energy error, namely $O(h^{2(q+1-m)})$ (see Section 16.4).

16.2   Mistakes

Both program and data must be properly prepared. A mistake in either can produce gross error or an error small enough to escape casual inspection. At the risk of recounting the obvious, we list some precautions that should be kept in mind.

Before use in new situations, a program should be known to produce good results in a variety of test cases. Even then the program should not be fully trusted, as the sample of test cases and mesh geometries could not have been exhaustive. Also, a program of moderate complexity may well contain computational paths never before exercised. Some trouble should be expected when running a program on a new machine—or even on the same machine, as systems programmers continually modify the operating system of large installations. Documentation (27) should be complete, readable, up to date, and contain references and some flow charts; the source deck should be liberally supplied with comment cards. In this way any newly discovered error may be traced, and modifications and improvements may be made with more ease and confidence. It is very helpful if statement numbers are assigned sequentially instead of at random. Security is needed to present casual users from modifying the coding. A set of test cases might be run occasionally to detect unexpected changes.

To properly prepare input data, the user should understand the purpose of the program, its capabilities and limitations, and something of the analysis method. Errors in measuring nodal coordinates, numbering nodes and elements, and keypunching are usually easy to see by inspection of a machine-generated plot of the mesh as understood by the program (Section 18.4). Input data (as understood by the program) should be output by the program, checked, and kept on file with the computed results, against the possibility that questions about the work may subsequently arise. The program itself can check for some data errors (program capacity exceeded, no boundary conditions specified, d.o.f. mismatched in adjacent elements, etc.) and stop execution when one is found. Indeed, coding for input, output, and error checking may comprise the bulk of a user-oriented program.

Computed output should be carefully examined. Comparison should be made with results obtained by other means—simplified analytical models, experiment, past experience with similar problems, and engineering judgment or "common sense." Some workers accept a computer analysis as valid only when substantially confirmed by an independent computer program, preferably based on a different analysis method (6).

16.3 Eigenvalue Test for Elements

We will describe the calculations first, then comment on what use may be made of them. Let nodal loads $\{\bar{r}\}$ on an element be made proportional to nodal displacements $\{d\}$:

$$[k]\{d\} = \{\bar{r}\} = \lambda\{d\} \tag{16.3.1}$$

This is an eigenvalue problem. If each eigenvector is normalized so that $\{d_i\}^T\{d_i\} = 1$, Eq. 16.3.1 yields

$$\lambda_i = \{d_i\}^T[k]\{d_i\} = 2U_i \tag{16.3.2}$$

That is, each eigenvalue λ_i equals twice the strain energy U_i of the element under normalized nodal displacements $\{d_i\}$ (2 of Chapter Six). Equation 16.3.1 may also be written as

$$([k] - \omega^2[I])\{d\} = 0 \tag{16.3.3}$$

where $\omega^2 = \lambda$. This is a free vibration problem with a unit mass matrix.

Sample results of an eigenvalue analysis, taken from Ref. 2 of Chapter Six, are shown in Fig. 16.3.1.

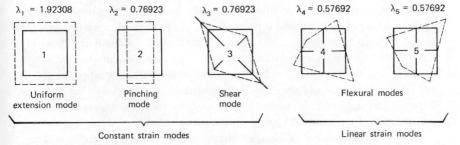

Figure 16.3.1. Nonzero eigenvalues and deformation modes of linear isoparametric element in plane strain. $E = 1.0$, $\nu = 0.3$, each side length = 1.0.

Zero eigenvalues arise when $\{d\}$ represents rigid body motion—translation or rotation, singly or in any combination (13 of Chapter Eight). Accordingly, there are three rigid body modes for the eight d.o.f. element of Fig. 16.3.1. There are six for a solid element, and only one for axisymmetric behavior of a solid or shell of revolution. Zero-energy deformation modes, mentioned in Sections 5.7, 6.3, and 9.5, also yield zero eigenvalues. The associated eigenvector $\{d\}$ may represent a combination of rigid body motion and the zero-energy deformation mode.

To check for error in element formulation by the eigenvalue test, one may first see if there are as many zero eigenvalues as there ought to be. Too few suggests a displacement field that does not permit rigid body motion (e.g., Table 8.2.1). Too many suggests zero-energy deformation modes (Section 5.7). Similar modes, for example, the flexural modes of Fig. 16.3.1, should have the same eigenvalue. Eigenvalues should not change when the element is displaced and rotated in global coordinates; if they do, the element is not invariant (Section 4.6). Nonzero eigenvalues should be real and positive.

In addition to its use in error checking, the eigenvalue test plays a role in assessing the relative merit of competing elements. A simplified argument is that since compatible elements based on assumed displacement fields are always too stiff, the least stiff element is best, and it is this element that has the smallest eigenvalues and smallest matrix trace. Apparently this argument may not be applied to elements having zero-energy deformation modes, elements softened by low order integration rules, and incompatible elements. Further arguments tend to be lengthy, and the reader is referred to the literature for more information (15 and references it cites; also 7 of Chapter Four).

16.4   Discretization Error and Convergence Rate

By "discretization error" we mean inaccuracy arising from the fact that the discretized model, which is what is actually analyzed, is rarely an exact representation of the physical structure. The element mesh may not exactly fit the structure geometry. The actual distribution of load and possibly variations of thickness and elastic properties may be approximated by simple interpolation functions. Boundary conditions may also be approximated. But even if these factors are exactly represented, it is unlikely that the true displacement field can be exactly represented by the piecewise interpolation field permitted by a model having only a finite number of d.o.f.

An unfortunate choice of element shape or size when discretizing a structure aggravates subsequent numerical error in computation, as described in Section 16.5. Computation error does not concern us here; in all but the last paragraph of this section we will assume that the computer represents and manipulates numbers exactly.

With regard to error in representing geometry, arguments for using curved elements to model curved structures are summarized in Section 8.2. Further theoretical studies appear in (7). In two-dimensional structures, an important source of geometric error is the use of elements with straight edges to model curved boundaries (8). Geometric error is reduced by using elements with curved edges or by arranging straight edges to approximate more correctly the actual volume of the structure (Fig. 16.4.1c).

Analysis of discretization error yields an expression that shows the convergence rate produced by refining the mesh. Consider, for example, a straight bar of cross-sectional area A that carries a uniformly distributed force Q per unit length (Fig. 16.4.2). From Fig. 16.4.2b we obtain the equilibrium equation

$$\frac{Q}{AE} + u_{,xx} = 0 \tag{16.4.1}$$

where E = elastic modulus. With a linear displacement field, the finite element model yields the nodal equilibrium equation (7)

$$AE(-u_{i-1} + 2u_i - u_{i+1}) = Qh^2 \tag{16.4.2}$$

Nodal displacements u_{i-1} and u_{i+1} may be expanded in Taylor series about node i, for example,

$$u_{i-1} = u_i - hu_{i,x} + \frac{h^2}{2} u_{i,xx} - \frac{h^3}{6} u_{i,xxx} + \cdots \tag{16.4.3}$$

Substitution into Eq. 16.4.2 yields

$$\frac{Q}{AE} + u_{i,xx} + \frac{h^2}{12} u_{i,xxxx} + \cdots = 0 \tag{16.4.4}$$

Therefore this finite element model approximates the differential equation with error $O(h^2)$ (read "error of order h^2"). The error vanishes as h approaches zero. If elements are halved in size, the error is quartered. Error of $O(h^n)$ is

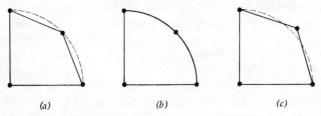

$$(a) \qquad\qquad (b) \qquad\qquad (c)$$

Figure 16.4.1. Quarter circle modeled by one quadrilateral element. (a) Straight edges, node on arc. (b) Curved edges. (c) Straight edges with good representation of volume.

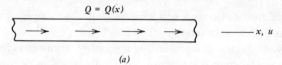

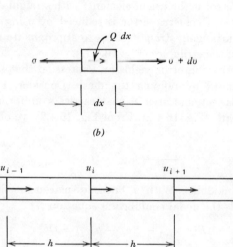

Figure 16.4.2. (*a*) Axial force member. (*b*) Equilibrium of differential element. (*c*) Finite element model.

said to be a discretization error and convergence rate of order h^n. For convergence to occur, n must be greater than zero. Several analysis have been done for one- and two-dimensional elements (7). The error for the standard four d.o.f. straight beam element is $O(h^4)$.

The foregoing order of error analysis is tedious to perform. Explicit algebraic expressions for stiffness are required, and these are not available for numerically integrated elements. Also, some shortcomings of the analysis method have been noted (10). With any given element, the order of convergence may be reduced if loading is changed from distributed to concentrated (7 of Chapter Nine). Also, convergence is not of the same order for differential equation, displacements, and stresses.

An argument for the convergence rate of strain energy has been given by Fried (11). His argument is paraphrased as follows. Again let h = a linear element dimension and let q = degree of the highest complete polynomial in the element displacement functions. For example, $q = 2$ for the element of Fig. 5.5.1, and h could represent the element width. Let $2m$ = order of the governing differential equation of the problem in terms of displacements (for plane problems, $m = 1$; in beam and thin plate theory, $m = 2$). An element can fit exactly the polynomial of degree q in a Taylor series expansion. Therefore, discretization error is due to the element's approximation of the next polynomial, of degree $q + 1$. Error in approximating true displacements is thus $O(h^{q+1})$; and by the same argument, error in the mth displacement derivative is $O(h^{q+1-m})$, where $q \geqq m$. Derivatives are squared in calculating strain energy. Therefore, because m is the highest order derivative in the strain energy expression, the strain energy error of the solution cannot exceed $O(h^{2(q+1-m)})$.

The effect of q and m on convergence rate is of interest. However, displacements and stresses may converge at different (and slower) rates than strain energy (7 of Chapter Nine). Also, the estimate is not valid for incompatible elements and for elements softened by low order numerical integration rules (9).

If the order of convergence of a particular quantity is known, a greatly improved estimate of the quantity can be obtained by extrapolating the results given by two different meshes (12).

Elements should not be severely distorted. For greatest accuracy, triangular elements should be equilateral, quadrilateral elements should be square, etc. Of course, even a severely distorted element should yield exact results in a field of constant strain. Several numerical examples of the effect of distorting certain plate elements have been given (11). Theoretical studies are offered by Fried (11, 13). He concludes that an isoparametric element of any order may be worsened to the point of converging no faster than a linear element by excessive curvature of its sides and by placing side nodes too close to corner nodes.

Computational error is increased by distorted elements, nonuniform meshes, and an increase in the number of d.o.f. used. The condition number $C(K)$ of the structure stiffness matrix, described in Section 16.6, is increased by uneven node spacing (14). A crude statement of the conclusion in Ref. 14 is that $C(K)$ is increased by the factor $(h_{max}/h_{min})^{2m-1}$, where h_{max} and h_{min} are the largest and smallest node point spacing and, as previously defined in this section, $2m$ is the differential equation order. Thus, by mixing beam elements of 10 to 1 length ratio, $C(K)$ is increased by a factor of about 1000. As more d.o.f. are used, the typical element dimension h decreases. Fried (22) finds that $C(K)$ is proportional to h^{-2m}, but that the condition number $C(M)$ of the mass matrix is independent of h.

16.5 Ill Conditioning. Computational Error

Equations $[K]\{D\} = \{R\}$ are termed "ill conditioned" if small changes in $[K]$ or $\{R\}$ produce large changes in the solution vector $\{D\}$. Ill conditioning may reflect physical reality, as in Fig. 16.5.1, where clearly a small change in $[K + K_\sigma]$ may greatly affect the lateral tip displacement. But ill conditioning may also arise even when the *physical* problem is strongly stable, because of the way numbers are represented and manipulated in the computer. It is this computational aspect that concerns us here. Pertinent references include Refs. 1, 16–20, and 23, from which the following comments are extracted.

"Computational error" is here defined as error that arises during computer operations such as generating stiffness coefficients and equation solving. A small loss of information is produced by rounding of the final digit during calculation. A larger and more sudden loss occurs when essential information is contained in the last few digits of a number, as in the familiar example of subtraction when the two numbers are almost equal. Actually, the "lost information" can be said to have disappeared when the structure stiffness matrix was generated; if word length limitations do not permit the original stiffness coefficients to be represented accurately enough, computed results will have few correct digits. In general, it is limited word length that limits the number of correct digits in computed results.

The structural equations $[K]\{D\} = \{R\}$ are equilibrium equations. These equations are written in terms of differences in the displacements of neighboring nodes. If these differences are quite small in comparison with the displacements themselves, the equations become ill conditioned (a familiar limiting case is that of an unsupported structure, which is free to move as a rigid body and has a singular stiffness matrix). Essential information regarding displacement differences is contained in the latter digits of the stiffness coefficients. Accordingly, the physical situation conducive to computational error is that of a stiff element surrounded by much more flexible elements. This includes the very large mesh (say, thousands of d.o.f.) where the total path between an interior element and a support is much less stiff than the element itself. If this physical situation is accompanied by truncation of essential information in stiffness coefficients so as to fit the finite computer word length, computed results may be worthless.

An example of information truncation may be helpful (17). For the

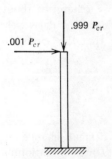

Figure 16.5.1. Column with near-buckling load.

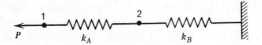

Figure 16.5.2. Two linear springs in series.

structure of Fig. 16.5.2, the structure stiffness matrix and its inverse are

$$[K] = \begin{bmatrix} k_A & -k_A \\ -k_A & k_A + k_B \end{bmatrix}, \qquad [K]^{-1} = \begin{bmatrix} \dfrac{1}{k_A} + \dfrac{1}{k_B} & \dfrac{1}{k_B} \\ \dfrac{1}{k_B} & \dfrac{1}{k_B} \end{bmatrix} \qquad (16.5.1)$$

If $k_A \gg k_B$, k_A dominates $[K]$, but k_B dominates $[K]^{-1}$ and hence the computed displacements. Numerical computation of $[K]^{-1}$ will be of satisfactory accuracy only if the term $k_A + k_B$ is represented to enough digits that k_B is not lost in comparison with k_A. If, for example, $k_A = 40$ and $k_B = 0.0015$, k_A must be represented in the computer as 40.0000 before forming $k_A + k_B$ if the last digit of k_B is to be retained, despite the fact that k_A is not really known to six significant digits.

Use of substructures alleviates the problem (16, 20). The rigid body motion of some area of a structure need not appear in the analysis if the area is treated as a substructure.

Summing up comments thus far, we remark that the major cause of ill conditioning is a large ratio of element stiffness to stiffness of supporting structure. All ill-conditioned matrix is sensitive to small changes in its coefficients, as when trailing digits are lost. Accordingly, the major cause of computational error is original truncation of significant information. It is therefore of marginal value at best to solve equations in double precision if they have been formulated in single precision. Use of partial double precision may even give worse answers than use of single precision throughout (18). Discretization error is reduced by using more elements. But using more elements or a poor element mesh promotes computational error.

16.6 The Condition Number

Causes of ill conditioning were discussed in Section 16.5. In this section a numerical measure of conditioning is introduced, known as the "condition number" and denoted by $C(K)$. The important result is that if coefficients in the structure stiffness matrix are represented with p decimal digits per computer word, then computed displacements will be correct to s decimal digits, where

$$s \geq p - \log_{10} C(K) \qquad (16.6.1)$$

Thus if $p = 7$ and $C(K) = 1000$, $s \geq 4$. The estimate of s accurate digits is often quite good, so that for a conservative rule of thumb we might replace the symbol $\geq$ in Eq. 16.6.1 by an equal sign (however, see also Section 16.7). The estimate of $\log_{10} C(K)$ digits lost does not include the effect of rounding the final digits during solution; instead, it represents the truncation error from having the really significant data in only the last few digits of computer words. Roy (18) believes that rounding and truncation errors often have roughly equal effects and may act on the solution in opposite directions. References pertinent to the present discussion include Refs. 1, 2, 16 18, 21, and 22; 17 and 27 of Chapter Two.

The condition number, defined below, is changed by scaling the matrix. Ideally, the minimum condition number $C(K)$ should be used in Eq. 16.6.1. However, optimum scaling is rather complicated. Accordingly, practitioners use near optimal scaling, which is produced by the following operation:

$$[K_s] = \lceil S \rfloor [K] \lceil S \rfloor \tag{16.6.2}$$

where $[K] = $ the given structure stiffness matrix, $[K_s] = $ the scaled matrix, and $\lceil S \rfloor = $ a diagonal matrix constructed from diagonal coefficients in $[K]$:

$$S_{ii} = 1/\sqrt{K_{ii}} \tag{16.6.3}$$

Diagonal coefficients of $[K_s]$ are therefore unity. The condition number of a symmetric matrix is defined as

$$C(K) = \lambda_{\max}/\lambda_{\min} \tag{16.6.4}$$

where $\lambda_{\max}$ and $\lambda_{\min}$ are, respectively, the maximum and minimum eigenvalues of the scaled matrix $[K_s]$. Conditioning can be no better than $C(K) = 1$. In large finite element problems, $C(K)$ may exceed 10^6.

The theory connecting Eqs. 16.6.1 and 16.6.4 is summarized as follows (16, 17, 18). Let $[K]$ have eigenvalues λ_i and eigenvectors $\{V_i\}$, normalized so that $\{V_i\}^T \{V_i\} = 1$. Then it is possible to write, for an n by n symmetric matrix,

$$[K] = \sum_{i=1}^{n} \lambda_i \{V_i\}\{V_i\}^T, \qquad [K]^{-1} = \sum_{i=1}^{n} \frac{1}{\lambda_i} \{V_i\}\{V_i\}^T \tag{16.6.5}$$

For each power of ten in the ratio $\lambda_{\max}/\lambda_{\min}$, the lowest mode (corresponding to $\lambda_{\min}$) loses about one digit when $[K]$ is represented in a computer (it is helpful here to imagine $\lambda_{\max}$ and $\lambda_{\min}$ represented with the same power of ten as a factor of each). Since the lowest mode dominates $[K]^{-1}$, we conclude that coefficients in computer representation of $[K]^{-1}$ may be *incorrect* in the *last N* digits, regardless of how many digits s are represented in a computer word and regardless of the accuracy with which inversion was carried out. The approximation for N is $N \leq \log_{10}(\lambda_{\max}/\lambda_{\min})$.

The scaling of Eq. 16.6.2 is done only in order to calculate $C(K)$. Accuracy is not affected by use of scaled equations in a direct equation-solving algorithm, provided that there is no change in the order of elimination or the position of pivot elements in the coefficient matrix (26). However, use of scaled equations greatly improves the convergence rate of energy-minimization solution methods (27 of Chapter Two; 64 of Chapter Twelve).

If $C(K)$ is calculated from the *unscaled* $[K]$, the expression $\log_{10} C(K)$ may greatly overestimate the number of digits lost in actual solution. In Fig. 16.5.2, for example, $C(K)$ of the unscaled matrix becomes large if $k_A \gg k_B$ *and* if $k_B \gg k_A$. However, $C(K)$ of the scaled matrix becomes large only if $k_A \gg k_B$. The case $k_B \gg k_A$ is an example of "artificial" ill conditioning.

Computation of λ_{max} and λ_{min} may be approached as follows (21). The eigenvalue problem might conveniently be viewed as a mechanical vibration problem with unit masses:

$$0 = ([K_s] - \lambda[I])\{D\} = ([K_s] - \lambda[M])\{D\} \qquad (16.6.6)$$

where $[M] = [I]$. Since the natural frequencies are not changed by use of different coordinates to define displacements, we may introduce the coordinate transformation $\{D\} = [T]\{D_1\}$ with $[T] = \lceil S \rfloor^{-1}$. Transformation of Eq. 16.6.6 according to the arguments of Chapter Eleven yields, since $\lceil S \rfloor^T = \lceil S \rfloor$,

$$(\lceil S \rfloor^{-1}[K_s]\lceil S \rfloor^{-1} - \lambda\lceil S \rfloor^{-1}\lceil S \rfloor^{-1})\{D_1\} = 0 \qquad (16.6.7)$$

$$([K] - \lambda\lceil K_{11} \quad K_{22} \quad \ldots \quad K_{nn}\rfloor)\{D_1\} = 0 \qquad (16.6.8)$$

The latter equation results from substitution of Eqs. 16.6.2 and 16.6.3 into Eq. 16.6.7. Thus the eigenvalue problem may be solved using the original unscaled $[K]$ with a diagonal "mass" matrix of the diagonal coefficients of $[K]$. This interpretation agrees with preceding remarks to the effect that isolated stiff members raise the condition number; isolated large "masses" K_{ii} reduce the lowest frequencies but have little effect on the highest, thus raising $\lambda_{max}/\lambda_{min}$ (21).

As $C(K)$ need not be computed exactly, λ_{max} and λ_{min} may be approximated. A close upper bound on λ_{max} of $[K_s]$ is found by summing coefficients in each row of the scaled matrix, and selecting the largest sum (17, 18):

$$\lambda_{max} \approx \max Q_i, \qquad \text{where} \quad Q_i = \sum_{j=1}^{n} \text{abs}\,(K_{sij}) \qquad (16.6.9)$$

and $n = $ order of (scaled) matrix $[K_s]$. Reference 21 finds $1.5 \le \lambda_{max} \le 3.2$ in a series of test cases and suggests that the approximation $\lambda_{max} \approx 2$ may be adequate for any size problem. Calculation of λ_{min} may be done by some method of eigenvalue extraction (Section 12.5). However, as even a crude estimate of λ_{min} may suffice, in some situations one might use the interpretation of Eq. 16.6.8 and estimate the "natural frequency" λ_{min} by hand calculation, using a single d.o.f. approximation of the system and an assumed mode shape (Rayleigh's method).

16.7 Equation Sequencing. Decay of Diagonal Coefficients

It is known that the accuracy of an elimination solution is affected by the choice of pivots (1). When diagonal coefficients are always used as pivots, good choice of pivots means good sequencing of equations.

To illustrate the point we take an example. Consider a train of linear springs, each of stiffness $k = 100$, with nodes constrained to move horizontally (Fig. 16.7.1).

Let a Gauss elimination solution proceed from right to left in Fig. 16.7.1a. After 100 eliminations, the stiffness of the condensed part of the structure is reduced to unity, leaving us with

$$\begin{bmatrix} 101 & -100 \\ -100 & 100 \end{bmatrix} \begin{Bmatrix} u_{101} \\ u_{102} \end{Bmatrix} = \begin{Bmatrix} 0 \\ -P \end{Bmatrix} \qquad (16.7.1)$$

Now suppose that elimination proceeds from left to right (Fig. 16.7.1b). The condensed part of the structure has zero stiffness because the left end is free. Elimination merely transfers load P from node to node. After 100 eliminations we have

$$\begin{bmatrix} 100 & -100 \\ -100 & 200 \end{bmatrix} \begin{Bmatrix} u_{101} \\ u_{102} \end{Bmatrix} = \begin{Bmatrix} -P \\ 0 \end{Bmatrix} \qquad (16.7.2)$$

The point to be made is that the problem of Eq. 16.7.1 is ill conditioned while that of Eq. 16.7.2 is well conditioned: this may be verified by calculating the condition numbers. The sequencing of operations from tip to root has preserved accuracy and made the estimate of $\log_{10} C(K)$ digits lost a very pessimistic one. Indeed, the condition number in Eq. 16.7.2 is much lower

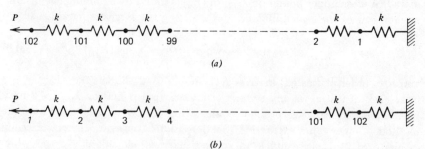

(a)

(b)

Figure 16.7.1. Trains of linear springs. Let $k = 100$.

than that of the original 102 by 102 structure stiffness matrix. As a general rule, for the sake of accuracy, it is desirable to number nodes in such a way that operations in a direct solution algorithm proceed from the more flexible part of a structure toward the stiffer part (19, 23).

The condition number $C(K)$ provides an estimate of solution accuracy and is available *before* solving for displacements. But, as seen above, the estimate may be pessimistic. Another test may be applied *during* solution. This simple test, described below, shows if the possible trouble actually materializes and permits automatic termination of the solution process if unacceptable error is indicated (19, 23, 24).

As each equation is treated in Gauss elimination (and in Choleski decomposition), remaining diagonal coefficients of $[K]$ are reduced by a subtraction operation. Equations 16.7.1 and 16.7.2 are examples of this behavior. Let D_{ii} represent the value of a diagonal coefficient just before it acts as a pivot in eliminating the ith unknown, and let K_{ii} be the original diagonal coefficient. If, for example, $K_{ii}/D_{ii} \approx 10^6$, then about six leading digits of K_{ii} have disappeared because of subtraction from K_{ii}. This is clearly unacceptable if numbers are represented to only seven digits. In the example of Fig. 16.7.1, the reader may check that the final elimination requires the subtraction $100 - (100^2/101) = 1$ in Eq. 16.7.1 and $200 - 100 = 100$ in Eq. 16.7.2. The test, then, is to store the K_{ii} in an auxiliary array and check K_{ii}/D_{ii} just before each equation is eliminated. A warning message can be printed if K_{ii}/D_{ii} is uncomfortably large, and execution terminated if K_{ii}/D_{ii} is unacceptably large. What is "large" depends on how many digits are carried in a machine word. If K_{ii}/D_{ii} is negative, the structure is unstable. Causes of large K_{ii}/D_{ii} include all causes of large condition number plus poor sequencing of equations.

Large error is *not guaranteed* even if K_{ii}/D_{ii} is large (24). The pertinent example is that of a structure without any supports but loaded by forces (or moments) that sum to zero. Computed displacements contain a random rigid body motion that may or may not be so large as to obliterate the deformations contained in the nodal displacement vector $\{D\}$.

16.8 Residuals and Iterative Improvement

After solving equations $[K]\{D\} = \{R\}$ for $\{D\}$, we may compute residuals $\{\text{Res}\}$ as

$$\{\text{Res}\} = \{R\} - [K]\{D\} \tag{16.8.1}$$

An *exact* solution yields $\{\text{Res}\} = 0$. But if we regard the solution as an approximation $\{D\}_1$ and assume the exact solution to be $\{D\}_1 + \{\Delta D\}_1$, we are led to the following iterative improvement scheme (1, 2):

$$[K]\{\Delta D\}_1 = \{R\} - [K]\{D\}_1 = \{\text{Res}\}_1$$

$$\{D\}_2 = \{D\}_1 + \{\Delta D\}_1 \tag{16.8.2}$$

$$[K]\{\Delta D\}_2 = \{R\} - [K]\{D\}_2 = \{\text{Res}\}_2, \text{ etc.}$$

After n iterations, presumably $\{\text{Res}\}_n$ and $\{\Delta D\}_n$ approach zero and $\{D\}_n$ is as exact as the original data permit. But we must question whether $\{\text{Res}\}$ is a valid measure of error and whether iteration really leads to improvement.

It is possible for a quite erroneous solution to have a very small residual, and possible that the more exact of two approximate solutions has the larger residual (1). A form of residual check is the "statics check," in which the integrated effect of stresses acting on a section is compared with the known force or moment on the section. Various authors suggest that the statics check is unreliable, since a satisfactory check may be obtained with stresses that are much in error. Similarly if support reaction R_i is computed by summing $K_{ij}D_j$, it may be nearly exact, but displacements D_j may have appreciable error.[1] If R_i is computed from element stresses, it is conceivable that stresses may be satisfactory while R_i is erroneous, since stresses at element corners are less reliable than elsewhere.

Nevertheless, it is suggested that residuals may be worth checking (1). The suggestion is that after a single precision solution, in any row i of the matrix the maximum value of $K_{ij}D_j$ in the row should be of order 10^t times the ith residual, where t is the number of decimal digits per single precision machine word. A residual greater than this may indicate trouble. Residuals should *always* be computed using double precision arithmetic (1; 11 of Chapter Two).

With regard to iterative improvement, no amount of iteration can recover essential information lost by original truncation. Iteration may not converge at all if the equations are very ill conditioned (1). A remedy is to use double precision throughout, or, if this has already been done, reformulate the problem to reduce the ill conditioning.

Roy (18) believes that the iterative improvement of Eqs. 16.8.2 is a waste of time and suggests the following method. After generating stiffness coefficients in double precision, truncate to single precision in forming the structure equations $[K]\{D\} = \{R\}$, but save the double precision coefficients in auxiliary storage. Solve for $\{D\}$ in single precision. Recall the double precision coefficients from auxiliary storage, blow $\{D\}$ up to double precision,

---

[1] This may happen even when computational error is negligible, as, for example, when a statically determinate structure is modeled by a coarse mesh of crude elements such as constant-strain triangles.

and solve for residuals in double precision. Finally, truncate residuals to single precision and solve for the next $\{\Delta D\}$ (in single precision) using the previously reduced single precision $[K]$. Whether this method is preferable to a full double precision solution depends on how many accurate digits are required and how fast the iteration converges—if it converges at all.

16.9 Concluding Remarks

There appears to be no test or group of tests that is fully reliable in checking for error. For every example there appears to be a counterexample: a large condition number does not guarantee error (1), small decay of diagonal coefficients does not guarantee accuracy (11 of Chapter One). With this reservation in mind let us summarize our remarks on errors.

Mistakes may enter the program by faulty programming logic, mispunched or missing cards, cards out of order, tape read and write error, change in operating system, and program modification that has unexpected results. Input data may have errors arising from similar causes, as well as from failure to follow data preparation instructions and ambiguous or incorrect instructions. Clearly it is helpful if the person who prepares data also understands how the program works.

Data preparation errors and badly shaped elements are easy to see on a computer-generated mesh plot. A trap built into the program may catch bad elements that otherwise escape notice; as suggested in Section 5.8, the Jacobian determinant should be positive at each sampling point (and elsewhere) in numerically integrated elements. Bad elements not numerically integrated may be trapped by a test for positive area. If a fictitious internal node 5 is assigned according to Eqs. 6.1.1, one or more component triangles in Fig. 16.9.1 will display negative area, as noted below Eq. 4.4.3.

A newly devised element may behave unexpectedly. The eigenvalue test of Section 16.3 should be applied to various shapes of element and various orientations of each shape as part of the search for defects.

One should reduce error in element formulation by use of local or centroidal coordinates (Section 4.4) and reduce error in single precision stress computation by initial subtraction of rigid body motion (Section 4.7).

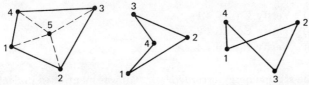

Figure 16.9.1. Quadrilateral element and two unacceptable distortions.

Computational error is increased by large element distortion, by great stiffness and size difference between adjacent elements, and by having an area of the structure whose displacements may have a large rigid body component. The possible error may not in fact occur, especially with node numbering which enforces good sequencing of equation-solving operations.

Possible computation error is forecast by a high condition number of the scaled structure stiffness matrix (Eq. 16.6.1). But the forecast may be overly pessimistic and expensive to compute. A very simple and quite valuable test is that for decay of diagonal coefficients (Section 16.7).

Error in the final solution may (and may not) be indicated by residual checks, and iterative improvement may (and may not) improve the solution.

The cantilever beam with equations sequenced from root to tip provides perhaps the most error-prone practical example (19, 23). Computational error becomes less as more supports are provided. Generalizing, we conclude that plates with clamped edges should have less computational error than plates with free edges, etc. It is well that this is so, as node numbering that begins (say) at the center of a clamped plate gives good equation sequencing but may be prohibited if the program requires a narrowly banded stiffness matrix.

Authors agree that if a set of equations is very ill conditioned it is better to reformulate the problem than to make heroic efforts to improve the solution. As stated in Ref. 1, "If a thing is not worth doing, it is not worth doing well." Use of double precision in all operations may be needed—if indeed this step has not already been taken—to retain enough significant information in the stiffness coefficients. There is little enthusiasm for use of partial double precision in equation solving.

After everything is apparently well debugged, well discretized, and well conditioned, it is safest to regard correct functioning of a program as something to be continually monitored instead of something established once and for all. Finally, recall that even correctly computed results are developed from a conceptual model and not from reality itself.

PROBLEMS

16.1. (a) Verify Eq. 16.4.1.
       (b) Verify Eq. 16.4.2.
       (c) Verify Eq. 16.4.4.

16.2. The strain energy error $O(h^{2(q+1-m)})$ was mentioned in Section 16.4. For the following elements, is the error measure valid, and if so what is the numerical value of $q + 1 - m$? The elements are described by:

(a) Eq. 4.4.1.   (e) Eqs. 8.4.8.
(b) Problem 4.4.   (f) Eq. 9.1.1.
(c) Fig. 5.5.3.   (g) Eq. 9.2.1.
(d) Eqs. 6.4.4.   (h) Fig. 10.6.1.

16.3. (a) Using the unscaled $[K]$ in Eq. 16.5.1, calculate $C(K)$ for $k_A = 10$, $k_B = 1$, and for $k_A = 1$, $k_B = 10$.

(b) Repeat Part a using the scaled matrix $[K_s]$.

(c) To illustrate true and artificial ill conditioning, in Fig. 16.5.2 let $k_A = 505$ and $k_B = 15$, then let $k_A = 15$ and $k_B = 505$. In each case compare exact displacements with those computed by dropping the final digit 5 from k_A and k_B.

16.4. Use a single beam element to model a cantilever beam of arbitrary length L. Evaluate the condition number of the scaled matrix $[K_s]$. Also approximate λ_{max} and λ_{min} as suggested in the latter part of Section 16.6 and compare the exact and approximate values of $C(K)$.

16.5. Let $k_1 = 1$ and $k_2 = 10$ in Problem 11.11. Suppose that $D_1 - D_2$ (or $D_2 - D_1$) replaces D_1 or D_2, but that $D_1 = D_2$ is not enforced so that $[K]$ remains 2 by 2. Is $C(K)$ changed by this transformation?

16.6. (a) Verify Eq. 16.7.1.

(b) Verify Eq. 16.7.2.

(c) Compute the condition numbers of the stiffness matrices in Eqs. 16.7.1 and 16.7.2.

(d) Use the spirit of Eq. 16.6.8 and rough physical approximation where necessary to estimate eigenvalues of the 102 by 102 scaled stiffness matrix of the structure of Fig. 16.7.1. Hence, approximate its condition number.

16.7. In Fig. 16.5.2 let $k_A = 0.30$, $k_B = 0.40$, and $P = 1$. Truncate each coefficient of $[K]^{-1}$ to a single digit. Calculate nodal displacements by the iterative improvement method of Section 16.8, and compare the results of three iterations with the exact displacements.

CHAPTER SEVENTEEN

Various Elements for Structural Mechanics

17.1 Equilibrium, Mixed, and Hybrid Elements

Most practical elements for structural mechanics have been formulated by use of assumed displacement fields and the potential energy principle. But there are other approaches, and some have scored notable successes. Pian (1 of Chapter Six) has classified various methods, some of which are quite briefly summarized in this section. It should be noted that a given element can sometimes be derived from several different variational principles (1.15, 1.25).[1]

Equilibrium elements are based on assumed stress fields and the complementary energy principle. Forces, not displacements, are the primary unknowns of the assembled structure. Displacements are obtained by use of the stress-strain relations and integration of the strain-displacement relations, a process that permits the displacement of a given point to have various values for various integration paths. A stiffness matrix may be obtained by inversion of the flexibility matrix produced by the basic formulation. Approximate solutions err by being too flexible. This suggests bounding the exact solution by use of a compatible displacement solution and an equilibrium solution. The bound is on strain energy, and while displacements and stresses may often follow the same trend, it cannot be said they will definitely be too low or too high in all problems. In the equilibrium model stresses are continuous across element boundaries but displacements are not,

---

[1] This notation is adopted for convenience in listing references by section instead of by chapter.

while in the compatible displacement model displacements are continuous but stresses are not. Sample references include Refs. 1.1 through 1.4.

Equilibrium models and compatible displacement models have a "dual" relationship. That is, for equations and principles in equilibrium analysis, there exist analogous equations and principles in displacement analysis. Perhaps the most familiar example of the duality is the "slab analogy," which is based on the fact that the Airy stress function and the lateral displacement of a thin plate both satisfy the biharmonic equation. A practical application of duality is the use of a program for plane stress analysis by the displacement method to solve problems in plate bending by the equilibrium method. Indeed, by invoking the duality concept it could be said that the library of equilibrium models includes the displacement models for plane stress, plates, and shells (1.18–1.22).

"Mixed" elements are based on a variational principle that admits both stress and displacement quantities as independent unknowns. It "mixes" potential and complementary energy formulations so that neither stresses nor displacements are given preferential treatment. A simple but effective plate bending element of this type is shown in Fig. 17.1.1. Nodal d.o.f. are lateral displacement at nodes 1, 3, 5 and bending moment normal to the edge at nodes 2, 4, 6, for a total of six d.o.f. Moments are computed with about the same accuracy as displacements. The structure matrix may have zeros on-diagonal even after boundary conditions have been imposed. Zeros fill in as an elimination solution proceeds, but clearly if diagonal coefficients are used as pivots the sequencing of eliminations cannot be arbitrary. Some awkwardness arises in branched shells, since a single moment cannot represent the edge moment in three or more elements that have the edge in common. A "mixed" element may be too stiff or too flexible; no bound can be guaranteed. References for "mixed" elements include Refs. 1.5 through 1.9 and 1.15 through 1.17.

"Hybrid" elements are founded on yet another variational principle. Actually a variety of hybrid elements is possible (1.23, 3.2; 1 of Chapter Six), but we will mention here only the assumed-stress hybrid, which has achieved greatest success. Elements of this type employ an assumed stress field within the element and assumed displacements on its boundaries. The two assumptions are independent. Thus, for example, in a plate element (where the hybrid approach has been most successful), the assumed moment field could be

$$M_x = \beta_1 + \beta_4 x + \beta_7 y + \beta_{10} x^2 + \cdots$$

$$M_y = \beta_2 + \beta_5 x + \beta_8 y + \beta_{11} x^2 + \cdots \qquad (17.1.1)$$

$$M_{xy} = \beta_3 + \beta_6 x + \beta_9 y + \beta_{12} x^2 + \cdots$$

while lateral displacement of any edge could be quadratic or cubic in the

Figure 17.1.1. "Mixed" plate element.

edge-parallel coordinate and governed by rotations and displacements of nodes on the edge. It is possible, but not necessary, to select the stress field so as to satisfy natural boundary conditions such as zero stress on a free edge. Use of these special edge-elements improves accuracy. Coefficients β_i in Eq. 17.1.1 play a role analogous to that of coefficients a_i in assumed displacement fields such as Eq. 4.4.1. Any number of β's may be used, but Eq. 17.1.1 must satisfy the differential equations of equilibrium. An element tends to become stiffer as more β's are used but more flexible as polynomials of higher degree are used for edge displacements. Thus no bound can be assured; computed results for displacement may be too low or too high. Nodal d.o.f. are displacement quantities. Thus, in contrast to equilibrium and mixed elements, hybrid elements are readily used in combination with assumed-displacement field elements. References for hybrid elements include Refs. 1.10 through 1.15 and 1.24; computer program statements in 16 and 18 of Chapter Nine; and 2.4 of Chapter Eighteen.

Hybrid plate elements may account for transverse shear deformation or may be restricted to thin plates. Triangular, rectangular, and general quadrilateral shapes are available. The versatility of formulation permits use of side nodes and corner nodes or use of corner nodes only. Published information suggests that for a unit of computational expense, hybrid elements are at least as effective as the eight-node isoparametric element discussed in Chapter Nine and are generally among the best plate elements. Details of the variational basis and of element formulation are lengthy, so we elect not to treat the matter further in this book. Those wishing only to use hybrid elements need not understand all details, provided that computer program statements for certain essential matrices can be obtained. Some troublesome zero-energy deformation modes appear in the rectangular elements of Ref. 1.14. A zero-energy mode is also possible in triangular elements (16, 18 of Chapter Nine), but it is of a type that is not possible if two or more elements are used.

Mass matrices $[m]$ and initial stress stiffness matrices $[k_\sigma]$ cannot be formulated as in Chapters Twelve and Thirteen if one demands a consistent formulation for equilibrium, mixed or hybrid elements. However, since approximate low order displacement fields are adequate for forming $[m]$ and $[k_\sigma]$ for use with assumed-displacement elements, it is reasonable to adopt similar displacement fields for use with other types of elements. Use of lumped masses is the most simple and obvious way to construct $[m]$.

17.2   Incompressible Elastic Media

As Poisson's ratio v approaches 0.5, a material becomes incompressible. Unless we are treating a case of plane stress, we cannot use $v = 0.5$, since to do so requires a division by zero in forming the material property matrix $[E]$ (see Eqs. 4.4.7 and 7.2.2). It is then tempting to seek an approximate solution by use of (say) $v = 0.49$, but there are dangers in this approach. For one thing, stresses in nearly incompressible materials are very sensitive to changes in v; a stress may double if v is changed from 0.48 to 0.50 (2.1). Also, structural equations become ill conditioned as v approaches 0.5. This is easily seen by noting that if an element is incompressible its nodal displacements cannot be independent.

Finite element models have been formulated that are valid for all values of Poisson's ratio, including $v = 0.5$ (2.2–2.6). Both isotropic and orthotropic materials have been considered. These elements have the usual displacement d.o.f. plus a mean pressure variable. The mean pressure away from nodes may be interpolated from nodal values in the same way as displacement variables (2.5). Most elements are based on the mixed variational principle, but the hybrid approach has also been used (2.3). Computer program statements appear in Ref. 2.7.

17.3   Beams, Stiffeners, and Related Members

Beams, either straight or curved, are useful as stiffening members for plates and shells and as structural members in their own right. Their stiffness, mass, and initial stress stiffness matrices are therefore of interest. An understandable development of these matrices for various members involves lengthy expressions that succeed more in consuming space than in providing enlightenment. Accordingly, we opt for the following summary instead.

Three methods of deriving element properties are in common use. In the first, governing differential equations of the problem are integrated. Particular solutions for various displacement or load conditions serve to establish individual coefficients of the stiffness or flexibility matrix. In the second, Castigliano's theorem (complementary energy) may be used with equilibrium equations to establish displacements for prescribed loads. The resulting flexibility matrix may be inverted to yield a stiffness matrix. The third method is the familiar one of assumed displacement fields. Use of internal d.o.f. may in some cases improve element properties (Table 8.2.1) but may also destroy displacement compatibility with a plate or shell the element is intended to stiffen. The question of compatibility may be of little consequence; indeed, the question might never arise when using the first and second methods of derivation mentioned above.

The effect of transverse shear deformation is significant for deep beams and sandwich beams. It is surprisingly awkward to include the shear effect by means of assumed displacement fields. For example, one might add to the expression for bending strain energy either of the following expressions for shear strain energy:

$$\frac{3(EI)^2}{5GA}\int_0^L \left(\frac{d^3v}{dx^3}\right)^2 dx, \quad \text{or} \quad \int_0^L \frac{GA}{2}\left(\theta - \frac{dv}{dx}\right)^2 dx \qquad (17.3.1)$$

The first expression leads to an incorrect result (3.1). The second expression is intended for use with separate assumptions for lateral displacement v and cross-section rotation θ. It leads to a stiffness matrix that is satisfactory only for very short elements (3.2). A satisfactory matrix is easily derived by the assumed-stress hybrid method; it is (3.1)

$$[k]\{d\} = \frac{2EI}{L(L^2+12g)}\begin{bmatrix} 6 & 3L & -6 & 3L \\ 3L & 2L^2+6g & -3L & L^2-6g \\ -6 & -3L & 6 & -3L \\ 3L & L^2-6g & -3L & 2L^2+6g \end{bmatrix}\begin{Bmatrix} v_1 \\ \theta_1 \\ v_2 \\ \theta_2 \end{Bmatrix} \qquad (17.3.2)$$

where the notation is as given in Section 3.9, and $g = nEI/AG$. Here $A =$ cross-sectional area, $G =$ shear modulus, and $n =$ shape factor ($n = 6/5$ for a rectangular cross section). As AG becomes large, Eq. 17.3.2 reduces to the conventional beam element stiffness matrix.

In vibration and dynamic response problems, both transverse shear effects and rotary inertia may become significant in the higher modes. Inclusion of these effects yields the "Timoshenko beam." Mass and stiffness matrices for Timoshenko beams are available (3.3, 3.4).

Stiffness, mass, and initial stress stiffness matrices have been given for straight members (3.7; 5 of Chapter Eleven) and for tapered members (3.6, 3.7; 53 of Chapter Twelve). Twisted members have also been considered (3.11, 3.22). Open and thin-walled sections, for which torque due to restraint of cross section warping may be significant, have been studied (3.5, 3.8–3.10; 53 of Chapter Twelve). It should be noted that to account for warping behavior, "extra" nodal d.o.f. are introduced. Thus there are more than the three rotations and three displacements that might be expected at each node of a member in three-dimensional space.

Stiffness matrices for curved bars under three-dimensional loading have been constructed (3.7, 3.12; 23 of Chapter Eleven). Here also, transverse shear deformation may be accounted for and mass matrices may include rotary inertia (3.13). If no displacements are permitted perpendicular to the plane of the element, we have a plane arch element. The comparative merit of various assumed displacement fields for arch elements has been studied (3.14–3.19); these studies are also pertinent to the formulation of cylindrical shell elements.

Thin-walled structures that are roughly beamlike, but too complicated to be analyzed as beams, may be built by combination of simpler elements such as plate or plane-stress elements (3.20, 3.21). However, idealization as a beam may suffice for preliminary design purposes (3.5).

17.4   Elements of Special Shape

Elements discussed in detail in Chapters Four to Ten are not restricted to any particular shape. This choice of elements was made in order to present elements with fewer restrictions and does *not* imply that more specialized elements are cumbersome, inaccurate, or obsolete. Where structures of a particular geometry—such as rectangular slabs, cylindrical shells, and building frames—must be frequently analyzed under various loadings and boundary conditions, use of a special purpose program based on a particular element may be convenient and economical. Common elements of restricted shape include rectangular elements for plane stress, rectangular and finite strip elements for plate bending, and elements for cylindrical shells. Some of these have been mentioned in preceding chapters. No attempt is made to provide a complete catalog of elements; however, some references for annular and skew elements (Fig. 17.4.1) should be mentioned. These include annular elements for plane stress (4.1–4.3, 4.13) annular elements for plate bending (4.3–4.6, 4.13) and skew elements for plate bending (4.7–4.12). A mass matrix is given in Ref. 4.4. The obvious advantage of annular elements is their ability to exactly represent a circular shape. The accuracy of skew elements declines somewhat with greater departure from a rectangular shape.

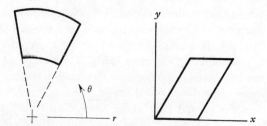

Figure 17.4.1.   Annular and skew elements.

17.5   Elastic Foundations

The simplest elastic foundation is provided by discrete springs, which resist either linear or angular displacement and are connected between node points and a fixed base. Each such spring is accounted for by adding its stiffness to the appropriate diagonal coefficient of the structure stiffness matrix.

The next more complicated model is the Winkler foundation, which idealizes reality by supposing that at any given point the foundation pressure is proportional to the displacement of that point only. Suppose, for example, that a plate lies in the xy plane and rests on a Winkler foundation whose modulus is β units of pressure per unit of displacement. Let z-direction displacement w be expressed in terms of nodal d.o.f. $\{d\}$: $w = [N]\{d\}$. Strain energy in the foundation is

$$U = \int \frac{\beta}{2} w^2 \, dx \, dy = \frac{1}{2} \{d\}^T \int \beta[N]^T[N] \, dx \, dy\{d\} = \frac{1}{2} \{d\}^T[k_f]\{d\}   \quad (17.5.1)$$

Here $[k_f]$ is an "element foundation stiffness" that may be added directly to the conventional element stiffness matrix. Note that $[k_f]$ and the "consistent" mass matrix $[m]$ of Eq. 12.2.3 have the same form. A simple "lumping," say of stiffness $\beta A/4$ to each node of a four-node element of area A, reduces the foundation to a set of discrete springs and makes $[k_f]$ a diagonal matrix.

A more realistic foundation model is that of an elastic half-space (5.1). Unfortunately the foundation stiffness matrix $[k_f]$ of this model is not sparse; the vertical displacements of *all* nodes in contact with the foundation are coupled in $[k_f]$. Extensions of this model have included horizontal contact pressure and the nonlinear effect of separation of contact surfaces where the structure displaces upward relative to the foundation (5.2). There is some indication that neither the Winkler foundation nor the elastic half-space is an accurate model for soils. Several other foundation models have been devised (5.3–5.5). Those that have attracted particular attention are the Pasternak models, which use two or three coefficients to describe foundation properties. To date there has been slight use of these models in finite element studies (5.6).

17.6   Triangular Elements and Area Coordinates

An arbitrary point P within a triangle serves to define three subareas A_1, A_2, A_3 in Fig. 17.6.1a and three altitudes s_1, s_2, s_3 in Fig. 17.6.1b. We define

$$L_1 = \frac{A_1}{A} = \frac{s_1}{h_1}, \qquad L_2 = \frac{A_2}{A} = \frac{s_2}{h_2}, \qquad L_3 = \frac{A_3}{A} = \frac{s_3}{h_3} \qquad (17.6.1)$$

where A = total area of the triangle. The ratios L_1, L_2, L_3 are known as "area coordinates" or "triangular coordinates." Since $A_1 + A_2 + A_3 = A$, the coordinates are not independent, but satisfy the relation $L_1 + L_2 + L_3 = 1$. As suggested by Fig. 17.6.1c, the equation L_i = constant defines a line parallel to side i. The location of a point P may be defined by any two of the three area coordinates. The centroid is at $L_1 = L_2 = L_3 = 1/3$.

Why should such unusual coordinates be used in formulating element properties? Area coordinates are "natural" for a triangle just as isoparametric coordinates $\xi\eta\zeta$ are natural for the elements in Chapter Five. An element formulated in natural coordinates is invariant (as defined in Section 4.6) even if the assumed displacement field is not a complete polynomial. In addition, assumed displacement fields may be written directly in terms of nodal displacements and area coordinate polynomials, just as we have written $\{f\} = [N]\{d\}$ for isoparametric elements. This direct approach leads to compact and computationally efficient expressions for element matrices. The alternative of working with undetermined coefficients $\{a\}$ until the last step, as described in Eqs. 4.2.14 to 4.2.16, makes for easier derivation but requires numerical inversion of matrix $[DA]$ for each different element.

Like the elements of Chapter Five, triangular elements with side nodes may be distorted so that the sides are curved. If the element shape and its displacements are defined by the same functions, the element may be called isoparametric, and its validity may be proved by the argument of Section 5.6.

To use area coordinates we must have expressions relating them to xy

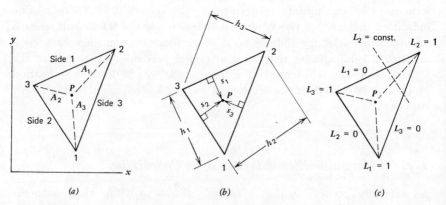

Figure 17.6.1. "Triangular" or "area" coordinates.

coordinates and rules for differentiation and integration. (With the exception of integration formulas, subsequent equations in this section are not difficult to obtain, and the reader is invited to verify them as an exercise.) For additional detail the reader is referred to the references cited. In particular, see Ref. 6 of Chapter Six, where area coordinates are exploited in many element formulations.

Triangular areas may be evaluated by a determinant, as noted in connection with Eq. 4.4.3. Thus we find an expression for A_1 in terms of x_2, y_2, x_3, y_3 and coordinates x and y of point P (Fig. 17.6.1). Similar treatment of the other subareas leads to the expression relating area coordinates to cartesian coordinates:

$$\begin{Bmatrix} L_1 \\ L_2 \\ L_3 \end{Bmatrix} = \frac{1}{2A} \begin{bmatrix} (x_2 y_3 - x_3 y_2) & (y_2 - y_3) & (x_3 - x_2) \\ (x_3 y_1 - x_1 y_3) & (y_3 - y_1) & (x_1 - x_3) \\ (x_1 y_2 - x_2 y_1) & (y_1 - y_2) & (x_2 - x_1) \end{bmatrix} \begin{Bmatrix} 1 \\ x \\ y \end{Bmatrix} \qquad (17.6.2)$$

where again A = total area of the triangle. From this relation we obtain expressions for derivatives, needed in calculating strains from displacements. Let $u = u(L_1, L_2, L_3)$; then

$$\frac{\partial u}{\partial x} = \frac{\partial u}{\partial L_1} \frac{\partial L_1}{\partial x} + \frac{\partial u}{\partial L_2} \frac{\partial L_2}{\partial x} + \frac{\partial u}{\partial L_3} \frac{\partial L_3}{\partial x}$$

$$(17.6.3)$$

where $\quad \dfrac{\partial L_1}{\partial x} = \dfrac{y_2 - y_3}{2A}, \quad \dfrac{\partial L_2}{\partial x} = \dfrac{y_3 - y_1}{2A}, \quad \dfrac{\partial L_3}{\partial x} = \dfrac{y_1 - y_2}{2A}$

and similarly for $\partial u/\partial y$, $\partial v/\partial x$, and $\partial v/\partial y$.

Integrals of polynomials in area coordinates are given by a remarkably simple formula involving factorials of the exponents (6.2; 5 of Chapter Five). It is

$$\int_{\text{area}} L_1^{m_1} L_2^{m_2} L_3^{m_3} \, dA = 2A \frac{m_1! \, m_2! \, m_3!}{(m_1 + m_2 + m_3 + 2)!} \qquad (17.6.4)$$

where m_1, m_2, m_3 are positive integers. If the integrand is expressed in terms of cartesian coordinates xy, compact closed form expressions for the integral are still possible, and some have been tabulated (6.1; 6 of Chapter Nine). In axisymmetric problems the radius appears in the denominator; the integrand is no longer a polynomial, and numerical integration is required. Several numerical integration formulas for triangles have been devised (6.3, 6.6, 6.8; 5 of Chapter Five). They have the form

$$\int_{\text{area}} f(x, y) \, dA = \sum_j f(x_j, y_j) W_j \qquad (17.6.5)$$

where W_j = a weight factor. One of them uses seven points within the triangle, integrates polynomials through the fifth degree exactly, and is used in the computer program of Ref. 2 of Chapter Seven. The sampling points and weight factors of the seven-point rule are as follows. Let x_i, y_i define corner coordinates of the triangle. The centroid is $x_c = (x_1 + x_2 + x_3)/3$, $y_c = (y_1 + y_2 + y_3)/3$. For the first three sampling points, $j = 1, 2, 3$, $i = j$,

$$\begin{Bmatrix} x_j \\ y_j \end{Bmatrix} = b \begin{Bmatrix} x_i \\ y_i \end{Bmatrix} + (1 - b) \begin{Bmatrix} x_c \\ y_c \end{Bmatrix}, \qquad b - \frac{1 + \sqrt{15}}{7}, \qquad W_j = \frac{155 - \sqrt{15}}{1200} A$$

(17.6.6)

For the next three sampling points, $j = 4, 5, 6$, $i = j - 3$,

$$\begin{Bmatrix} x_j \\ y_j \end{Bmatrix} = c \begin{Bmatrix} x_i \\ y_i \end{Bmatrix} + (1 - c) \begin{Bmatrix} x_c \\ y_c \end{Bmatrix}, \qquad c = \frac{1 - \sqrt{15}}{7}, \qquad W_j = \frac{155 + \sqrt{15}}{1200} A$$

(17.6.7)

The seventh sampling point, $j = 7$, is at the centroid and has weight $9A/40$. If desired, the sampling point locations may be expressed in terms of area coordinates by use of Eq. 17.6.2. These locations are as follows. With $i = 1, 2, 3$, let $L_i = (9 + 2g\sqrt{15})/21$ and $L_{i+1} = L_{i+2} = (6 - g\sqrt{15})/21$. For points $j = 1, 2, 3$ with $i = j$, set $g = 1$. For points $j = 4, 5, 6$ with $i = j - 3$, set $g = -1$.

As an example of use of area coordinates, consider the plane element of Fig. 17.6.2 (6 of Chapter Six). Nodes 4, 5, and 6 are at the midsides. In *cartesian* coordinates the displacement field is a complete quadratic; hence

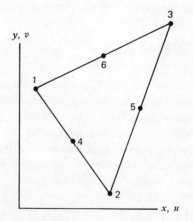

Figure 17.6.2. Linear-strain triangle.

formulation according to Eqs. 4.2.14 through 4.2.16 produces a satisfactory and invariant stiffness matrix but requires numerical inversion of a 6 by 6 matrix in forming $[DA]^{-1}$. In area coordinates the quadratic displacement field, formulated by inspection and trial, is

$$u = [L_1(2L_1 - 1) \quad L_2(2L_2 - 1) \quad L_3(2L_3 - 1)$$

$$4L_1L_2 \quad 4L_2L_3 \quad 4L_3L_1]\{u_1\,u_2\,u_3\,u_4\,u_5\,u_6\} \quad (17.6.8)$$

and similarly for displacement v. The reader may check that Eq. 17.6.8 satisfies interelement compatibility and, by the same arguments as used for isoparametric elements in Section 5.6, show that rigid body and constant strain modes are included. By differentiation, the strain-displacement relation $\{\epsilon\} = [B]\{d\}$ is constructed:

$$\begin{Bmatrix} \epsilon_x \\ \epsilon_y \\ \gamma_{xy} \end{Bmatrix} = \begin{bmatrix} B_x & 0 \\ 0 & B_y \\ B_y & B_x \end{bmatrix} \{u_1\,u_2\,\ldots\,u_6\,v_1\,v_2\,\ldots\,v_6\} \quad (17.6.9)$$
$$\underset{12\times 1}{}$$

where, for example,

$$[B_x] = \frac{1}{2A} [(4L_1 - 1)(y_2 - y_3) \quad (4L_2 - 1)(y_3 - y_1) \quad \cdots \quad \cdots$$
$$\underset{1\times 6}{}$$

$$4L_2(y_1 - y_2) + 4L_3(y_3 - y_1) \qquad 4L_3(y_2 - y_3) + 4L_1(y_1 - y_2)]$$
$$(17.6.10)$$

Equation 17.6.9 may be evaluated at nodes 1, 2, and 3 to yield an expression for strains at the corners in terms of nodal displacements. We write for this

$$\{\epsilon 123\} = [B123]\{d\}$$
$$\underset{9\times 1}{} \quad \underset{9\times 12}{} \; \underset{12\times 1}{}$$

where $\qquad\qquad\qquad\qquad\qquad\qquad\qquad\qquad\qquad\qquad (17.6.11)$

$$\{\epsilon 123\} = \{\epsilon_{x1}\,\epsilon_{x2}\,\epsilon_{x3}\,\epsilon_{y1}\,\epsilon_{y2}\,\epsilon_{y3}\,\gamma_{xy1}\,\gamma_{xy2}\,\gamma_{xy3}\}$$

Now because the strain field is a complete linear polynomial, strains at any point may be linearly interpolated from the corner values:

$$\begin{Bmatrix} \epsilon_x \\ \epsilon_y \\ \gamma_{xy} \end{Bmatrix} = \begin{bmatrix} L_1 & L_2 & L_3 & 0 & 0 & 0 & 0 & 0 & 0 \\ 0 & 0 & 0 & L_1 & L_2 & L_3 & 0 & 0 & 0 \\ 0 & 0 & 0 & 0 & 0 & 0 & L_1 & L_2 & L_3 \end{bmatrix} \underset{9\times 1}{\{\epsilon 123\}} \quad (17.6.12)$$

or symbolically, $\{\epsilon\} = [L123]\{\epsilon 123\}$. Therefore $\{\epsilon\} = [L123][B123]\{d\}$, and the stiffness matrix is

$$[k] = [B123]^T \int_{\text{area}} [L123]^T[E][L123]\,t\,dA[B123] \quad (17.6.13)$$

If material properties $[E]$ and element thickness t are constant, the integral is easily evaluated explicitly by use of Eq. 17.6.4, and a purely algebraic expression for $[k]$ is produced.

Area coordinates have been most frequently used in the derivation of plate elements. References include Refs. 6.4 and 6.5; 2 of Chapter Four; and 6 and 8 of Chapter Nine. Analogous coordinates have been used for tetrahedral solid elements (12 of Chapter Five).

Finally it may be noted that area coordinates are the two-space case of formulas for a "simplex" having $n + 1$ vertices in n-dimensional space (6.2). An n-dimensional "simplex" is a region having $n + 1$ vertices and bounded by $n + 1$ surfaces of dimensionality $n - 1$. Let the space be spanned by coordinates $x_1, x_2, x_3, \ldots, x_n$. The size S of the simplex is defined by a determinant whose rows are vertex coordinates:

$$S = \frac{1}{n!} \begin{vmatrix} 1 & {}_1x_1 & {}_1x_2 & \cdots & {}_1x_n \\ 1 & {}_2x_1 & {}_2x_2 & \cdots & {}_2x_n \\ \cdot & \cdot & \cdot & & \cdot \\ \cdot & \cdot & \cdot & & \cdot \\ \cdot & \cdot & \cdot & & \cdot \\ 1 & {}_{n+1}x_1 & {}_{n+1}x_2 & \cdots & {}_{n+1}x_n \end{vmatrix} \qquad (17.6.14)$$

Thus in one dimension ($n = 1$), $S =$ length; in two dimensions ($n = 2$), $S =$ area; in three dimensions ($n = 3$), $S =$ volume. Coordinates L_i may be defined analogously to Eq. 17.6.1; in three-space they are volume ratios and could be called "tetrahedron" coordinates. For any n, $L_1 + L_2 + \cdots + L_{n+1} = 1$. The integration formula for polynomials is

$$\int_S L_1^{m_1} L_2^{m_2} \ldots L_{n+1}^{m_{n+1}} \, dS = S \frac{m_1! \, m_2! \ldots m_n! \, n!}{(m_1 + m_2 + \cdots + m_n + n)!} \qquad (17.6.15)$$

In the mathematics of topology, coordinates L_i are known as *barycentric coordinates*.

17.7   Computer Generation of Element Matrices

The topic of the present section is that of transferring, from the analyst to the computer, the labor of element derivation and algebraic formulation. An appalling amount of algebra is needed to produce the stiffness matrix of, say, a plate element based on a high order polynomial in area coordinates. Also, if the element is not isoparametric it is sometimes easier to formulate matrices in terms of generalized coordinates $\{a\}$, leaving the transformation to nodal d.o.f. $\{d\}$ until last as in Eqs. 4.2.14 through 4.2.16. Computer

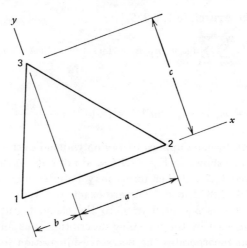

Figure 17.7.1. Triangular element with local coordinates xy.

time is increased by assigning these tasks to the machine, but the trade may be worthwhile. Whether it is or not depends on the type and complexity of elements used, the number of *different* element matrices needed in a typical run, the availability of useful programs, and the costs of manpower and computer time.

Let us consider first the matter of formulation in terms of generalized coordinates $\{a\}$, as in Eqs. 4.2.14 through 4.2.16. We will consider only enough detail to indicate the overall approach. Consider the triangular element 123 of Fig. 17.7.1. A convenient set of local cartesian coordinates xy is attached. If this is a thin plate element, the displacement field is entirely defined by z direction displacement w:

$$w = \sum_{i=1}^{NT} a_i x^{m_i} y^{n_i} \tag{17.7.1}$$

where m_i, n_i are integers. The number of terms NT depends on the number of nodes and the number of d.o.f. assigned to each. As an example, Table 17.7.1 lists the exponents used for the complete fifth-degree polynomial used in Ref. 6 of Chapter Nine.

In forming the stiffness matrix of a constant-thickness element, a typical term that must be integrated is the curvature-squared term $w_{,xx}^2$. The

Table 17.7.1. Exponents in Eq. 17.7.1 for a Plate Element

i	1	2	3	4	5	6	7	8	9	10	11	12	13	14	15	16	17	18	19	20	21
m_i	0	1	0	2	1	0	3	2	1	0	4	3	2	1	0	5	4	3	2	1	0
n_i	0	0	1	0	1	2	0	1	2	3	0	1	2	3	4	0	1	2	3	4	5

integral may be written (6.1, 6.2, 7.1):

$$\int_{\text{area}} w^2_{,xx}\, dA = \sum_{i=1}^{NT} \sum_{j=1}^{NT} a_i a_j m_i m_j (m_i - 1)(m_j - 1) F(m_i + m_j - 4, n_i + n_j)$$

where   (17.7.2)

$$F(m, n) = c^{n+1}[a^{m+1} - (-b)^{m+1}] \frac{m!\, n!}{(m + n + 2)!}$$

This integration formula applies to *any* polynomial over a triangle with the dimensions a, b, c shown in Fig. 17.7.1, and not only to the particular polynomial of Table 17.7.1. Other terms may be integrated by use of similar expressions (7.1, 7.2). Integrations are exact, not approximate. The coefficient of each $a_i a_j$ is a contribution to the element stiffness matrix. The conclusion suggested by the foregoing description is as follows: a computer subroutine that incorporates the necessary integration formulas need only be supplied with elastic constants, element dimensions, and the polynomial to be used; the subroutine can then generate the element stiffness matrix $[k_a]$ of Eq. 4.2.15.

In this way matrices for a variety of elements can be produced by a single and rather simple program. For example, Ref. 7.2 contains two subroutines of about 65 Fortran statements each, which formulate nine different plate bending elements and eight different plane elements. Mass and initial stress stiffness matrices are also generated for the plate elements, which may be anisotropic and of variable thickness. The matrix $[DA]$, needed to convert from coordinates $\{a\}$ to nodal d.o.f. $\{d\}$, is not so easily established and is different for each different type of element; however, formulation of $[DA]$ apparently presents no substantial difficulty. This method of forming element matrices may be extended, say to cylindrical elements, by use of the appropriate energy expressions.

The foregoing generation of element matrices relies largely on closed-form expressions for integrals. An alternate approach is that of having the computer perform symbolically (not numerically) the algebraic operations of element derivation (7.3, 7.4). These operations include multiplication, differentiation, and integration of polynomials. Input to such a program consists of element shape (triangular, rectangular, etc.), type of problem (plane, shell, etc.), the assumed polynomial field, and the nodal d.o.f. to be used. Output consists of algebraic expressions or program statements, by means of which element matrices are evaluated numerically. Such a symbol-processing program would be most helpful in assessing the quality of a newly proposed element. A derivation may require several minutes of computer time, but derivation by hand may require months (7.4).

The expressed concern about long computer run times prompts two concluding remarks. First, it is obvious that if successive elements in a mesh are identical, element matrices need be generated only once. Second, on some machines time may be saved by coding the innermost loop of a nest of DO loops in assembly language. Other time-saving rules are probably better known (avoid BACKSPACE and large amounts of input data, do not include arithmetic operations in a subscript, define constants by DATA statements, transfer data to auxiliary storage in large blocks instead of small, etc.).

CHAPTER EIGHTEEN

Various Problems and Techniques

18.1 *Other Formulations and Other Applications*

Structural elements mentioned so far have been based on the stationary potential energy principle and other variational principles cited in Section 17.1. There appears to be an indefinite number of additional variational principles on which finite elements might be based. Some attention has been given to the use of Lagrange multipliers (1.1–1.6). An application involves formulation of displacement elements without regard to interelement compatibility, then enforcing some amount of compatibility by introduction of Lagrange multipliers. The multipliers become additional unknowns that increase the size of the problem.

Although it is helpful to have a governing functional for the physical problem at hand, finite elements may be formulated without reference to variational principles (1.26). An alternate approach is through weighted residual methods, which include point matching, subdomain, least squares, and the Galerkin method (13 of Chapter Twelve). Of these the Galerkin method has received particular attention in finite element literature (1.7–1.11). The approximating function for the Galerkin method need not satisfy *all* boundary conditions. If the problem is suitably formulated, only the forced boundary conditions need be explicitly met (1.11).

The finite element approach has been used in many physical problems. No variational statement is available for some of them, and here the Galerkin method finds application. The list of problems that have been treated by finite elements is a long one. It includes but is not limited to the following: electric and magnetic fields, viscoelasticity, steady-state and transient heat conduction, steady-state analysis of temperature distribution in flowing bodies of water, frequencies and modes in acoustic enclosures, seepage flow in porous media, diffusion, convection, lubrication, slow deformation of geologic structures, compressible and incompressible fluid flow, fluid sloshing in tanks and oscillation in harbors, and fluid-structure interaction. Not all the various problems are independent in a mathematical sense, since some are represented by differential equations or functionals of the same form. For example, Stokes flow (where inertia is neglected and viscous action predominates) has the same fundamental equations as plate bending and is therefore readily treated by a finite element program for plates. While there are fundamental differences between thermal analysis and stress analysis, the two may operate on the same finite element mesh. Thus, node point temperatures from thermal analysis may be directly transferred to a subsequent stress analysis, and the labor of mesh preparation need not be repeated.

For further information the reader is referred to the literature. A small sampling is Refs. 1.7, 1.10 through 1.33 and 16 and 21 of Chapter Eleven. Several papers on fluids appear in *USJap1 Conf.* (See Notation for References.)

18.2   Stress Concentrations. Peak Stress Under Live Load

It is, of course, the highest stresses that are of greatest interest, and these are likely to be found at holes and notches where stress gradients are also high. An attempt to model a stress concentration by a profusion of tiny elements is objectionable for at least two reasons. First, an exceptionally large mesh may exceed storage capacity, and if not will have a long computation time. Second, very small elements raise the condition number of the structure stiffness matrix and thus may lead to poor results. These two drawbacks may be avoided by carrying out the analysis in two stages. In the first stage the entire structure is analyzed, using a coarse mesh near the stress concentration. In the second stage a separate analysis is performed on the region containing the stress concentration, the region now being modeled by a fine mesh. This "regional" structure is loaded on its boundaries by either the displacements or forces calculated in the first stage.

The approach is shown in Fig. 18.2.1. The "regional" mesh in Fig. 18.2.1b may use directly the nodal forces or displacements along $ABCD$ as calculated from the mesh of Fig. 18.2.1a. The mesh of Fig. 18.2.1b is rather awkward and mixes elements of considerably different size. In Fig. 18.2.1c new nodes are introduced along $ABCD$; their prescribed displacements must be interpolated from the displacements of the original nodes A, A', A'', B, B', and B''. The reader may recognize this technique as similar to that of the "reduced substructure," described in connection with Fig. 11.9.2.

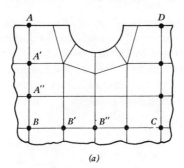

(a)

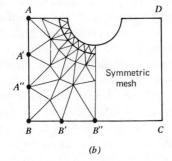

(b)

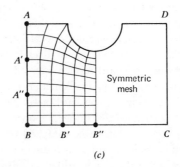

(c)

Figure 18.2.1. (a) Coarse mesh near stress concentration. (b) Separate "regional" structure, graded from nodes of the coarse mesh. (c) More regular mesh requiring interpolation of displacements on boundary *ABCD*.

An alternate approach to the stress concentration problem has been proposed (2.1). In this method the region containing a stress concentration is modeled by a single element that incorporates the "exact" solution for the stress concentration from theory of elasticity. Such an element could replace Fig. 18.2.1b or 18.2.1c, and might have as its nodes only *A*, *A'*, *A''*, *B*, etc. The same approach has been used in fracture mechanics, where the object is to compute stress intensity factors for cracks of arbitrary orientation in arbitrary stress fields (2.1–2.6, 2.10). The special "crack elements" incorporate the appropriate singularity functions. The method has been extended into the elastic-plastic range (2.2, 2.5). Others (2.7, 2.8) have suggested that stress intensity factors of engineering accuracy may be obtained without resorting to the special elements.

The analyst may wish to know where a group of movable loads should be placed in order to produce maximum effect at some location on the structure. The "effect" may be a moment, a reaction, a transverse shear, etc. It is helpful to have an influence line for the effect in question. In beam theory, according to the Müller-Breslau principle, an influence line is the same as the displaced shape of the beam when a displacement corresponding to the effect is imposed. More generally we will have an influence surface, and such a surface may be generated by the displacements of a finite element structure (2.9). For example, the influence surface for a support reaction on a shell is the displaced shape (minus the original shape) when the shell is loaded by a unit displacement at the desired reaction.

18.3   Incremental Construction

Analysis commonly proceeds under the assumption that the structure is assembled in a stress-free condition, and that only on completion of construction are loads "switched on." Actually, under body force loading, stresses often accumulate as the structure is built, with each additional piece applying more load to an already stressed structure (35 of Chapter Fifteen).

Proper accounting for the effects of incremental construction may be made as follows (3.1). Consider, for example, a structure built layer by layer, such as an earthen embankment. The first layer is modeled (by finite elements, say) and analyzed under its own weight. A second layer is added, and the two-layer structure is analyzed under the weight of the second layer only. Displacements and stresses computed in the two analyses are superposed. Then a third layer is added and its weight alone is used in a third analysis, etc. In an n-layer structure, n analyses are superposed to obtain conditions in the first layer, $n - 1$ analyses are superposed for the second layer, and so on. The number of layers needed for analysis may be markedly less than the number used in actual construction.

As the final solution must be obtained by a series of steps, material non-linearity is easily accommodated (3.1). This is done by adopting new material properties each time a layer is added and presuming that these properties remain constant during the step. The cost of this method lies in having to completely reform the structure stiffness matrix at the beginning of each step, instead of merely extending the matrix which was there already.

In comparison with conventional analysis, which presumes complete construction followed by gravity "switch-on," analysis of incremental construction may show similar stresses but rather different displacements (3.1). However, another incremental analysis (3.2) was prompted by the observation that conventional analysis gave negative backfill pressures.

18.4   Mesh Generation and Computer Graphics

Communication with the computer is a major problem in finite element or finite difference work. The volume of necessary input data can be formidable, and the amount of output data overwhelming. Of the total cost of a computer-based analysis, 50% may be used for preparation and checking of input data, 35 to 40% for interpretation of results, and 10 to 15% for computer time (4.1, 4.12). These percentages, of course, vary with the problem and the computer program. It is worth noting that most academic effort goes toward better elements, better equation solvers, nonlinear analysis, etc., while practitioners must devote most effort toward management of data and communication with the computer (4.2, 4.12).

"Mesh generation" refers to automatic generation of node locations and numbering of nodes and elements, proceeding from a minimal amount of user-supplied data, or perhaps from Fortran statements alone. The obvious advantage is in reducing long and tedious human effort. There should be fewer errors simply because there is less data to prepare. Solutions may become more accurate; a computer program may generate a more regular mesh than the user would, and elements that are not of greatly distorted shape give the best answers.

A survey (4.3) and other papers (4.4, 4.5; 22 of Chapter Eleven) describe many methods of mesh generation. Some easily understood concepts may be explained with reference to Fig. 18.4.1 (2 of Chapter Seven). If numbers and coordinates are supplied for nodes 1 and 5, a suitable program can easily recognize that 1 and 5 are not consecutive numbers and proceed to assign coordinates to nodes 2, 3, and 4 by spacing them uniformly along a straight line. With a little more sophistication, nodes 10, 15, and 20 can be generated from data regarding nodes 5 and 25 and the curvature of the arc. Indeed,

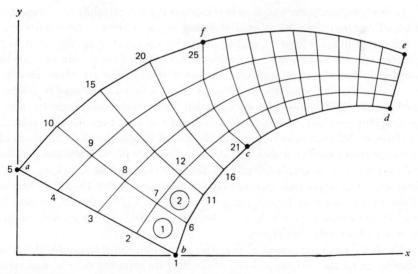

Figure 18.4.1. Example mesh of elements.

from data regarding nodes a through f, all remaining boundary nodes may be automatically located and numbered. Coordinates of each interior node may be computed as the average of coordinates of surrounding nodes, for example,

$$x_7 = \frac{(x_2 + x_6 + x_8 + x_{12})}{4}$$

$$y_7 = \frac{(y_2 + y_6 + y_8 + y_{12})}{4}$$

(18.4.1)

As coordinates of all interior nodes are initially unknown, Eqs. 18.4.1 must be applied iteratively (to all internal nodes) until the coordinates "steady down." Readers may recognize Eq. 18.4.1 as the finite-difference form of Laplace's equation. It tends to produce a uniformly spaced mesh, as seen in Fig. 18.4.1. An alternative way of locating internal nodes is to model the structure, or a large part of it, by a single isoparametric element (Chapter Five) and let intersections of lines $\xi =$ constant and $\eta =$ constant define nodal locations. As for element data, we note that node numbers in element 2 may be obtained by adding 5 to each node number in element 1 (Fig. 18.4.1). Remaining elements are similarly numbered, and it is clear that these steps may be automated.

A rather different form of node generation relies on electromechanical devices (4.3). A mesh is drawn on paper, or perhaps on a three-dimensional object or model. A probe is manually positioned at a node, then at the push of a button the coordinates of the probe are recorded on some storage device.

In analysis, computer graphics aids in checking of input data and presentation of computed results. Visual checking of input data is quite effective, since errors in node location and connectivity are much more apparent in a plotted mesh than in a table of numbers. A variety of things can be done in display of the structure and the computed results: the structure may be rotated so that it is seen from a different angle, selected areas may be magnified for detailed inspection, an exploded view shows how separate components fit together, the deformed state of the structure may be shown, surfaces of different orientation may be shaded differently, contour maps of stress or strain may be drawn and stress trajectories plotted, stresses may be plotted on various user-selected cuts through the structure, at some point of interest the strain-time history may be shown, etc. (4.6–4.10; 23 of Chapter Twelve). Results may be displayed on paper, on CRT (cathode ray tube), and to some extent on the line printer (4.11). Paper copy plots may be incorporated directly into reports.

The CRT terminal gives the user the most immediate response, the best way to grasp the "whole picture" instead of its separate details, and thus permits the best control over computed information. Initially, to avoid submerging the user in output, results of an analysis may be written on tape instead of printed. Then, after some study at the CRT terminal, the results of particular interest may be printed or plotted on paper. As another application, motion pictures of a vibrating structure could be made by sequencing separate exposures of the structure in varying amplitudes of deflection.

The user need not wait until a solution is completed to use the graphics terminal. In "interactive graphics" he participates in the solution or design process by changing thicknesses, constraints, mesh layout, loading, etc., whenever it seems necessary to do so. Iterative procedures may be terminated when near convergence or when not converging. While the terminal is idle the computer does its "number crunching" in the background. Apparently interactive graphics is of more potential value in design than in analysis.

There is, of course, a great deal of information regarding computer handling of graphical data in publications other than the structural mechanics literature.

18.5   Large-Scale Programs and Projects

Many programs, both large and small, have become widely available (6.1, 6.2). Users must decide which type is best suited to their particular needs. In this section some characteristics of large general purpose programs are summarized (4.1, 4.12, 5.3–5.6, 5.10; 17, 19 of Chapter Two; 23 of Chapter Twelve). Large programs require particular foresight in organization, development, and maintenance because of their size and complexity. The same might be said of large-scale projects that depend on these programs (5.5–5.9).

One expects a "large program" to be large in physical size (there are some 150,000 Fortran statements in NASTRAN). Beyond this, a general purpose program makes it possible to analyze a structure of almost any size; the number of d.o.f. is more likely to be limited by the expense of analysis instead of by the storage capacity of available computers. To accomplish this the program and data are segmented, so that segments that are momentarily inactive reside in auxiliary storage. Storage allocation is dynamic, meaning that the storage format for data is not fixed but adapts to handle efficiently the problem presented. The user may choose the method of data input or the scheme of mesh generation, select from a variety of element types and use them in combination, call for a static or dynamic analysis, select the quantity and mode of output, etc. Restart capability is provided. That is, the program, either automatically or as directed, writes intermediate results on an auxiliary storage device for possible future use. Hence, after scheduled exits, detection of an error or machine failure, analysis may be resumed without having to start again at the very beginning. Restart capability reduces the cost of examination of the mesh prior to analysis, examination of displacements prior to stress computation, application of different loads to a given structure, etc.

Modular design is a common feature of large programs. A modular program is one in which computation is done by a quantity of subroutines under the control of an executive routine. For example, the executive routine may call Level 1 subroutines, which in turn call Level 2 subroutines, etc.; however, subroutines on a given level do not call one another. Subroutines must share data, which underlines the need for effective organization and handling of data files. With some large programs the user must write his own executive routine in order to solve a problem. This executive routine consists of calls to the Level 1 subroutines that are needed. Such an arrangement has the beneficial effect of forcing the user to understand the analysis method, at least to a greater extent than a user's manual that directs him to place certain data in certain columns of punched cards.

Advantages of a large general purpose program include its capacity and versatility. Its modular design permits modification and correction of errors in one module without having to change all the others. Modularity also facilitates addition of new features, and these may make use of various utility routines already contained in the program (5.3). Of course, there are also disadvantages. Considerable expense is involved in the production and maintenance of a large general purpose program. Because of its versatility some time is required to learn to use it properly, and because of its complexity the average user is probably not capable of modifying it (5.3). Small special purpose programs will probably be faster in execution and can be maintained and run in-house by the organization with a small computer.

18.6   Finding Additional Information

The more useful sources of information about matrix methods of structural mechanics include textbooks, technical journals, proceedings of conferences and symposia, and reports available from N.T.I.S. (National Technical Information Service, Springfield, Va. 22151). A bibliography (6.3), which does not claim to be complete, lists some 1100 references as of early 1972, with author and key work indexes. Reports of government agencies and other research groups are indexed and abstracted in *STAR* (*Scientific and Technical Aerospace Reports*) and *GRA/GRI* (*Government Reports Announcements* and *Government Reports Index*). Most of the reports cited are available from N.T.I.S. Journal papers are abstracted in *IAA* (*International Aerospace Abstracts*). Currently *STAR*, *GRA/GRI*, and *IAA* are published once every two weeks.

Doctoral dissertations are occasionally valuable; they are indexed in

Dissertation Abstracts International, which also describes how copies may be purchased. For pursuit of ongoing developments in a particular speciality, the *Science Citation Index* may be helpful; for a specific paper it cites subsequent publications that have used that paper as a reference. Literature searches, done by computer and of either broad or highly specific coverage, are offered by N.T.I.S., the National Aeronautics and Space Administration, and the Defense Documentation Center. Fortunately for speakers of English, relatively little finite element work has been published in other languages. At present the number of papers in Japanese is rapidly increasing.

Listings of computer programs are frequently contained in reports cited in *STAR* and *GRA/GRI* and are occasionally contained in doctoral dissertations. The National Aeronautics and Space Administration has begun the quarterly journal *Computer Program Abstracts* to announce documented computer programs offered for public sale (journal available from Superintendent of Documents, U.S. Government Printing Office, Washington, D.C. 20402). Program decks, tapes, and documentation are sold by COSMIC (Barrow Hall, University of Georgia, Athens, Ga. 30601). This organization was established to collect and distribute software developed by government agencies. As an example, COSMIC sells NASTRAN. It has not been fashionable for regularly published journals to include programs or even useful subroutines; but there are exceptions, such as the *International Journal for Numerical Methods in Engineering* and *Computers and Structures.* Surveys of available software appear occasionally (6.1, 6.2). Commercial organizations that maintain large computers and sell their services usually have many programs available for use, including large general purpose programs. These organizations also offer consulting services. Other sources of software include manufacturers of digital computers and various users groups, such as the Integrated Civil Engineering System (ICES) users group (6.4).[1]

---

[1] Several papers directed toward the practitioner appear in the same journal issue as this reference.

REFERENCES

Notation for References

PERIODICALS

ActaM	Acta Mechanica
AeroJ	Aeronautical Journal of the Royal Aeronautical Society
AeroQu	The Aeronautical Quarterly
AIAAJ	Journal, American Institute of Aeronautics and Astronautics
AircrE	Aircraft Engineering
APRev	Applied Mechanics Reviews
BAMSoc	Bulletin of the American Mathematical Society
CivEng	Civil Engineering (ASCE)
CMAME	Computer Methods in Applied Mechanics and Engineering
CompAD	Computer Aided Design
CompFl	Computers and Fluids, An International Journal
CompJ	The Computer Journal
CompSt	Computers and Structures, An International Journal
EEStDy	International Journal of Earthquake Engineering and Structural Dynamics
Engin	The Engineer (London)
ExperM	Experimental Mechanics (Society for Experimental Stress Analysis)
Geotec	Geotechnique
IJESci	International Journal of Engineering Science
IJFraM	International Journal of Fracture Mechanics

IJMSci	International Journal of Mechanical Sciences
IJNME	International Journal for Numerical Methods in Engineering
IJSoSt	International Journal of Solids and Structures
JAeroS	Journal of the Aeronautical (*or* Aerospace) Sciences
JAircr	Journal of Aircraft (AIAA)
JAmCI	Journal of the American Concrete Institute
JApplM	Journal of Applied Mechanics, Transactions of ASME
JEInd	Journal of Engineering for Industry, Transactions of ASME
JEMDiv	Journal of the Engineering Mechanics Division, Proceedings of ASCE
JFlMec	Journal of Fluid Mechanics
JFrIns	Journal of the Franklin Institute
JGeoRe	Journal of Geophysical Research
JHeTra	Journal of Heat Transfer, Transactions of ASME
JHYDiv	Journal of the Hydraulics Division, Proceedings of ASCE
JMESci	Journal of Mechanical Engineering Science
JSMDiv	Journal of the Soil Mechanics and Foundations Division, Proceedings of ASCE
JSouVi	Journal of Sound and Vibration
JSpRoc	Journal of Spacecraft and Rockets (AIAA)
JSTDiv	Journal of the Structural Division, Proceedings of ASCE
JStMec	Journal of Structural Mechanics
JStrAn	Journal of Strain Analysis
MathCo	Mathematics of Computation
MoMath	Monatshefte für Mathematik
MTAC	Mathematical Tables and Other Aids to Computation
NucEDe	Nuclear Engineering and Design
PInsCE	Proceedings of the Institution of Civil Engineers
RocMec	Rock Mechanics
WaterR	Water Resources Research
ZAMP	Zeitschrift für Angewandte Mathematik und Physik

CONFERENCES AND SYMPOSIA

CompAE Sympos. "Computer Aided Engineering," Proceedings of a Symposium at Waterloo, Ontario, May 1971; G. M. L. Gladwell, ed.; University of Waterloo Press, 1971.

CompSh Conf. "Computer Oriented Analysis of Shell Structures," Proceedings of a Conference held at Palo Alto, California, August 1971 (AFFDL-TR-71-79, June 1972; AD-740-547, N.T.I.S.)

FElCE1 Sympos. "Application of Finite Element Methods in Civil Engineering," Proceedings of a Symposium held at Nashville, Tennessee, November 1969; W. H. Rowan Jr. and R. M. Hackett, eds.; American Society of Civil Engineers, 1969.

FElCE2 Conf. Proceedings of the Speciality Conference on Finite Element Method in Civil Engineering, Montreal, Quebec, June 1972; McGill University Printing Services, Montreal, 1972.

IUTAM Sympos. "High Speed Computing of Elastic Structures," Proceedings of the Symposium of International Union of Theoretical and Applied Mechanics, University of Liege, Belgium, August 1970; B. Fraeijs de Veubeke, ed.; University of Liege Press, 1971.

NCoACM Conf. National Conference of the Association for Computing Machinery.

ONR Sympos. "Numerical and Computer Methods in Structural Mechanics," Proceedings of a Symposium held at Urbana, Illinois, September 1971; Sponsored by the Office of Naval Research; S. J. Fenves, N. Perrone, A. R. Robinson, and W. C. Schnobrich, eds.; Academic Press, New York, 1973.

SMRT1 Conf. Proceedings of the First International Conference on Structural Mechanics in Reactor Technology, Berlin, September 1971; published by Commission of the European Communities, Brussels, 1972.

Sparse1 Conf. "Large Sparse Sets of Linear Equations," Proceedings of the Oxford Conference of the Institute of Mathematics and Its Applications held in April 1970, J. K. Reid, ed.; Academic Press, London, 1971.

Sparse2 Conf. "Sparse Matrices and Their Applications," Proceedings of a Symposium held September 1971; D. J. Rose and R. A. Willoughby, eds.; Plenum Press, New York, 1972.

SSDM Conf. AIAA/ASME/SAE Structures, Structural Dynamics and Materials Conference.

UCCE Sympos. Symposium on the Use of Computers in Civil Engineering, Lisbon, Portugal, Oct. 1962.

USJap1 Conf. "Recent Advances in Matrix Methods of Structural Analysis and Design," papers presented at the U.S.-Japan Seminar, Tokyo, Japan, 1969; R. H. Gallagher, Y. Yamada and J. T. Oden, eds.; University of Alabama Press, University, Alabama, 1971.

USJap2 Conf. "Advances in Computational Methods in Structural Mechanics and Design," papers presented at the Second U.S.-Japan Seminar, Berkeley, California, August 1972; J. T. Oden, R. W. Clough and Y. Yamamoto, eds.; UAH Press, Huntsville, Alabama, 1972.

WPAFB1 Conf. Proceedings of the (First) Conference on Matrix Methods in Structural Mechanics, Wright-Patterson Air Force Base, Ohio, October 1965 (AFFDL-TR-66-80, November 1966; AD-646-300, N.T.I.S.).

WPAFB2 Conf. Proceedings of the Second Conference on Matrix Methods in Structural Mechanics, Wright-Patterson Air Force Base, Ohio, October 1968 (AFFDL-TR-68-150, December 1969; AD-703-685, N.T.I.S.).

WPAFB3 Conf. Proceedings of the Third Conference on Matrix Methods in Structural Mechanics, Wright-Patterson Air Force Base, Ohio, October 1971.

References

CHAPTER ONE

1. A. L. Yettram and H. M. Husain, "Plane Framework Models for Plates in Extension," *JEMDiv*, Vol. 92, No. EM1, 1966, pp. 157–168.

2. S. G. Lekhnitskii, *Theory of Elasticity of an Anisotropic Elastic Body*, Holden-Day, San Francisco, 1963.

3. S. Timoshenko and J. N. Goodier, *Theory of Elasticity*, 2nd Ed., McGraw-Hill, New York, 1951.

CHAPTER TWO

1. M. J. Turner, R. W. Clough, H. C. Martin, and L. J. Topp, "Stiffness and Deflection Analysis of Complex Structures," *JAeroS*, Vol. 23, No. 9, 1956, pp. 805–823.

2. L. Fox, *An Introduction to Numerical Linear Algebra*, Oxford University Press, New York, 1966.

3. See Ref. 4 of Chapter Twelve.

4. J. H. Argyris and S. Kelsey, *Energy Theorems and Structural Analysis*, Butterworths, London, 1960 (collection of papers published in *Aircraft Engineering* in 1954 and 1955).

5. K. J. Bathe, E. L. Wilson, and F. E. Peterson, *SAP IV: A Structural Analysis Program for Static and Dynamic Response of Linear Systems*, Report EERC-73-11, Earthquake Engineering Research Center, University of California, Berkeley, June 1973 (PB-221-967/3, N.T.I.S.).

6. H. R. Grooms, "Algorithm for Matrix Bandwidth Reduction," *JSTDiv*, Vol. 98, No. ST1, 1972, pp. 203–214 (discussion: Vol. 98, No. ST12, 1972 pp. 2820–2821).

7. E. Cuthill and J. McKee, "Reducing the Bandwidth of Sparse Symmetric Matrices," *24th NCoACM Conf.*, 1969, pp. 157–172.

8. J. Barlow and C. G. Marples, Comment on "An Automatic Node-Relabeling Scheme for Bandwidth Minimization of Stiffness Matrices," *AIAAJ*, Vol. 7, No. 2, 1969, pp. 380–382.

9. G. Cantin, "An Equation Solver of Very Large Capacity," *IJNME*, Vol. 3, No. 3, 1971, pp. 379–388.

10. W. Weaver, *Computer Programs for Structural Analysis*, D. Van Nostrand Co., New York, 1967.

11. Discussion and closure on "Efficient Solution of Load-Deflection Equations," *JSTDiv*, Vol. 96, No. ST2, 1970, pp. 421–426; Vol. 97, No. ST2, 1971, pp. 713–717.

12. D. J. Rose and R. A. Willoughby, eds., *Sparse Matrices and Their Applications*, Plenum Press, New York, 1972.

13. J. K. Reid, ed., *Large Sparse Sets of Linear Equations*, Academic Press, New York, 1971.

14. H. G. Jensen and G. A. Parks, "Efficient Solutions for Linear Matrix Equations," *JSTDiv*, Vol. 96, No. ST1, 1970, pp. 49–64.

15. R. J. Melosh and R. M. Bamford, "Efficient Solution of Load-Deflection Equations," *JSTDiv*, Vol. 95, No. ST4, 1969, pp. 661–676.

16. A. Jennings, "A Compact Storage Scheme for the Solution of Symmetric Linear Simultaneous Equations," *CompJ*, Vol. 9, No. 3, 1966, pp. 281–285.

17. E. Schrem, "Computer Implementation of the Finite Element Procedure," *ONR Sympos.*, pp. 79–121.

18. I. P. King, "An Automatic Reordering Scheme for Simultaneous Equations Derived from Network Systems," *IJNME*, Vol. 2, No. 4, 1970, pp. 523–533.

19. S. J. Fenves, "Design Philosophy of Large Interactive Systems," *ONR Sympos.*, pp. 403–414.

20. E. L. Wilson, *Finite Element Analysis of Two Dimensional Structures*, Report 63–2, Civil Engineering Dept., University of California, Berkeley, June 1963.

21. H. A. Kamel, D. Liu, and E. I. White, "The Computer in Ship Structure Design," *ONR Sympos.*, pp. 643–668.

22. Y. R. Rashid, "Three Dimensional Analysis of Elastic Solids," *IJSoSt*, "I—Analysis Procedure," Vol. 5, No. 12, 1969, pp. 1311–1331; "II—The Computational Problem," Vol. 6, No. 1, 1970, pp. 195–207.

23. P. B. Lindley, "Plane-Stress Analysis of Rubber at High Strains Using Finite Elements," *JStrAn*, Vol. 6, No. 1, 1971, pp. 45–52.

24. I. Fried, "A Gradient Computational Procedure for the Solution of Large Problems Arising from the Finite Element Discretization Method," *IJNME*, Vol. 2, No. 4, 1970, pp. 477–494.

25. A. L. Yettram and M. J. S. Hirst, "The Solution of Structural Equilibrium Equations by the Conjugate Gradient Method with Particular Reference to Plane Stress Analysis," *IJNME*, Vol. 3, No. 3, 1971, pp. 349–360.

26. R. H. Mallett and L. A. Schmit, "Nonlinear Structural Analysis by Energy Search," *JSTDiv*, Vol. 93, No. ST3, 1967, pp. 221–234.

27. R. L. Fox and E. L. Stanton, "Developments in Structural Analysis by Direct Energy Minimization," *AIAAJ*, Vol. 6, No. 6, 1968, pp. 1036–1042.

28. B. M. Irons and D. K. Y. Yan, "Equation-Solving Algorithms for the Finite Element Method," *ONR Sympos.*, pp. 497–511.

29. R. A. Willoughby, "A Survey of Sparse Matrix Technology," *CompSh Conf.*, pp. 65–171.

30. G. von Fuchs, J. R. Roy, and E. Schrem, "Hypermatrix Solution of Large Sets of Symmetric Positive-Definite Linear Equations," *CMAME*, Vol. 1, No. 2, 1972, pp. 197–216.

31. J. Lestingi and S. Prachuktam, "A Blocking Technique for Large Scale Structural Analysis," *CompSt*, Vol. 3, No. 3, 1973, pp. 669–714.

32. R. J. Collins, "Bandwidth Reduction by Automatic Renumbering," *IJNME*, Vol. 6, No. 3, 1973, pp. 345–356.

33. J. E. Key, "Computer Program for Solution of Large, Sparse, Unsymmetric Systems of Linear Equations, *IJNME*, Vol. 6, No. 4, 1973, pp. 497–509.

34. T. S. Chow and J. S. Kowalik, "Computing with Sparse Matrices," *IJNME*, Vol. 7, No. 2, 1973, pp. 211–223.

35. C. Meyer, "Solution of Linear Equations—State-of-the-Art," *JSTDiv*, Vol. 99, No. ST7, 1973, pp. 1507–1526.

CHAPTER THREE

1. See Ref. 4 of Chapter Two.

2. Y. C. Fung, *Foundations of Solid Mechanics*, Prentice-Hall, Englewood Cliffs, 1965.

3. R. E. Jones, "A Generalization of the Direct-Stiffness Method of Structural Analysis," *AIAAJ*, Vol. 2, No. 5, 1964, pp. 821–826.

4. H. L. Langhaar, *Energy Methods in Applied Mechanics*, John Wiley & Sons, New York, 1962.

5. J. T. Oden, *Mechanics of Elastic Structures*, McGraw-Hill, New York, 1967.

6. J. S. Przemieniecki, *Theory of Matrix Structural Analysis*, McGraw-Hill, New York, 1968.

7. I. S. Sokolnikoff, *Mathematical Theory of Elasticity*, McGraw-Hill, New York, 1956.

8. K. Washizu, *Variational Methods in Elasticity and Plasticity*, Pergamon Press, Oxford, 1968.

9. R. W. McLay, "On Natural Boundary Conditions in the Finite Element Method," *JApplM*, Vol. 39, No. 4, 1972, pp. 1149–1150.

CHAPTER FOUR

1. R. J. Melosh, "Basis for Derivation of Matrices for the Direct Stiffness Method," *AIAAJ*, Vol. 1, No. 7, 1963, pp. 1631–1637.

2. G. P. Bazeley, Y. K. Cheung, B. M. Irons, and O. C. Zienkiewicz, "Triangular Elements in Plate Bending—Conforming and Non-Conforming Solutions," *WPAFB1 Conf.*, pp. 547–576.

3. B. M. Irons, O. C. Zienkiewicz, and E. R. de Arantes e Oliveira, "Comments on the Paper: Theoretical Foundations of the Finite Element Method," *IJSoSt*, Vol. 6, No. 5, 1970, pp. 695–697.

4. F. K. Bogner, R. L. Fox, and L. A. Schmit, Jr., "The Generation of Inter-element-Compatible Stiffness and Mass Matrices by the Use of Interpolation Formulas," *WPAFB1 Conf.*, pp. 397–443.

5. See Ref. 3 of Chapter Nine.

6. See Ref. 4 of Chapter Eight.

7. R. H. Gallagher, "Analysis of Plate and Shell Structures," *FElCE1 Sympos.*, pp. 155–205.

8. P. C. Dunne, "Complete Polynomial Displacement Fields for Finite Element Method," *AeroJ*, Vol. 72, No. 687, 1968, pp. 245–246 (discussion: Vol. 72, No. 692, 1968, pp. 709–711).

9. K. T. Kavanagh and S. W. Key, "A Note on Selective and Reduced Integration Techniques in the Finite Element Method," *IJNME*, Vol. 4, No. 1, 1972, pp. 148–150.

10. S. F. Pawsey, *The Analysis of Moderately Thick to Thin Shells by the Finite Element Method*, Report UC-SESM 70-12, Civil Engineering Dept., University of California, Berkeley, 1970.

11. E. C. Pestel, "Dynamic Stiffness Matrix Formulation by Means of Hermitian Polynomials," *WPAFB1 Conf.*, pp. 479–502.

12. E. R. Arantes e Oliveira, "Theoretical Foundations of the Finite Element Method," *IJSoSt*, Vol. 4, No. 10, 1968, pp. 929–952.

13. D. R. Navaratna, "Computation of Stress Resultants in Finite Element Analysis," *AIAAJ*, Vol. 4, No. 11, 1966, pp. 2058–2060.

14. B. M. Irons and A. Razzaque, "Introduction of Shear Deformations into a Thin Plate Displacement Formulation," *AIAAJ*, Vol. 11, No. 10, 1973, pp. 1438–1439.

15. D. Bushnell, "Finite Difference Energy Models versus Finite Element Models: Two Variational Approaches in One Computer Program," *ONR Sympos.*, pp. 291–336.

16. T. H. H. Pian, "Variational Formulations of Numerical Methods in Solid Continua," *CompAE Sympos.*, pp. 421–448.

17. "Finite Elements versus Finite Differences," panel discussion at *CompSh Conf.*, pp. 798–824.

18. K. J. Forsberg, "General Summary of the Conference," *CompSh Conf.*, pp. 1250–1284.

19. O. C. Zienkiewicz, "The Finite Element Method: From Intuition to Generality," *APRev*, Vol. 23, No. 3, 1970, pp. 249–256.

CHAPTER FIVE

1. B. M. Irons, "Engineering Applications of Numerical Integration in Stiffness Methods," *AIAAJ*, Vol. 4, No. 11, 1966, pp. 2035–2037.

2. I. Ergatoudis, B. M. Irons, and O. C. Zienkiewicz, "Curved, Isoparametric, 'Quadrilateral' Elements for Finite Element Analysis," *IJSoSt*, Vol. 4, No. 1, 1968, pp. 31–42.

3. W. Kaplan, *Advanced Calculus*, Addison-Wesley, Reading, 1953.

4. Z. Kopal, *Numerical Analysis*, John Wiley & Sons, New York, 1955.

5. A. H. Stroud and D. Secrest, *Gaussian Quadrature Formulas*, Prentice-Hall, Englewood Cliffs, 1966.

6. B. M. Irons, "Economical Computer Techniques for Numerically Integrated Finite Elements," *IJNME*, Vol. 1, No. 2, 1969, pp. 201–203.

7. R. L. Taylor, "On Completeness of Shape Functions for Finite Element Analysis," *IJNME*, Vol. 4, No. 1, 1972, pp. 17–22.

8. O. C. Zienkiewicz, "Isoparametric and Allied Numerically Integrated Elements—A Review," *ONR Sympos.*, pp. 13–41.

9. B. M. Irons, "Comment on 'Stiffness Matrices for Sector Elements,'" *AIAAJ*, Vol. 8, No. 1, 1970, pp. 191–192.

10. See Ref. 14 of Chapter Nine.

11. See Ref. 19 of Chapter Nine.

12. R. W. Clough, "Comparison of Three Dimensional Finite Elements," *FEICE1 Sympos.*, pp. 1–26.

13. B. M. Irons, "Quadrature Rules for Brick Based Finite Elements," *IJNME*, Vol. 3, No. 2, 1971, pp. 293–294.

14. T. K. Hellen, "Effective Quadrature Rules for Quadratic Solid Isoparametric Finite Elements," *IJNME*, Vol. 4, No. 4, 1972, pp. 597–599.

15. A. K. Gupta and B. Mohraz, "A Method of Computing Numerically Integrated Stiffness Matrices," *IJNME*, Vol. 5, No. 1, 1972, pp. 83–89.

16. P. C. Hammer and A. H. Stroud, "Numerical Evaluation of Multiple Integrals II," *MTAC*, Vol. 12, No. 64, 1958, pp. 272–280.

CHAPTER SIX

1. T. H. H. Pian and P. Tong, "Basis of Finite Element Methods for Solid Continua," *IJNME*, Vol. 1, No. 1, pp. 3–28.

2. W. P. Doherty, E. L. Wilson, and R. L. Taylor, *Stress Analysis of Axisymmetric Solids Utilizing Higher-Order Quadrilateral Finite Elements*, Report UC-SESM-69-3, Civil Engineering Dept., University of California, Berkeley, 1969 (PB-190-321, N.T.I.S.).

3. J. L. Krahula and J. F. Polhemus, "Use of Fourier Series in the Finite Element Method," *AIAAJ*, Vol. 6, No. 4, 1968, pp. 726–728.

4. W. Carnegie, J. Thomas, and E. Dokumaci, "An Improved Method of Matrix Displacement Analysis in Vibration Problems," *AeroQu*, Vol. 20, Part 4, 1969, pp. 321–332.

5. E. L. Wilson, R. L. Taylor, W. P. Doherty, and J. Ghaboussi, "Incompatible Displacement Models," *ONR Sympos.*, pp. 43–57.

6. C. A. Felippa, *Refined Finite Element Analysis of Linear and Nonlinear Two-Dimensional Structures*, Ph.D. Dissertation, University of California, Berkeley, 1966 (also available as PB-178-418 and PB-178-419, N.T.I.S.; computer programs in the latter).

7. I. Holand and K. Bell, eds., *Finite Element Methods in Stress Analysis*, Tapir Press, Trondheim, 1969.

8   L. R. Horrmann, computer program for "Finite Element Bending Analysis for Plates," *JEMDiv*, Vol. 93, No. EM5, 1967, pp. 13–26.

9. R. D. Cook, "Strain Resultants in Certain Finite Elements," *AIAAJ*, Vol. 7, No. 3, 1969, p. 535.

CHAPTER SEVEN

1. E. L. Wilson, "Structural Analysis of Axisymmetric Solids," *AIAAJ*, Vol. 3, No. 12, 1965, pp. 2269–2274.

2. J. G. Crose and R. M. Jones, *SAAS III: Finite Element Analysis of Axisymmetric and Plane Solids with Different Orthotropic, Temperature-Dependent Material Properties in Tension and Compression*, Aerospace Corp., San Bernardino, Calif., 1971 (AD-729-188, N.T.I.S.).

3. R. S. Dunham and R. E. Nickell, *Finite Element Analysis of Axisymmetric Solids with Arbitrary Loadings*, Report 67-6, Civil Engineering Dept., University of California, Berkeley, 1967 (AD-655-253, N.T.I.S.).

4. S. Timoshenko and J. N. Goodier, *Theory of Elasticity*, 2nd Ed., McGraw-Hill, New York, 1951.

5. G. A. Greenbaum, L. D. Hofmeister, and D. A. Evenson, "Pure Moment Loading of Axisymmetric Finite Element Models," *IJNME*, Vol. 5, No. 4, 1973, pp. 459–463.

6. E. L. Wilson and P. C. Pretorius, *A Computer Program for the Analysis of Prismatic Solids*, Report UC-SESM-70-21, Civil Engineering Dept., University of California, Berkeley, 1970 (PB-196-462, N.T.I.S.).

7. R. D. Cook, "A Note on Certain Incompatible Elements," *IJNME*, Vol. 6, No. 1, 1973, pp. 146–147.

8. J. G. Crose, "Stress Analysis of Axisymmetric Solids with Asymmetric Properties," *AIAAJ*, Vol. 10, No. 7, 1972, pp. 866–871.

9. O. C. Zienkiewicz and J. M. Too, "The Finite Prism in Analysis of Thick Simply Supported Bridge Boxes," *PInsCE*, Vol. 53, Part 2, 1972, pp. 147–172.

10. T. Belytschko, "Finite Elements for Axisymmetric Solids under Arbitrary Loadings with Nodes on Origin," *AIAAJ*, Vol. 10, No. 11, 1972, pp. 1532–1533 (discussion and closure: Vol. 11, No. 9, 1973, pp. 1357–1358).

11. O. C. Zienkiewicz and Y. K. Cheung, "Stresses in Shafts," *Engin*, Vol. 224, No. 5835, 1967, pp. 696–697.

12. O. C. Zienkiewicz and B. Schimming, "Torsion of Non-Homogeneous Bars with Axial Symmetry," *IJMSci*, Vol. 4, No. 1, 1962, pp. 15–23.

CHAPTER EIGHT

1. R. E. Jones and D. R. Strome, "Direct Stiffness Method of Analysis of Shells of Revolution Utilizing Curved Elements," *AIAAJ*, Vol. 4, No. 9, 1966, pp. 1519–1525.

2. R. E. Jones and D. R. Strome, "A Survey of Analysis of Shells by the Displacement Method," *WPAFB1 Conf.*, pp. 205–229.

3. W. E. Haisler and J. A. Stricklin, "Rigid-Body Displacements of Curved Elements in the Analysis of Shells by the Matrix Displacement Method," *AIAAJ*, Vol. 5, No. 8, 1967, pp. 1525–1527.

4. P. M. Mebane and J. A. Stricklin, "Implicit Rigid Body Motion in Curved Finite Elements," *AIAAJ*, Vol. 9, No. 2, 1971, pp. 344–345.

5. E. P. Popov and P. Sharifi, "A Refined Curved Element for Thin Shells of Revolution," *IJNME*, Vol. 3, No. 4, 1971, pp. 495–508.

6. L. J. Brombolich and P. L. Gould, "A High-Precision Curved Shell Finite Element," *12th SSDM Conf.*, April 1971.

7. I. Fried, "Basic Computational Problems in the Finite Element Analysis of Shells," *IJSoSt*, Vol. 7, No. 12, 1971, pp. 1705–1715.

8. See Ref. 35 of Chapter Twelve.

9. E. A. Witmer and J. J. Kotanchik, "Progress Report on Discrete-Element Elastic and Elastic-Plastic Analyses of Shells of Revolution Subjected to Axisymmetric and Asymmetric Loading," *WPAFB2 Conf.*, pp. 1341–1453.

10. J. H. Percy, T. H. H. Pian, S. Klein, and D. R. Navaratna, "Application of Matrix Displacement Method to Linear Elastic Analysis of Shells of Revolution," *AIAAJ*, Vol. 3, No. 11, 1965, pp. 2138–2145.

11. J. A. Stricklin, D. R. Navaratna, and T. H. H. Pian, "Improvements on the Analysis of Shells by the Matrix Displacement Method," *AIAAJ*, Vol. 4, No. 11, 1966, pp. 2069–2071.

12. D. Bushnell, "Stress, Buckling and Vibration of Prismatic Shells," *AIAAJ*, Vol. 9, No. 10, 1971, pp. 2004–2013.

13. G. A. Fonder and R. W. Clough, "Explicit Addition of Rigid-Body Motions in Curved Finite Elements," *AIAAJ*, Vol. 11, No. 3, 1973, pp. 305–312.

14. D. Bushnell, "Stress, Stability and Vibration of Complex, Branched Shells of Revolution," *14th SSDM Conf.*, March 1973.

CHAPTER NINE

1. S. Timoshenko and S. Woinowsky-Krieger, *Theory of Plates and Shells*, 2nd Ed., McGraw-Hill, New York, 1959.

2. R. J. Melosh, "A Stiffness Matrix for the Analysis of Thin Plates in Bending," *JAeroS*, Vol. 28, No. 1, 1961, pp. 34–42.

3. R. W. Clough and J. L. Tocher, "Finite Element Stiffness Matrices for Analysis of Plate Bending," *WPAFB1 Conf.*, pp. 515–545.

4. B. M. Irons and K. J. Draper, "Inadequacy of Nodal Connections in a Stiffness Solution for Plate Bending," *AIAAJ*, Vol. 3, No. 5, 1965, p. 961.

5. F. K. Bogner, R. L. Fox, and L. A. Schmit, Jr., "The Generation of Inter-Element-Compatible Stiffness and Mass Matrices by the Use of Interpolation Formulas," *WPAFB1 Conf.*, pp. 397–443.

6. K. Bell, "A Refined Triangular Plate Bending Finite Element," *IJNME*, Vol. 1, No. 1, 1969, pp. 101–122 (discussion: Vol. 1, No. 4, p. 395; Vol. 2, No. 1, pp. 146–147).

7. G. R. Cowper, E. Kosko, G. M. Lindberg, and M. D. Olson, "Static and Dynamic Applications of a High-Precision Triangular Plate Element," *AIAAJ*, Vol. 7, No. 10, 1969, pp. 1957–1965.

8. J. A. Stricklin, W. E. Haisler, P. R. Tisdale, and R. Gunderson, "A Rapidly Converging Triangular Plate Element," *AIAAJ*, Vol. 7, No. 1, 1969, pp. 180–181.

9. Y. K. Cheung, "Folded Plate Structures by the Finite Strip Method," *JSTDiv*, Vol. 95, No. ST12, 1969, pp. 2963–2979 (discussion: Vol. 96, No. ST7, pp. 1622–1624; Vol. 96, No. ST8, pp. 1848–1851; closure: Vol. 97, No. ST4, pp. 1354–1355).

10. G. M. Folie, "Stiffness Matrix for Sandwich Folded Plates," *JSTDiv*, Vol. 97, No. ST2, 1971, pp. 603–617.

11. C. Meyer and A. C. Scordelis, "Analysis of Curved Folded Plate Structures," *JSTDiv*, Vol. 97, No. ST10, 1971, pp. 2459–2480.

12. Y. K. Cheung and M. S. Cheung, "Flexural Vibrations of Polygonal Plates," *JEMDiv*, Vol. 97, No. EM2, 1971, pp. 391–411.

13. S. Ahmad, B. M. Irons, and O. C. Zienkiewicz, "Analysis of Thick and Thin Shell Structures by Curved Finite Elements," *IJNME*, Vol. 2, No. 3, 1970, pp. 419–451.

14. O. C. Zienkiewicz, R. L. Taylor, and J. M. Too, "Reduced Integration Technique in General Analysis of Plates and Shells," *IJNME*, Vol. 3, No. 2, 1971, pp. 275–290.

15. S. F. Pawsey and R. W. Clough, "Improved Numerical Integration of Thick Shell Finite Elements," *IJNME*, Vol. 3, No. 4, 1971, pp. 575–586.

16. R. D. Cook, "Two Hybrid Elements for Analysis of Thick, Thin and Sandwich Plates," *IJNME*, Vol. 5, No. 2, 1972, pp. 277–288.

17. F. J. Plantema, *Sandwich Construction*, John Wiley & Sons, New York, 1966.

18. R. D. Cook, "Some Elements for Analysis of Plate Bending," *JEMDiv*, Vol. 98, No. EM6, 1972, pp. 1453–1470.

19. S. F. Pawsey, "Discussion of Papers by O. C. Zienkiewicz, R. L. Taylor, and J. M. Too and S. F. Pawsey and R. W. Clough," *IJNME*, Vol. 4, No. 3, 1972, pp. 449–450.

20. R. D. Cook, "More on Reduced Integration and Isoparametric Elements," *IJNME*, Vol. 5, No. 1, 1972, pp. 141–142.

21. J. F. Abel and C. S. Desai, "Comparison of Finite Elements for Plate Bending," *JSTDiv*, Vol. 98, No. ST9, 1972, pp. 2143–2148.

CHAPTER TEN

1. R. W. Clough and C. P. Johnson, "A Finite Element Approximation for the Analysis of Thin Shells," *IJSoSt*, Vol. 4, No. 1, 1968, pp. 43–60.

2. See Ref. 35 of Chapter Twelve.

3. A. B. Sabir and D. G. Ashwell, "A Stiffness Matrix for Shallow Shell Finite Elements," *IJMSci*, Vol. 11, No. 3, 1969, pp. 269–279.

4. G. R. Cowper, G. M. Lindberg, and M. D. Olson, "A Shallow Shell Finite Element of Triangular Shape," *IJSoSt*, Vol. 6, No. 8, 1970, pp. 1133–1156.

5. G. Cantin and R. W. Clough, "A Curved, Cylindrical Shell, Finite Element," *AIAAJ*, Vol. 6, No. 6, 1968, pp. 1057–1062.

6. F. K. Bogner, R. L. Fox, and L. A. Schmit, "A Cylindrical Shell Discrete Element," *AIAAJ*, Vol. 5, No. 4, 1967, pp. 745–751.

7. D. G. Ashwell and A. B. Sabir, "A New Cylindrical Shell Finite Element Based on Simple Independent Strain Functions," *IJMSci*, Vol. 14, No. 3, 1972, pp. 171–183.

8. G. Cantin, "Strain Displacement Relationships for Cylindrical Shells," *AIAAJ*, Vol. 6, No. 9, 1968, pp. 1787–1788.

9. G. Cantin, "Rigid Body Motions in Curved Finite Elements," *AIAAJ*, Vol. 8, No. 7, 1970, pp. 1252–1255.

10. G. Dupuis and J. Goël, "A Curved Finite Element for Thin Elastic Shells," *IJSoSt*, Vol. 6, No. 11, 1970, pp. 1413–1428.

11. S. Ahmad, B. M. Irons, and O. C. Zienkiewicz, "A Simple Matrix-Vector Handling Scheme for Three-Dimensional and Shell Analysis," *IJNME*, Vol. 2, No 4, 1970, pp. 500 522.

12. S. Ahmad, B. M. Irons, and O. C. Zienkiewicz, "Curved Thick Shell and Membrane Elements with Particular Reference to Axisymmetric Problems," *WPAFB2 Conf.*, pp. 539–572.

13. B. M. Irons and A. Razzaque, "A Further Modification to Ahmad's Shell Element," *IJNME*, Vol. 5, No. 4, 1973, pp. 588–589.

14. A. C. Scordelis and K. S. Lo, "Computer Analysis of Cylindrical Shells," *JAmCI*, Vol. 61, No. 5, 1964, pp. 539–561.

15. I. Fried, "Shear in C^0 and C^1 Bending Finite Elements," *IJSoSt*, Vol. 9, No. 4, 1973, pp. 449–460.

16. H. Takemoto and R. D. Cook, "Some Modifications of an Isoparametric Shell Element," *IJNME*, Vol. 7, No. 3, 1973, pp. 401–405.

CHAPTER ELEVEN

1. J. S. Przemieniecki and L. Berke, *Digital Computer Program for the Analysis of Aerospace Structures by the Matrix Displacement Method*, Report FDL TDR 64-18, Wright-Patterson Air Force Base, Flight Dynamics Lab., April 1964 (AD-600-518, N.T.I.S.).

2. D. J. Dawe, "Shell Analysis Using a Simple Facet Element," *JStrAn*, Vol. 7, No. 4, 1972, pp. 266–270.

3. L. R. Herrmann and D. M. Campbell, "A Finite-Element Analysis for Thin Shells," *AIAAJ*, Vol. 6, No. 10, 1968, pp. 1842–1847.

4. E. Spira and Y. Sokal, discussion of "New Rectangular Finite Element for Shear Wall Analysis," *JSTDiv*, Vol. 96, No. ST8, 1970, pp. 1799–1802.

5. S. S. Tezcan, "Computer Analysis of Plane and Space Structures," *JSTDiv*, Vol. 92, No. ST2, 1966, pp. 143–173.

6. W. Weaver, Jr., *Computer Programs for Structural Analysis*, D. Van Nostrand Co., New York, 1967.

7. R. K. Livesley, *Matrix Methods of Structural Analysis*, Pergamon Press, Oxford, 1964.

8. F. W. Beaufait, "Stiffness Analysis Using Multiglobal Axes System," *AIAAJ*, Vol. 9, No. 7, 1971, pp. 1400–1402.

9. P. A. Iversen, "Some Aspects of the Finite Element Method in Two Dimensional Analysis," in *Finite Element Methods in Stress Analysis*, ed. by I. Holand and K. Bell, Tapir Press, Trondheim, 1969, pp. 93–114.

10. O. C. Zienkiewicz and F. C. Scott, "On the Principle of Repeatability and Its Application in Analysis of Turbine and Pump Impellers," *IJNME*, Vol. 4, No. 3, 1972, pp. 445–448.

11. J. S. Przemieniecki, "Matrix Structural Analysis of Substructures," *AIAAJ*, Vol. 1, No. 1, 1963, pp. 138–147.

12. I. C. Taig, "Automated Stress Analysis Using Substructures," *WPAFB1 Conf.*, pp. 255–274.

13. R. Rosen and M. F. Rubinstein, "Substructure Analysis by Matrix Decomposition," *JSTDiv*, Vol. 96, No. ST3, 1970, pp. 663–670 (discussion and closure: Vol. 96, No. ST10, pp. 2249–2251; Vol. 97, No. ST2, pp. 724–726; Vol. 97, No. ST9, p. 2425).

14. R. H. Gallagher, "An Overview and Some Projections," *USJap1 Conf.*, pp. 3–22.

15. C. J. Meissner, "A Multiple Coupling Algorithm for the Stiffness Method of Structural Analysis," *AIAAJ*, Vol. 6, No. 11, 1968, pp. 2184–2185.

16. R. H. Gallagher and R. H. Mallett, "Efficient Solution Processes for Finite Element Analysis of Transient Heat Conduction," *JHeTra*, Vol. 93, No. 3, 1971, pp. 257–263.

17. R. R. Craig, Jr. and M. C. Bampton, "Coupling of Substructures for Dynamic Analyses," *AIAAJ*, Vol. 6, No. 7, 1968, pp. 1313–1319.

18. W. A. Benfield and R. F. Hruda, "Vibration Analysis of Structures by Component Mode Substitution," *AIAAJ*, Vol. 9, No. 7, 1971, pp. 1255–1261.

19. W. C. Hurty, J. D. Collins, and G. C. Hart, "Dynamic Analysis of Large Structures by Modal Synthesis Techniques," *CompSt*, Vol. 1, No. 4, 1971, pp. 535–563.

20. R. H. MacNeal, "A Hybrid Method of Component Mode Synthesis, *CompSt*, Vol. 1, No. 4, 1971, pp. 581–601.

21. D. A. Hunt, "Discrete Element Structural Theory of Fluids," *AIAAJ*, Vol. 9, No. 3, 1971, pp. 457–461.

22. H. A. Kamel, D. Liu, W. McCabe, and V. Philippopoulos, "Some Developments in the Analysis of Complex Ship Structures," *USJap2 Conf.*, pp. 703–726.

23. D. L. Morris, "Curved Beam Stiffness Coefficients," *JSTDiv*, Vol. 94, No. ST5, 1968, pp. 1165–1174.

24. C. W. Coale and W. A. Loden, "The Role of Component Modal Techniques in Dynamic Analysis of Engineering Structures," *CompSh Conf.*, pp. 1032–1062.

25. F. W. Williams, "Comparison Between Sparse Stiffness Matrix and Sub-Structure Methods, *IJNME*, Vol. 5, No. 3, 1973, pp. 383–394.

26. I. J. Somervaille, "A Technique for Mesh Grading Applied to Conforming Plate Bending Finite Elements," *IJNME*, Vol. 6, No. 2, 1973, pp. 310–312.

27. T. Furike, "Computerized Multiple Level Substructuring Analysis," *CompSt*, Vol. 2, Nos. 5/6, 1972, pp. 1063–1073.

CHAPTER TWELVE

1. J. W. S. Rayleigh, *Theory of Sound*, Vols. I and II, Dover Publications, New York, 1945 (2nd Ed., originally published in 1894).

2. W. C. Hurty and M. F. Rubinstein, *Dynamics of Structures*, Prentice-Hall, Englewood Cliffs, 1964.

3. A. L. Kimball, "Vibration Damping, Including the Case of Solid Friction," *JApplM*, Vol. 51, 1929, pp. 227–236.

4. E. L. Wilson, *A Computer Program for the Dynamic Analysis of Underground Structures*, Report 68-1, Civil Engineering Dept., University of California, Berkeley, January 1968 (AD-832-681, N.T.I.S.).

5. R. W. Clough, "Analysis of Structural Vibrations and Dynamic Response," *USJap1 Conf.*, pp. 441–486.

6. E. L. Wilson and J. Penzien, "Evaluation of Orthogonal Damping Matrices," *IJNME*, Vol. 4, No. 1, 1972, pp. 5–10.

7. J. S. Archer, "Consistent Mass Matrix for Distributed Mass Systems," *JSTDiv*, Vol. 89, No. ST4, 1963, pp. 161–178.

8. J. F. Abel and E. P. Popov, "Static and Dynamic Finite Element Analysis of Sandwich Structures," *WPAFB2 Conf.*, pp. 213–245.

9. R. W. Clough and C. A. Felippa, "A Refined Quadrilateral Element for Analysis of Plate Bending," *WPAFB2 Conf.*, pp. 399–440.

10. P. Tong, T. H. H. Pian, and L. L. Bucciarelli, "Mode Shapes and Frequencies by Finite Element Method Using Consistent and Lumped Masses," *CompSt*, Vol. 1, No. 4, 1971, pp. 623–638.

11. J. S. Archer, "Consistent Matrix Formulations for Structural Analysis Using Finite-Element Techniques," *AIAAJ*, Vol. 3, No. 10, 1965, pp. 1910–1918.

12. D. M. Egle and J. L. Sewall, "An Analysis of Free Vibration of Orthogonally Stiffened Cylindrical Shells with Stiffeners Treated as Discrete Elements," *AIAAJ*, Vol. 6, No. 3, 1968, pp. 518–526.

13. S. H. Crandall, *Engineering Analysis*, McGraw-Hill, New York, 1956.

14. See Ref. 6 of Chapter Six.

15. B. Noble, *Applied Linear Algebra*, Prentice-Hall, Englewood Cliffs, 1969.

16. R. J. Guyan, "Reduction of Stiffness and Mass Matrices," *AIAAJ*, Vol. 3, No. 2, 1965, p. 380.

17. J. N. Ramsden and J. R. Stoker, "Mass Condensation: a Semi-automatic Method for Reducing the Size of Vibration Problems," *IJNME*, Vol. 1, No. 4, 1969, pp. 333–349.

18. R. G. Anderson, B. M. Irons, and O. C. Zienkiewicz, "Vibration and Stability of Plates Using Finite Elements," *IJSoSt*, Vol. 4, No. 10, 1968, pp. 1031–1055.

19. K. Appa, G. C. C. Smith, and J. T. Hughes, "Rational Reduction of Large-Scale Eigenvalue Problems," *AIAAJ*, Vol. 10, No. 7, 1972, pp. 964–965.

20. J. H. Wilkinson, *The Algebraic Eigenvalue Problem*, Clarendon Press, Oxford, 1965.

21. A. Jennings, "Natural Vibrations of a Free Structure," *AircrE*, Vol. 24, No. 397, 1962, pp. 81–83.

22. R. Craig, Jr. and M. C. C. Bampton, "On the Iterative Solution of Semi-definite Eigenvalue Problems," *AeroJ*, Vol. 75, No. 724, 1971, pp. 287–290.

23. R. H. MacNeal, ed., *The NASTRAN Theoretical Manual*, National Aeronautics and Space Administration, Washington, D.C., 1970 (available from COSMIC, University of Georgia, Athens, Ga.).

24. R. Rosen and M. F. Rubinstein, "Dynamic Analysis by Matrix Decomposition," *JEMDiv*, Vol. 94, No. EM2, 1968, pp. 385–395.

25. L. Fox, *An Introduction to Numerical Linear Algebra*, Oxford University Press, New York, 1965.

26. A. Jennings and D. R. L. Orr, "Application of the Simultaneous Iteration Method to Undamped Vibration Problems," *IJNME*, Vol. 3, No. 1, 1971, pp. 13–24.

27. G. Peters and J. H. Wilkinson, "Eigenvalues of $Ax = \lambda Bx$ with Band Symmetric A and B," *CompJ*, Vol. 12, No. 4, 1969, pp. 398–404.

28. K. K. Gupta, "Vibration of Frames and Other Structures with Banded Stiffness Matrix," *IJNME*, Vol. 2, No. 2, 1970, pp. 221–228.

29. K. K. Gupta, "Solution of Eigenvalue Problems by Sturm Sequence Method," *IJNME*, Vol. 4, No. 3, 1972, pp. 379–404 (addendum: Vol. 6, No. 3, 1973, p. 456).

30. R. L. Fox and M. P. Kapoor, "A Minimization Method for the Solution of the Eigenproblem Arising in Structural Dynamics," *WPAFB2 Conf.*, pp. 271–306.

31. I. Fried, "Gradient Methods for Finite-Element Eigenproblems," *AIAAJ*, Vol. 7, No. 4, 1969, pp. 739–741.

32. L. W. Rehfield, "Higher Vibration Modes by Matrix Iteration," *AIAAJ*, Vol. 9, No. 7, 1972, pp. 505–506.

33. H. Rutishauser, "Deflation bei Bandmatrizen," *ZAMP*, Vol. 10, No. 3, 1959, pp. 314–319.

34. R. F. Hartung, "An Assessment of Current Capability for Computer Analysis of Shell Structures," *CompSt*, Vol. 1, Nos. 1/2, 1971, pp. 3–32.

35. R. W. Clough and E. L. Wilson, "Dynamic Finite Element Analysis of Arbitrary Thin Shells," *CompSt*, Vol. 1, Nos. 1/2, 1971, pp. 33–56.

36. See Ref. 12 of Chapter Fourteen.

37. W. C. Carpenter, "Viscoelastic Stress Analysis," *IJNME*, Vol. 4, No. 3, 1972, pp. 357–366.

38. R. E. Nickell, "On the Stability of Approximation Operators in Problems of Structural Dynamics," *IJSoSt*, Vol. 7, No. 3, 1971, pp. 301–319.

39. R. S. Dunham, R. E. Nickell, and D. C. Stickler, "Integration Operators for Transient Structural Response," *CompSt*, Vol. 2, Nos. 1/2, 1972, pp. 1–15.

40. N. M. Newmark, "A Method of Computation for Structural Dynamics," *JEMDiv*, Vol. 85, No. EM3, 1959, pp. 67–94.

41. G. L. Goudreau and R. L. Taylor, "Evaluation of Numerical Integration Methods in Elastodynamics," *CMAME*, Vol. 2, No. 1, 1973, pp. 69–97.

42. E. L. Wilson and R. W. Clough, "Dynamic Response by Step-by-Step Matrix Analysis," *UCCE Sympos.*, pp. 45.1–45.14.

43. R. W. H. Wu and E. A. Witmer, "Stability of the De Vogelaere Method for Timewise Numerical Integration," *AIAAJ*, Vol. 11, No. 10, 1973, pp. 1432–1436.

44. K. D. Willmert, "Numerical Determination of the Response of a Linear Vibration System With a Singular Mass Matrix," *JEInd*, Vol. 94, No. 1, 1972, pp. 64–69.

45. J. F. McNamara and P. V. Marcal, "Incremental Stiffness Method for Finite Element Analysis of the Nonlinar Dynamic Problem," *ONR Sympos.*, pp. 353–376.

46. R. D. Krieg and S. W. Key, "Transient Shell Response by Numerical Time Integration," *USJap2 Conf.*, pp. 237–258.

47. Y. K. Cheung and S. Chakrabarti, "Free Vibrations of Thick, Layered Rectangular Plates by a Finite Layer Method," *JSouVi*, Vol. 21, No. 3, 1972, pp. 277–284.

48. M. A. Dokainish and S. Rawtani, "Vibration Analysis of Rotating Cantilever Plates," *IJNME*, Vol. 3, No. 2, 1971, pp. 233–248.

49. C. Mei, "Nonlinear Vibrations of Beams by Matrix Displacement Method," *AIAAJ*, Vol. 10, No. 3, 1972, pp. 355–357.

50. S. Klein and R. J. Sylvester, "The Linear Elastic Dynamic Analysis of Shells of Revolution by the Matrix Displacement Method," *WPAFB1 Conf.*, pp. 299–328.

51. I. Fried, "Finite Element Analysis of Time-Dependent Phenomena," *AIAAJ*, Vol. 7, No. 6, 1969, pp. 1170–1173.

52. J. H. Argyris and D. W. Scharpf, "Finite Elements in Time and Space," *NucEDe*, Vol. 10, No. 4, 1969, pp. 456–464.

53. R. S. Barsoum, "Finite Element Method Applied to the Problem of Stability of a Non-Conservative System," *IJNME*, Vol. 3, No. 1, 1971, pp. 63–87.

54. C. D. Mote and G. Y. Matsumoto, "Coupled, Nonconservative Stability-Finite Element," *JEMDiv*, Vol. 98, No. EM3, 1972, pp. 595–608.

55. V. Kariappa and B. R. Somashekar, "Flutter of Skew Panels by the Matrix Displacement Approach," *AeroJ*, Vol. 74, No. 716, 1970, pp. 672–675.

56. M. D. Olson, "Some Flutter Solutions Using Finite Elements," *AIAAJ*, Vol. 8, No. 4, 1970, pp. 747–752.

57. K. J. Bathe, *Solution Methods for Large Generalized Eigenvalue Problems in Structural Engineering*, Report UC-SESM-71-20, Civil Engineering Dept., University of California, Berkeley, November 1971 (PB-208-853, N.T.I.S.).

58. C. V. Girija Vallabhan, "Numerical Method for Elastic Stability Problems," *JSTDiv*, Vol. 97, No. ST11, 1971, pp. 2691–2706.

59. S. T. Mau and R. H. Gallagher, "A Finite-Element Procedure for Nonlinear Prebuckling and Initial Postbuckling Analysis," *NASA Contractor Report*, January 1972 (NASA CR-1936, N.T.I.S.).

60. G. Weeks, "Temporal Operators for Nonlinear Structural Dynamics Problems," *JEMDiv*, Vol. 98, No. EM5, 1972, pp. 1087–1104.

61. M. Geradin, "Error Bounds for Eigenvalue Analysis by Elimination of Variables," *JSouVi*, Vol. 19, No. 2, 1971, pp. 111–132.

62. K. J. Bathe and E. L. Wilson, "Large Eigenvalue Problems in Dynamic Analysis," *JEMDiv*, Vol. 98, No. EM6, 1972, pp. 1471–1485.

63. S. W. Key and Z. E. Beisinger, "The Transient Dynamic Analysis of Thin Shells by the Finite Element Method," *WPAFB3 Conf.*

64. E. L. Stanton and D. J. McGovern, "The Application of Gradient Minimization Methods and Higher Order Discrete Elements to Shell Buckling and Vibration Eigen-Problems," *CompSh Conf.*, pp. 1138–1172.

65. H. A. Kamel and R. L. Lambert, "Solution of Structural Eigenvalue Problems Using Sparsely Populated Matrices," *CompSh Conf.*, pp. 746–797.

66. K. K. Gupta, "Free Vibration Analysis of Spinning Structural Systems," *IJNME*, Vol. 5, No. 3, 1973, pp. 395–418.

67. E. L. Wilson, I. Farhoomand, and K. J. Bathe, "Nonlinear Dynamic Analysis of Complex Structures, *EEStDy*, Vol. 1, No. 3, 1973, pp. 241–252.

68. K. J. Bathe and E. L. Wilson, "Stability and Accuracy Analysis of Direct Integration Methods," *EEStDy*, Vol. 1, No. 3, 1973, pp. 283–291.

69. C. Mei, "Finite Element Analysis of Nonlinear Vibration of Beam Columns," *AIAAJ*, Vol. 11, No. 1, 1973, pp.115–117.

70. C. Mei, "Finite Element Displacement Method for Large Amplitude Free Flexural Vibrations of Beams and Plates," *CompSt*, Vol. 3, No. 1, 1973, pp. 163–174.

71. R. E. Nickell, "Direct Integration Methods in Structural Dynamics," *JEMDiv*, Vol. 99, No. EM2, 1973, pp. 303–317.

72. K. J. Bathe and E. L. Wilson, "Solution Methods for Eigenvalue Problems in Structural Mechanics," *IJNME*, Vol. 6, No. 2, 1973, pp. 213–226.

73. R. L. Kidder, "Reduction of Structural Frequency Equations," *AIAAJ*, Vol. 11, No. 6, 1973, p. 892.

74. B. R. Corr and A. Jennings, "Implementation of Simultaneous Iteration for Vibration Analysis," *CompSt*, Vol. 3, No. 3, 1973, pp. 497–507.

75. A. Jennings, "Mass Condensation and Simultaneous Iteration for Vibration Problems," *IJNME*, Vol. 6, No. 4, 1973, pp. 543–552.

76. J. H. Argyris, P. C. Dunne, and T. Angelopoulos, "Non-linear Oscillations Using the Finite Element Technique," *CMAME*, Vol. 2, No. 2, 1973, pp. 203–250.

77. K. J. Bathe and E. L. Wilson, "Eigensolution of Large Structural Systems with Small Bandwidth," *JEMDiv*, Vol. 99, No. EM3, 1973, pp. 467–479.

78. K. K. Gupta, "Eigenproblem Solution by a Combined Sturm Sequence and Inverse Iteration Technique," *IJNME*, Vol. 7, No. 1, 1973, pp. 17–42.

79. I. B. Alpay and S. Utku, "On Response of Initially Stressed Structures to Random Excitations," *CompSt*, Vol. 3, No. 5, 1973, pp. 1079–1097.

CHAPTER THIRTEEN

1. R. H. Gallagher and J. Padlog, "Discrete Element Approach to Structural Stability Analysis," *AIAAJ*, Vol. 1, No. 6, 1963, pp. 1437–1439.

2. B. J. Hartz, "Matrix Formulation of Structural Stability Problems," *JSTDiv*, Vol. 91, No. ST6, 1965, pp. 141–157.

3. K. K. Kapur and B. J. Hartz, "Stability of Plates Using the Finite Element Method," *JEMDiv*, Vol. 92, No. EM2, 1966, pp. 177–195.

4. H. C. Martin, "On the Derivation of Stiffness Matrices for the Analysis of Large Displacement and Stability Problems," *WPAFB1 Conf.*, pp. 697–716.

5. D. R. Navaratna, T. H. H. Pian, and E. A. Witmer, "Analysis of Elastic Stability of Shells of Revolution by the Finite Element Method," *AIAAJ*, Vol. 6, No. 2, 1968, pp. 355–361.

6. J. A. Stricklin, "Geometrically Nonlinear Static and Dynamic Analysis of Shells of Revolution," *IUTAM Sympos.*, pp. 383–412.

7. G. Powell and R. Klingner, "Elastic Lateral Buckling of Steel Beams," *JSTDiv*, Vol. 96, No. ST9, 1970, pp. 1919–1932.

8. R. S. Barsoum and R. H. Gallagher, "Finite Element Analysis of Torsional and Torsional-Flexural Stability Problems," *IJNME*, Vol. 2, No. 3, 1970, pp. 335–352.

9. D. Bushnell, "Analysis of Ring-Stiffened Shells of Revolution Under Combined Thermal and Mechanical Loading," *AIAAJ*, Vol. 9, No. 3, 1971, pp. 401–410.

10. J. S. Przemieniecki, "Matrix Analysis of Local Instability in Plates, Stiffened Panels and Columns," *IJNME*, Vol. 5, No. 2, 1972, pp. 209–216.

11. D. V. Wallerstein, "A General Linear Geometric Matrix for a Fully Compatible Finite Element," *AIAAJ*, Vol. 10, No. 4, 1972, pp. 545–546.

12. R. H. Gallagher, R. A. Gellatly, J. Padlog, and R. H. Mallett, "A Discrete Element Procedure for Thin-Shell Instability Analysis," *AIAAJ*, Vol. 5, No. 1, 1967, pp. 138–145.

13. A. D. Kerr and M. T. Soifer, "The Linearization of the Prebuckling State and Its Effect on the Determined Stability Loads," *JApplM*, Vol. 36, No. 4, 1969, pp. 775–783.

14. B. O. Almroth and F. A. Brogan, "Bifurcation Buckling as an Approximation of the Collapse Load for General Shells," *AIAAJ*, Vol. 10, No. 4, 1972, pp. 463–467.

15. J. S. Przemieniecki, "Finite Element Structural Analysis of Local Instability," *AIAAJ*, Vol. 11, No. 1, 1973, pp. 33–39.

CHAPTER FOURTEEN

1. J. T. Oden, *Finite Elements of Nonlinear Continua*, McGraw-Hill, New York, 1972.

2. D. W. Murray, *Large Deflection Analysis of Plates*, Ph.D. Dissertation, University of California, Berkeley, 1967.

3. D. W. Murray and E. L. Wilson, "Finite-Element Large Deflection Analysis of Plates," *JEMDiv*, Vol. 95, No. EM1, 1969, pp. 143–165.

4. A. Jennings, "Frame Analysis Including Change of Geometry," *JSTDiv*, Vol. 94, No. ST3, 1968, pp. 627–644.

5. G. H. Powell, "Theory of Nonlinear Elastic Structures," *JSTDiv*, Vol. 95, No. ST12, 1969, pp. 2687–2701.

6. R. H. Mallett and P. V. Marcal, "Finite-Element Analysis of Nonlinear Structures," *JSTDiv*, Vol. 94, No. ST9, 1968, pp. 2081–2105.

7. J. A. Stricklin, W. E. Haisler, and W. A. Von Riesemann, "Geometrically Nonlinear Analysis by Stiffness Method," *JSTDiv*, Vol. 97, No. ST9, 1971, pp. 2299–2314.

8. P. G. Bergan and R. W. Clough, "Convergence Criteria for Iterative Processes," *AIAAJ*, Vol. 10, No. 8, 1972, pp. 1107–1108.

9. W. E. Haisler, J. A. Stricklin, and F. J. Stebbins, "Development and Evaluation of Solution Procedures for Geometrically Nonlinear Structural Analysis by the Direct Stiffness Method," *AIAAJ*, Vol. 10, No. 3, 1972, pp. 264–272.

10. L. D. Hofmeister, G. A. Greenbaum, and D. A. Evenson, "Large Strain, Elasto-Plastic Finite Element Analysis," *AIAAJ*, Vol. 9, No. 7, 1971. pp. 1248–1254.

11. J. A. Stricklin, W. E. Haisler, H. R. MacDougall, and F. J. Stebbins, "Nonlinear Analysis of Shells of Revolution by the Matrix Displacement Method," *AIAAJ*, Vol. 6, No. 12, 1968, pp. 2306–2312.

12. J. A. Stricklin, J. E. Martinez, J. R. Tillerson, J. H. Hong, and W. E. Haisler, "Nonlinear Dynamic Analysis of Shells of Revolution by Matrix Displacement Method," *AIAAJ*, Vol. 9, No. 4, 1971, pp. 629–636.

13. T. Kawai and N. Yoshimura, "Analysis of Large Deflection of Plates by the Finite Element Method," *IJNME*, Vol. 1, No. 1, 1969, pp. 123–133.

14. J. A. Stricklin, W. E. Haisler, and W. A. Von Riesemann, "Evaluation of Solution Procedures for Material and/or Geometrically Nonlinear Structural Analysis," *AIAAJ*, Vol. 11, No. 3, 1973, pp. 292–299.

15. J. A. Stricklin, W. A. Von Riesemann, J. R. Tillerson, and W. E. Haisler, "Survey of Static Geometric and Material Nonlinear Analysis by the Finite Element Method," *USJap2 Conf.*, pp. 301–324.

16. P. Sharifi and E. P. Popov, "Nonlinear Buckling Analysis of Sandwich Arches," *JEMDiv*, Vol. 97, No. EM5, 1971, pp. 1397–1412.

17. F. Baron and M. S. Venkatesan, "Nonlinear Analysis of Cable and Truss Structures," *JSTDiv*, Vol. 97, No. ST2, 1971, pp. 679–710.

18. J. H. Argyris and D. W. Scharpf, "Large Deflection Analysis of Prestressed Networks," *JSTDiv*, Vol. 98, No. ST3, 1972, pp. 633–654.

19. J. T. Oden and T. Sato, "Finite Strains and Displacements of Elastic Membranes by the Finite Element Method," *IJSoSt*, Vol. 3, No. 4, 1967, pp. 471–488.

20. J. T. Oden, "Note on an Approximate Method for Computing Nonconservative Generalized Forces on Finitely Deformed Finite Elements," *AIAAJ*, Vol. 8, No. 11, 1970, pp. 2088–2090.

21. P. G. Smith and E. L. Wilson, "Automatic Design of Shell Structures," *JSTDiv*, Vol. 97, No. ST1, 1971, pp. 191–201.

22. D. J. White and L. R. Enderby, "Finite-Element Stress Analysis of a Non-Linear Problem: A Connecting-Rod Eye Loaded by Means of a Pin," *JStrAn*, Vol. 5, No. 1, 1970, pp. 41–48.

23. S. K. Chan and I. S. Tuba, "A Finite Element Method for Contact Problems of Solid Bodies—Part I. Theory and Validation," *IJMSci*, Vol. 13, No. 7, 1971, pp. 615–625.

24. A. Scholes and E. M. Strover, "The Piecewise-Linear Analysis of Two Connecting Structures Including the Effect of Clearance at the Connections," *IJNME*, Vol. 3, No. 1, 1971, pp. 45–51.

25. E. A. Wilson and B. Parsons, "Finite Element Analysis of Elastic Contact Problems Using Differential Displacements," *IJNME*, Vol. 2, No. 3, 1970, pp. 387–395.

26. J. Connor and N. Morin, "Perturbation Techniques in the Analysis of Geometrically Nonlinear Shells," *IUTAM Sympos.*, pp. 683–705.

27. R. T. Haftka, R. H. Mallett, and W. Nachbar, "Adaption of Koiter's Method to Finite Element Analysis of Snap-Through Buckling Behavior," *IJSoSt*, Vol. 7, No. 10, 1971, pp. 1427–1445.

28. T. Y. Yang, "A Finite Element Procedure for Large Deflection Analysis of Plates with Initial Imperfections," *AIAAJ*, Vol. 9, No. 8, 1971, pp. 1468–1473.

29. T. Y. Yang, "Elastic Snap-Through Analysis of Curved Plates Using Discrete Elements," *AIAAJ*, Vol. 10, No. 4, 1972, pp. 371–372.

30. B. M. Irons and R. C. Tuck, "A Version of the Aitken Accelerator for Computer Iteration," *IJNME*, Vol. 1, No. 3, 1969, pp. 275–277.

31. O. C. Zienkiewicz and B. M. Irons, "Matrix Iteration and Acceleration Processes in Finite Element Problems of Structural Mechanics," in *Methods for Nonlinear Algebraic Equations*, P. Rabinowitz, ed., Gordon & Breach, London, 1970, pp. 183–194.

32. G. C. Nayak and O. C. Zienkiewicz, "Note on the 'Alpha'-Constant Stiffness Method for the Analysis of Non-Linear Problems," *IJNME*, Vol. 4, No. 4, 1972, pp. 579–582.

33. Y. Hangai and S. Kawamata, "Purturbation Method in the Analysis of Geometrically Nonlinear and Stability Problems," *USJap2 Conf.*, pp. 473–492.

34. G. Cohen, "Imperfection Sensitivity of Elastic Structures: The Second Approximation for Unique Mode Buckling," *CompSh Conf.*, pp. 1107–1137.

35. R. G. Vos and W. P. Vann, "Finite Element Tensor Approach to Plate Buckling and Postbuckling," *IJNME*, Vol. 5, No. 3, 1973, pp. 351–365.

36. C. K. Mak and D. W. Kao, "Finite Element Analysis of Buckling and Postbuckling Behaviors of Arches with Geometric Imperfections," *CompSt*, Vol. 3, No. 1, 1973, pp. 149–162.

37. P. G. Bergan and R. W. Clough, "Large Deflection Analysis of Plates and Shallow Shells Using the Finite Element Method," *IJNME*, Vol. 5, No. 4, 1973, pp. 543–556 (a more detailed account is available as AD-735-937, N.T.I.S.).

38. R. T. Haftka and J. C. Robinson, "Effect of Out-of-Planeness of Membrane Quadrilateral Finite Elements," *AIAAJ*, Vol. 11, No. 5, 1973, pp. 742–744.

39. T. Belytschko and B. J. Hsieh, "Nonlinear Transient Finite Element Analysis with Convected Coordinates," *IJNME*, Vol. 7, No. 3, 1973, pp. 255–271.

40. A. K. Kar and C. Y. Okazaki, "Convergence in Highly Nonlinear Cable Net Problems," *JSTDiv*, Vol. 99, No. ST3, 1973, pp. 321–334.

41. A. M. Ebner and J. J. Ucciferro, "A Theoretical and Numerical Comparison of Elastic Nonlinear Finite Element Methods," *CompSt*, Vol. 2, Nos. 5/6, 1972, pp. 1043–1061.

42. R. H. Gallagher, "The Finite Element Method in Shell Stability Analysis," *CompSt*, Vol. 3, No. 3, 1973, pp. 543–557.

CHAPTER FIFTEEN

1. A. Mendelson, *Plasticity: Theory and Application*, The Macmillan Co., New York, 1968.

2. O. C. Zienkiewicz, S. Valliappan, and I. P. King, "Elastoplastic Solutions of Engineering Problems—'Initial Stress,' Finite Element Approach," *IJNME*, Vol. 1, No. 1, 1969, pp. 75–100.

3. P. V. Marcal, "Large Deflection Analysis of Elastic-Plastic Shells of Revolution," *AIAAJ*, Vol. 8, No. 9, 1970, pp. 1627–1633.

4. Y. Yamada, N. Yoshimura, and T. Sakurai, "Plastic Stress-Strain Matrix and Its Application for the Solution of Elastic-Plastic Problems by the Finite Element Method," *IJMSci*, Vol. 10, No. 5, 1968, pp. 343–354.

5. J. A. Stricklin, W. E. Haisler, and W. A. Von Riesemann, "Computation and Solution Procedures for Nonlinear Analysis by Combined Finite Element—Finite Difference Methods," *CompSt*, Vol. 2, Nos. 5/6, 1972, pp. 955–974.

6. J. A. Stricklin, W. E. Haisler, and W. A. Von Riesemann, *Formulation, Computation and Solution Procedures for Material and/or Geometric Nonlinear Structural Analysis by the Finite Element Method*, Report SC-CR-72 3102, Sandia Laboratories, Albuquerque, New Mexico, July 1972.

7. G. Venkateswara Rao and A. V. Krishna Murty, "An Alternate Form of the Ramberg-Osgood Formula for Matrix Displacement Analysis," *NucEDe*, Vol. 17, No. 3, 1971, pp. 297–308.

8. Y. Yamada, T. Kawai, N. Yoshimura, and T. Sakurai, "Analysis of the Elastic-Plastic Problems by the Matrix Displacement Method," *WPAFB2 Conf.*, pp. 1271–1299.

9. Y. Yamada, "Recent Developments in Matrix Displacement Method for Elastic-Plastic Problems in Japan," *USJap1 Conf.*, pp. 283–316.

10. B. Whang, "Elasto-Plastic Orthotropic Plates and Shells," *FElCE1 Sympos.*, pp. 481–515.

11. G. C. Nayak and O. C. Zienkiewicz, "Elasto-Plastic Stress Analysis. A Generalization for Various Constitutive Relations Including Strain Softening," *IJNME*, Vol. 5, No. 1, 1972, pp. 113–135.

12. P. V. Marcal and I. P. King, "Elastic-Plastic Analysis of Two-Dimensional Stress Systems by the Finite Element Method," *IJMSci*, Vol. 9, No. 3, 1967, pp. 143–155.

13. H. G. Harris and A. B. Pifko, "Elastic-Plastic Buckling of Stiffened Rectangular Plates," *FElCE1 Sympos.*, pp. 207–253.

14. P. V. Marcal, "Finite Element Analysis with Material Nonlinearities— Theory and Practice," *USJap1 Conf.*, pp. 257–282.

15. D. W. Murray and E. L. Wilson, "An Approximate Nonlinear Analysis of Plates," *WPAFB2 Conf.*, pp. 1207–1230.

16. K. S. Havner, "A Discretized Variational Formulation of Anisotropic Small Strain Plasticity Problems," *NucEDe*, Vol. 11, No. 2, 1969, pp. 308–322.

17. J. O. Smith and O. M. Sidebottom, *"Inelastic Behavior of Load-Carrying Members,"* John Wiley & Sons, New York, 1965.

18. O. C. Zienkiewicz and M. Watson, "Some Creep Effects in Stress Analysis with Particular Reference to Concrete Pressure Vessels," *NucEDe*, Vol. 4, No. 4, 1966, pp. 406–412.

19. G. A. Greenbaum and M. F. Rubinstein, "Creep Analysis of Axisymmetric Bodies Using Finite Elements," *NucEDe*, Vol. 7, No. 4, 1968, pp. 379–397.

20. A. P. Boresi and O. M. Sidebottom, "Creep of Metals Under Multiaxial States of Stress," *NucEDe*, Vol. 18, No. 3, 1972, pp. 415–456.

21. N. A. Cyr, R. D. Teter, and B. B. Stocks, "Finite Element Thermoelasto-plastic Analysis," *JSTDiv*, Vol. 98, No. ST7, 1972, pp. 1585–1603.

22. O. C. Zienkiewicz, S. Valliappan, and I. P. King, "Stress Analysis of Rock as a 'No Tension' Material," *Geotec*, Vol. 18, No. 1, 1968, pp. 56–66.

23. J. C. Jofriet and G. M. McNiece, "Finite Element Analysis of Reinforced Concrete Slabs," *JSTDiv*, Vol. 97, No. ST3, 1971, pp. 784–806.

24. S. Valliappan and T. F. Doolan, "Nonlinear Stress Analysis of Reinforced Concrete," *JSTDiv*, Vol. 98, No. ST4, 1972, pp. 885–898.

25. A. C. Scordelis, "Finite Element Analysis of Reinforced Concrete Structures," *FElCE2 Conf.*, pp. 71–114.

26. C. H. Lee and S. Kobayashi, "Elastoplastic Analysis of Plane-Strain and Axisymmetric Flat Punch Indentation by the Finite-Element Method," *IJMSci*, Vol. 12, No. 4, 1970, pp. 349–370.

27. R. W. H. Wu and E. A. Witmer, "Finite-Element Analysis of Large Elastic-Plastic Transient Deformations of Simple Structures," *AIAAJ*, Vol. 9, No. 9, 1971, pp. 1719–1724.

28. P. B. Lindley, "Plane-Stress Analysis of Rubber at High Strains Using Finite Elements," *JStrAn*, Vol. 6, No. 1, 1971, pp. 45–52.

29. J. T. Oden, "Finite Element Applications in Nonlinear Structural Analysis," *FElCE1 Sympos.*, pp. 419–456.

30. O. Egeland, "Application of Finite Element Techniques to Plasticity Problems," in *Finite Element Methods in Stress Analysis*, ed. by I. Holand and K. Bell, Tapir Press, Trondheim, 1969, pp. 435–450.

31. Y. Ueda and T. Yamakawa, "Thermal Nonlinear Behavior of Structures," *USJap2 Conf.*, pp. 375–392.

32. J. Ghaboussi, E. L. Wilson, and J. Isenberg, "Finite Element for Rock Joints and Interfaces," *JSMDiv*, Vol. 99, No. SM10, 1973, pp. 833–848.

33. K. I. Majid, *Non-Linear Structures*, Butterworths, London, 1972.

34. D. N. Trikha and A. D. Edwards, "Analysis of Concrete Box Girders Before and After Cracking," *PInsCE*, Vol. 53, Part 2, 1972, pp. 515–528.

35. C. S. Desai, ed., *Applications of the Finite Element Method in Geotechnical Engineering*, Proceedings of the Symposium held at Vicksburg, Miss., May 1972, U. S. Army Engineer Waterways Experiment Station, Corps of Engineers, Vicksburg, Miss., 1972.

36. R. S. Barsoum, "A Convergent Method for Cyclic Plasticity Analysis with Application to Nuclear Components," *IJNME*, Vol. 6, No. 2, 1973, pp. 227–236.

37. H. S. Levine, H. Armen, R. Winter, and A. Pifko, "Nonlinear Behavior of Shells of Revolution under Cyclic Loading," *CompSt*, Vol. 3, No. 3, 1973, pp. 589–617.

38. N. A. Cyr and R. D. Teter, "Finite Element Elastic-Plastic-Creep Analysis of Two-Dimensional Continuum with Temperature Dependent Material Properties," *CompSt*, Vol. 3, No. 4, 1973, pp. 849–863.

39. E. F. Boyle and A. Jennings, "Accelerating the Convergence of Elastic-Plastic Stress Analysis," *IJNME*, Vol. 7, No. 2, 1973, pp. 232–235.

40. M. Suidan and W. C. Schnobrich, "Finite Element Analysis of Reinforced Concrete," *JSTDiv*, Vol. 99, No. ST10, 1973, pp. 2109–2122.

CHAPTER SIXTEEN

1. B. Noble, *Applied Linear Algebra*, Prentice-Hall, Inc., Englewood Cliffs, 1969.

2. L. Fox, *An Introduction to Numerical Linear Algebra*, Oxford University Press, New York, 1965.

3. M. Geradin, "Error Bounds for Eigenvalue Analysis by Elimination of Variables," *JSouVi*, Vol. 19, No. 2, 1971, pp. 111–132.

4. I. Fried, "Condensation of Finite Element Eigenproblems," *AIAAJ*, Vol. 10, No. 11, 1972, pp. 1529–1530.

5. I. Fried, "Accuracy of Finite Element Eigenproblems," *JSouVi*, Vol. 18, No. 2, 1971, pp. 289–295.

6. C. F. Beck, "Responsibilities of Computer-Aided Design," *CivEng*, June 1972, pp. 58–60.

7. J. E. Walz, R. E. Fulton, and N. J. Cyrus, "Accuracy and Convergence of Finite Element Approximations," *WPAFB2 Conf.*, pp. 995–1027.

8. M. W. Chernuka, G. R. Cowper, G. M. Lindberg, and M. D. Olson, "Finite Element Analysis of Plates with Curved Edges," *IJNME*, Vol. 4, No. 1, 1972, pp. 49–65.

9. G. R. Cowper, "Variational Procedures and Convergence of Finite Element Methods," *ONR Sympos.*, pp. 1–12.

10. J. T. Oden and H. J. Brauchli, "A Note on Accuracy and Convergence of Finite Element Approximations," *IJNME*, Vol. 3, No. 2, 1971, pp. 291–292.

11. R. D. Henshell, D. Walters, and G. B. Warburton, "A New Family of Curvilinear Plate Bending Elements for Vibration and Stability," *JSouVi*, Vol. 20, No. 3, 1972, pp. 381–397 (discussion by I. Fried and authors' closure: Vol. 23, No. 4, 1972, pp. 507–513).

12. S. H. Crandall, *Engineering Analysis*, McGraw-Hill, New York, 1956, pp. 171–173.

13. I. Fried, "Accuracy of Complex Finite Elements," *AIAAJ*, Vol. 10, No. 3, 1972, pp. 347–349.

14. I. Fried, "Condition of Finite Element Matrices Generated from Nonuniform Meshes," *AIAAJ*, Vol. 10, No. 2, 1972, pp. 219–221.

15. G. L. Rigby and G. M. McNeice, "A Strain Energy Basis for Studies of Element Stiffness Matrices," *AIAAJ*, Vol. 10, No. 11, 1972, pp. 1490–1493.

16. R. A. Rosanoff and T. A. Ginsburg, "Matrix Error Analysis for Engineers," *WPAFB1 Conf.*, pp. 887–910.

17. R. A. Rosanoff, J. F. Gloudemann, and S. Levy, "Numerical Conditioning of Stiffness Matrix Formulations for Frame Structures," *WPAFB2 Conf.*, pp. 1029–1060.

18. J. R. Roy, "Numerical Error in Structural Solutions," *JSTDiv*, Vol. 97, No. ST4, 1971, pp. 1039–1054 (author's closure to discussion: Vol. 98, No. ST7, 1972, pp. 1663–1666).

19. R. J. Melosh, "Manipulation Errors in Finite Element Analysis," *USJap1 Conf.*, pp. 857–877.

20. Y. Yamamoto, "Some Considerations on Round-Off Errors of the Finite Element Method," *USJap2 Conf.*, pp. 69–86.

21. S. Kelsey, K. N. Lee, and C. K. K. Mak, "The Condition of Some Finite Element Coefficient Matrices," *CompAE Sympos.*, pp. 267–283.

22. I. Fried, "Bounds on the Extremal Eigenvalues of the Finite Element Stiffness and Mass Matrices and Their Spectral Condition Number," *JSouVi*, Vol. 22, No. 4, 1972, pp. 407–418.

23. R. J. Melosh and E. L. Palacol, *Manipulation Errors in Finite Element Analysis of Structures*, Report NASA CR-1385, Philco-Ford Corp., August 1969 (N69-34360, N.T.I.S.).

24. B. M. Irons, "Roundoff Criteria in Direct Stiffness Solutions," *AIAAJ*, Vol. 6, No. 7, 1968, pp. 1308–1312.

25. W. C. Hurty, "Truncation Errors in Natural Frequencies as Computed by the Method of Component Mode Synthesis," *WPAFB1 Conf.*, pp. 803–821.

26. G. Forsythe and C. B. Moler, *Computer Solution of Linear Algebraic Systems*, Prentice-Hall, Englewood Cliffs, 1967.

27. R. M. Jones, "Effective Development, Documentation, and Distribution of Computer Programs," *CompSt*, Vol. 2, Nos. 5/6, 1972, pp. 1089–1095.

CHAPTER SEVENTEEN

(Section 17.1)

1.1. E. Anderheggen, "Finite Element Plate Bending Equilibrium Analysis," *JEMDiv*, Vol. 95, No. EM4, 1969, pp. 841–857.

1.2. B. Fraejis de Veubeke and G. Sander, "An Equilibrium Model for Plate Bending," *IJSoSt*, Vol. 4, No. 4, 1968, pp. 447–468.

1.3. B. Fraejis de Veubeke, "Displacement and Equilibrium Models in the Finite Element Method," Chapter 9 of *Stress Analysis*, ed. by O. C. Zienkiewicz and G. S. Holister, John Wiley & Sons, New York, 1965, pp. 145–197.

1.4. L. S. D. Morley, "A Triangular Equilibrium Element with Linearly Varying Bending Moments for Plate Bending Problems," *AeroJ*, Vol. 71, No. 682, 1967, pp. 715–719.

1.5. L. R. Herrmann, "Finite-Element Bending Analysis for Plates," *JEMDiv*, Vol. 93, No. EM5, 1967, pp. 13–26.

1.6. L. R. Herrmann and D. M. Campbell, "A Finite-Element Analysis for Thin Shells," *AIAAJ*, Vol. 6, No. 10, 1968, pp. 1842–1847.

1.7. W. Visser, "A Refined Mixed-Type Plate Bending Element," *AIAAJ*, Vol. 7, No. 9, 1969, pp. 1801–1803.

1.8. C. A. Prato, "Shell Finite Element Method via Reissner's Principle," *IJSoSt*, Vol. 5, No. 10, 1969, pp. 1119–1133.

1.9. R. D. Cook, "Eigenvalue Problems with a 'Mixed' Plate Element," *AIAAJ*, Vol. 7, No. 5, 1969, pp. 982–983.

1.10. T. H. H. Pian, "Derivation of Element Stiffness Matrices by Assumed Stress Functions," *AIAAJ*, Vol. 2, No. 7, 1964, pp. 1333–1336.

1.11. T. H. H. Pian and P. Tong, "Rationalization in Deriving Element Stiffness Matrix by Assumed Stress Approach," *WPAFB2 Conf.*, pp. 441–469.

1.12.   P. Tong and T. H. H. Pian, "A Variational Principle and the Convergence of a Finite-Element Method Based on Assumed Stress Distribution," *IJSoSt*, Vol. 5, No. 5, 1969, pp. 463–472.

1.13.   R. D. Henshell, B. K. Neale, and G. B. Warburton, "A New Hybrid Cylindrical Shell Finite Element," *JSouVi*, Vol. 16, No. 4, 1971, pp. 519–531 (discussion and closure: Vol. 21, No. 3, 1972, pp. 369–378).

1.14.   T. H. H. Pian and S. T. Mau, "Some Recent Studies in Assumed Stress Hybrid Models," *USJap2 Conf.*, pp. 87–106.

1.15.   Y. Yoshida, "Equivalent Finite Elements of Different Bases," *USJap2 Conf.*, pp. 133–149.

1.16.   P. L. Gould and S. K. Sen, "Refined Mixed Method Finite Elements for Shells of Revolution," *WPAFB3 Conf.*

1.17.   L. R. Herrmann and W. E. Mason, "Mixed Formulations for Finite Element Shell Analysis," *CompSh Conf.*, pp. 290–336.

1.18.   B. Fraejis de Veubeke and O. C. Zienkiewicz, "Strain Energy Bounds in Finite Element Analysis by Slab Analogy," *JStrAn*, Vol. 2, No. 4, 1967, pp. 265–271.

1.19.   Z. M. Elias, "Duality in Finite Element Methods," *JEMDiv*, Vol. 94, No. EM4, 1968, pp. 931–946.

1.20.   N. D. Hung, "Duality in the Analysis of Shells by the Finite Element Method," *IJSoSt*, Vol. 7, No. 3, 1971, pp. 281–299.

1.21.   E. R. Arantes e Oliveira, "The Convergence Theorems and Their Role in the Theory of Structures," *IUTAM Sympos.*, pp. 1–42.

1.22.   G. Sander, "Application of the Dual Analysis Principle," *IUTAM Sympos.*, pp. 167–207.

1.23.   F. Kikuchi and Y. Ando, "Some Finite Element Solutions for Plate Bending Problems by Simplified Hybrid Displacement Method," *NucEDe*, Vol. 23, No. 2, 1972, pp. 155–178.

1.24.   J. P. Wolf, "Systematic Enforcement of Stress Boundary Conditions in the Assumed Stress Hybrid Model Based on the Deformation Method," *SMRT1 Conf.*, Vol. 6, Part M, pp. 463–484.

1.25.   K. Hellan, "On the Unity of the Constant Strain/Constant Moment Finite Element Methods," *IJNME*, Vol. 6, No. 2, 1973, pp. 191–200.

(Section 17.2)

2.1.   M. E. Fourney, "A Pseudo Two-Dimensional Photoelastic Method of Testing Axisymmetric Geometries," *ExperM*, Vol. 11, No. 1, 1971, pp. 19–25.

2.2. L. R. Herrmann, "Elasticity Equations for Incompressible and Nearly Incompressible Materials by a Variational Theorem," *AIAAJ*, Vol. 3, No. 10, 1965, pp. 1896–1900.

2.3. P. Tong, "An Assumed Stress Hybrid Finite Element Method for an Incompressible and Near-Incompressible Material," *IJSoSt*, Vol. 5, No. 5, 1969, pp. 455–461.

2.4. R. L. Taylor, K. S. Pister, and L. R. Herrmann, "On a Variational Theorem for Incompressible and Nearly Incompressible Orthotropic Elasticity," *IJSoSt*, Vol. 4, No. 9, 1968, pp. 875–883.

2.5. T. J. R. Hughes and H. Allik, "Finite Elements for Compressible and Incompressible Continua," *FElCE1 Sympos.*, pp. 27–62.

2.6. S. W. Key, "A Variational Principle for Incompressible and Nearly-Incompressible Anisotropic Elasticity," *IJSoSt*, Vol. 5, No. 9, 1969, pp. 951–964.

2.7. S. W. Key, *A Finite Element Program for Orthotropic Incompressible Axisymmetric Elasticity Problems*, Report SC-R-68-1868, Sandia Laboratories, Albuquerque, New Mexico, December 1968.

(Section 17.3)

3.1. R. T. Severn, "Inclusion of Shear Deflection in the Stiffness Matrix for a Beam Element," *JStrAn*, Vol. 5, No. 4, 1970, pp. 239–241.

3.2. P. Tong, "New Displacement Hybrid Finite Element Models for Solid Continua," *IJNME*, Vol. 2, No. 1, 1970, pp. 73–83.

3.3 D. L. Thomas, J. M. Wilson, and R. R. Wilson, "Timoshenko Beam Finite Elements," *JSouVi*, Vol. 31, No. 3, 1973, pp. 315–330.

3.4. R. E. Nickel and G. A. Secor, "Convergence of Consistently Derived Timoshenko Beam Finite Elements," *IJNME*, Vol. 5, No. 2, 1972, pp. 243–253.

3.5. T. Kawai, "The Application of Finite Element Methods to Ship Structures," *CompSt*, Vol. 3, No. 5, 1973, pp. 1175–1194.

3.6. R. H. Gallagher and C. H. Lee, "Matrix Dynamic and Instability Analysis with Non-Uniform Elements," *IJNME*, Vol. 2, No. 2, 1970, pp. 265–275.

3.7. L. R. L. Wang, "Parametric Matrices of Some Structural Members," *JSTDiv*, Vol. 96, No. ST8, 1970, pp. 1735–1759.

3.8. R. S. Barsoum and R. H. Gallagher, "Finite Element Analysis of Torsional and Torsional-Flexural Stability Problems," *IJNME*, Vol. 2, No. 3, 1970, pp. 335–352.

3.9.   D. Krajcinovic, "A Consistent Discrete Elements Technique for Thin-walled Assemblages," *IJSoSt*, Vol. 5, No. 7, 1969, pp. 639–662.

3.10. R. J. Reilly, "Stiffness Analysis of Grids Including Warping," *JSTDiv*, Vol. 98, No. ST7, 1972, pp. 1511–1523.

3.11. K. L. Lawrence, "Twisted Beam Element Matrices for Bending," *AIAAJ*, Vol. 8, No. 6, 1970, pp. 1160–1162.

3.12. H. P. Lee, "Generalized Stiffness Matrix of a Curved-Beam Element," *AIAAJ*, Vol. 7, No. 10, 1969, pp. 2043–2045.

3.13. R. Davis, R. D. Henshell, and G. B. Warburton, "Curved Beam Finite Elements for Coupled Bending and Vibration," *EEStDy*, Vol. 1, No. 2, 1972, pp. 165–175.

3.14. D. G. Ashwell and A. B. Sabir, "Limitations of Certain Curved Finite Elements when Applied to Arches," *IJMSci*, Vol. 13, No. 2, 1971, pp. 133–139.

3.15. D. G. Ashwell, A. B. Sabir, and T. M. Roberts, "Further Studies in the Application of Curved Finite Elements to Circular Arches," *IJMSci*, Vol. 13, No. 6, 1971, pp. 507–517 (errata: Vol. 15, No. 4, 1973, pp. 325–327).

3.16. G. Hommel, discussion of Refs. 3.14 and 3.15, *IJMSci*, Vol. 14, No. 4, 1972, pp. 275–277.

3.17. M. Petyt and C. C. Fleischer, "Free Vibration of a Curved Beam," *JSouVi*, Vol. 18, No. 1, 1971, pp. 17–30.

3.18. A. B. Sabir and D. G. Ashwell, "A Comparison of Curved Beam Finite Elements When Used in Vibration Problems," *JSouVi*, Vol. 18, No. 4, 1971, pp. 555–563.

3.19. K. M. Ahmed, "Free Vibration of Curved Sandwich Beams by the Method of Finite Elements," *JSouVi*, Vol. 18, No. 1, 1971, pp. 61–74.

3.20. M. A. Crisfield, "Finite Element Methods for the Analysis of Multi-cellular Structures," *PInsCE*, Vol. 48, 1971, pp. 413–437.

3.21. K. J. Willam and A. C. Scordelis, "Cellular Structures of Arbitrary Plan Geometry," *JSTDiv*, Vol. 98, No. ST7, 1972, pp. 1377–1394.

3.22. E. Dokumaci, J. Thomas, and W. Carnegie, "Matrix Displacement Analysis of Coupled Bending-Bending Vibrations of Pretwisted Blading," *JMESci*, Vol. 9, No. 4, 1967, pp. 247–254.

(Section 17.4)

4.1.   I. S. Raju and A. K. Rao, "Stiffness Matrices for Sector Elements," *AIAAJ*, Vol. 7, No. 1, 1969, pp. 156–157 (discussion and closure: Vol. 8, No. 1, 1970, p. 192).

4.2. A. Fam and C. J. Turkstra, "An Annular In-Plane Finite Element," *FElCE2 Conf.*, pp. 263–286.

4.3. C. R. Lansberry and S. Shore, "A Fully Compatible Annular Segment Finite Element," *FElCE2 Conf.*, pp. 233–262.

4.4. M. D. Olson and G. M. Lindberg, "Annular and Circular Sector Finite Elements for Plate Bending," *IJMSci*, Vol. 12, No. 1, 1970, pp. 17–33.

4.5. F. Sawko and P. A. Merriman, "An Annular Segment Finite Element for Plate Bending," *IJNME*, Vol. 3, No. 1, 1971, pp. 119–129 (discussion and closure: Vol. 3, No. 4, 1971, pp. 589–591).

4.6. S. Singh and G. S. Ramaswamy, "A Sector Element for Thin Plate Flexure," *IJNME*, Vol. 4, No. 1, 1972, pp. 133–142.

4.7. T. G. Brown and A. Ghali, "Finite Strip Analysis of Skew Slabs," *FElCE2 Conf.*, pp. 1141–1158.

4.8. P. V. T. Babu and D. V. Reddy, "Frequency Analysis of Skew Orthotropic Plates by the Finite Strip Method," *JSouVi*, Vol. 18, No. 4, 1971, pp. 465–474.

4.9. G. R. Monforton and L. A. Schmit Jr., "Finite Element Analysis of Skew Plates in Bending," *AIAAJ*, Vol. 6, No. 6, 1968, pp. 1150–1152.

4.10. G. R. Monforton and M. G. Michail, "Finite Element Analysis of Skew Sandwich Plates," *JEMDiv*, Vol. 98, No. EM3, 1972, pp. 763–769.

4.11. G. R. Monforton, "Some Orthotropic Skew Plate Finite Element Results," *JSTDiv*, Vol. 98, No. ST4, 1972, pp. 955–959.

4.12. D. J. Dawe, "Parallelogrammic Elements in the Solution of Rhombic Cantilever Plate Problems," *JStrAn*, Vol. 1, No. 3, 1966, pp. 223–230.

4.13. I. S. Raju, A. V. Krishna Murty, and A. K. Rao, "Sector Elements for Matrix Displacement Analysis," *IJNME*, Vol. 6, No. 4, 1973, pp. 553–563.

(Section 17.5)

5.1. Y. K. Cheung and O. C. Zienkiewicz, "Plates and Tanks on Elastic Foundations—An Application of Finite Element Method," *IJSoSt*, Vol. 1, No. 4, 1965, pp. 451–461.

5.2. Y. K. Cheung and D. K. Nag, "Plates and Beams on Elastic Foundations—Linear and Non-Linear Behavior," *Geotec*, Vol. 18, No. 2, 1968, pp. 250–260.

5.3. A. D. Kerr, "Elastic and Viscoelastic Foundation Models," *JApplM*, Vol. 31, No. 3, 1964, pp. 491–498.

5.4. A. D. Kerr, "A Study of a New Foundation Model," *ActaM*, Vol. 1, No. 2, 1965, pp. 135–147.

5.5. D. Q. Fletcher and L. R. Herrmann, "Elastic Foundation Representation of Continuum," *JEMDiv*, Vol. 97, No. EM1, 1971, pp. 95–107.

5.6. T. Y. Yang, "A Finite Element Analysis of Plates on a Two Parameter Foundation Model," *CompSt*, Vol. 2, No. 4, 1972, pp. 593–614.

(Section 17.6)

6.1. G. C. Best, "Helpful Formulas for Integrating Polynomials in Three Dimensions," *MathCo*, Vol. 18, No. 86, 1964, pp. 310–312.

6.2. I. Fried, "Some Aspects of the Natural Coordinate System in the Finite-Element Method," *AIAAJ*, Vol. 7, No. 7, 1969, pp. 1366–1368.

6.3. P. C. Hammer, O. J. Marlow, and A. H. Stroud, "Numerical Integration Over Simplexes and Cones," *MTAC*, Vol. 10, No. 3, 1956, pp. 130–137.

6.4. B. M. Irons, "A Conforming Quartic Triangular Element for Plate Bending," *IJNME*, Vol. 1, No. 1, 1969, pp. 29–45.

6.5. R. W. Clough and C. A. Felippa, "A Refined Quadrilateral Element for Analysis of Plate Bending," *WPAFB2 Conf.*, pp. 399–440.

6.6. J. Radon, "Zur Mechanischen Kubatur," *MoMath*, Vol. 52, No. 4, 1948, pp. 286–300.

6.7. L. S. D. Morley and B. C. Merrifield, "On the Conforming Cubic Triangular Element for Plate Bending," *CompSt*, Vol. 2, Nos. 5/6, 1972, pp. 875–892.

6.8. G. R. Cowper, "Gaussian Quadrature Formulas for Triangles," *IJNME*, Vol. 7, No. 3, 1973, pp. 405–408.

(Section 17.7)

7.1. G. R. Cowper, E. Kosko, G. M. Lindberg, and M. D. Olson, *A High Precision Triangular Plate Bending Element*, Aeronautical Report LR-514, National Research Council of Canada, Ottawa, December 1968 (AD-685-576, N.T.I.S.).

7.2. R. A. Tinawi, "Anisotropic Tapered Elements Using Displacement Models," *IJNME*, Vol. 4, No. 4, 1972, pp. 475–489.

7.3. R. W. Luft, J. M. Roesset, and J. J. Connor, "Automatic Generation of Finite Element Matrices," *JSTDiv*, Vol. 97, No. ST1, 1971, pp. 349–362.

7.4. R. H. Gunderson and A. Cetiner, "Element Stiffness Matrix Generator," *JSTDiv*, Vol. 97, No. ST1, 1971, pp. 363–375.

CHAPTER EIGHTEEN

(Section 18.1)

1.1. R. E. Jones, "A Generalization of the Direct Stiffness Method of Structural Analysis," *AIAAJ*, Vol. 2, No. 5, 1964, pp. 821–826.

1.2. B. E. Greene, R. E. Jones, R. W. McLay, and D. R. Strome, "Generalized Variational Principles in the Finite Element Method," *AIAAJ*, Vol. 7, No. 7, 1969, pp. 1254–1260.

1.3. R. W. McLay, "A Special Variational Principle for the Finite Element Method," *AIAAJ*, Vol. 7, No. 3, 1969, pp. 533–534.

1.4. S. W. Key, "A Specialization of Jones' Generalization of the Direct Stiffness Method of Structural Analysis," *AIAAJ*, Vol. 9, No. 5, 1971, pp. 984–985.

1.5. K. Washizu, *Variational Methods in Elasticity and Plasticity*, Pergamon Press, Oxford, 1968.

1.6. F. Kikuchi and Y. Ando, "A New Variational Functional for the Finite Element Method and Its Application to Plate and Shell Problems" *NucEDe*, Vol. 21, No. 1, 1972, pp. 95–113.

1.7. O. C. Zienkiewicz, *The Finite Element Method in Engineering Science*, McGraw-Hill, London, 1971.

1.8. O. C. Zienkiewicz and C. J. Parekh, "Transient Field Problems: Two Dimensional and Three Dimensional Analysis by Isoparametric Finite Elements," *IJNME*, Vol. 2, No. 1, 1970, pp. 61–71.

1.9. B. A. Szabo and G. C. Lee, "Stiffness Matrix for Plates by Galerkin's Method," *JEMDiv*, Vol. 95, No. EM3, 1969, pp. 571–585.

1.10. J. W. Leonard and T. T. Bramlette, "Finite Element Solutions for Differential Equations," *JEMDiv*, Vol. 96, No. EM6, 1970, pp. 1277–1283.

1.11. S. G. Hutton and D. L. Anderson, "Finite Element Method: A Galerkin Approach," *JEMDiv*, Vol. 97, No. EM5, 1971, pp. 1503–1520.

1.12. J. W. Leonard, "Finite Element Analysis of Perturbed Compressible Flow," *IJNME*, Vol. 4, No. 1, 1972, pp. 123–132.

1.13. J. W. Leonard and D. Melfi, "Three-Dimensional Finite Element Model for Lake Circulation," *WPAFB3 Conf.*

1.14.   J. H. Dieterich and E. T. Onat, "Slow Finite Deformations of Viscous Solids," *JGeoRe*, Vol. 74, No. 8, 1969, pp. 2081–2088.

1.15.   A. F. Emery and W. W. Carson, "An Evaluation of the Use of the Finite Element Method in the Computation of Temperature," *JHeTra*, Vol. 93, No. 2, 1971, pp. 136–145.

1.16.   R. D. Cook, "Comment on 'Discrete Element Idealization of an Incompressible Liquid for Vibration Analysis' and 'Discrete Element Structural Theory of Fluids'," *AIAAJ*, Vol. 11, No. 5, 1973, pp. 766–767.

1.17.   D. W. Malone and J. J. Connor, "Finite Elements and Dynamic Viscoelasticity," *JEMDiv*, Vol. 97, No. EM4, 1971, pp. 1145–1158.

1.18.   G. de Vries and D. H. Norrie, "The Application of the Finite Element Technique to Potential Flow Problems," *JApplM*, Vol. 38, No. 4, 1971, pp. 798–802.

1.19.   P. Tong and Y. C. Fung, "Slow Particulate Viscous Flow in Channels and Tubes—Application to Biomechanics," *JApplM*, Vol. 38, No. 4, 1971, pp. 721–728.

1.20.   P. Tong and D. Vawter, "An Analysis of Peristaltic Pumping," *JApplM*, Vol. 39, No. 4, 1972, pp. 857–862.

1.21.   G. L. Guymon, V. H. Scott, and L. R. Herrmann, "A General Numerical Solution of the Two-Dimensional Diffusion-Convection Equation by the Finite Element Method," *WaterR*, Vol. 6, No. 6, 1970, pp. 1611–1617.

1.22.   C. Taylor and J. F. O'Callaghan, "A Numerical Solution of the Elastohydrodynamic Lubrication Problem Using Finite Elements," *JMESci*, Vol. 14, No. 4, 1972, pp. 229–237.

1.23.   A Craggs, "The Use of Simple Three-Dimensional Acoustic Finite Elements for Determining the Natural Modes and Frequencies of Complex Shaped Enclosures," *JSouVi*, Vol. 23, No. 3, 1972, pp. 331–339.

1.24.   R. S. Sandhu and E. L. Wilson, "Finite Element Analysis of Seepage in Elastic Media," *JEMDiv*, Vol. 95, No. EM3, 1969, pp. 641–652.

1.25.   R. H. Gallagher, "Applications of Finite Element Analysis," *USJap2 Conf.*, pp. 641–678.

1.26.   J. T. Oden, "A General Theory of Finite Elements: I. Topological Considerations. II. Applications," *IJNME*, Vol. 1, 1969: No. 2, pp. 205–221; No. 3, pp. 247–259.

1.27.   T. T. Bramlette and R. H. Mallett, "A Finite Element Solution Technique for the Boltzmann Equation," *JFlMech*, Vol. 42, Part 1, 1970, pp. 177–191.

1.28.   J. Ghaboussi and E. L. Wilson, "Flow of Compressible Fluid in Porous Elastic Media," *IJNME*, Vol. 5, No. 3, 1973, pp. 419–442.

1.29. C. Taylor and P. Hood, "A Numerical Solution of the Navier-Stokes Equations Using the Finite Element Technique," *CompFl*, Vol. 1, No. 1, 1973, pp. 73–100.

1.30. L. A. Loziuk, J. C. Anderson, and T. Belytschko, "Hydrothermal Analysis by Finite Element Method," *JHYDiv*, Vol. 98, No. HY11, 1972, pp. 1983–1998.

1.31. O. C. Zienkiewicz, P. Mayer, and Y. K. Cheung, "Solution of Anisotropic Seepage by Finite Elements," *JEMDiv*, Vol. 92, No. EM1, 1966, pp. 111–120.

1.32. E. L. Wilson and R. E. Nickell, "Application of the Finite Element Method to Heat Conduction Analysis," *NucEDe*, Vol. 4, No. 3, 1966, pp. 276–286.

1.33. O. C. Zienkiewicz, A. K. Bahrani, and P. L. Arlett, "Solution of Three-Dimensional Field Problems by the Finite Element Method," *Engin*, Vol. 224, No. 5831, 1967, pp. 547–550.

(Section 18.2)

2.1. A. K. Rao, I. S. Raju, and A. V. Krishna Murty, "A Powerful Hybrid Method in Finite Element Analysis," *IJNME*, Vol. 3, No. 3, 1971, pp. 389–403.

2.2. N. Levy, P. V. Marcal, W. J. Ostergren, and J. R. Rice, "Small Scale Yielding Near a Crack Tip in Plane Strain: A Finite Element Analysis," *IJFraM*, Vol. 7, No. 2, 1971, pp. 143–156.

2.3. E. Byskov, "The Calculation of Stress Intensity Factors Using the Finite Element Method With Cracked Elements," *IJFraM*, Vol. 6, No. 2, 1970, pp. 159–167.

2.4. P. Tong, T. H. H. Pian, and S. J. Lasry, "A Hybrid-Element Approach to Crack Problems in Plane Elasticity," *IJNME*, Vol. 7, No. 3, 1973, pp. 297–308.

2.5. J. R. Rice and D. M. Tracey, "Computational Fracture Mechanics," *ONR Sympos.*, pp. 585–623.

2.6. P. F. Walsh, "The Computation of Stress Intensity Factors by a Special Finite Element Technique," *IJSoSt*, Vol. 7, No. 10, 1971, pp. 1333–1342.

2.7. G. P. Anderson, V. L. Ruggles, and G. S. Stibor, "Use of Finite Element Computer Programs in Fracture Mechanics," *IJFraM*, Vol. 7, No. 1, 1971, pp. 63–76.

2.8. R. K. Leverenz, "A Finite Element Stress Analysis of a Crack in a Bi-Material Plate," *IJFraM*, Vol. 8, No. 3, 1972, pp. 311–324.

2.9. H. C. Fu, "Indirect Structural Analysis by Finite Element Method," *JSTDiv*, Vol. 99, No. ST1, 1973, pp. 91–11 1.

2.10. Y. Yamamoto and N. Tokuda, "Determination of Stress Intensity Factors in Cracked Plates by the Finite Element Method," *IJNME*, Vol. 6, No. 3, 1973, pp. 427–439.

(Section 18.3)

3.1. R. W. Clough and R. J. Woodward III, "Analysis of Embankment Stresses and Deformations," *JSMDiv*, Vol. 93, No. SM4, 1967, pp. 529–549.

3.2. J. M. Duncan and G. W. Clough, "Finite Element Analysis of Port Allen Lock," *JSMDiv*, Vol. 97, No. SM8, 1971, pp. 1053–1068.

(Section 18.4)

4.1. D. N. Yates, T. J. Vinson, and W. W. Sable, "The Development of Large Scale Digital Computer Codes for Production Structural Analysis," *CompSh Conf.*, pp. 172–222.

4.2. J. L. Tocher and C. A. Felippa, "Computer Graphics Applied to Production Structural Analysis," *IUTAM Sympos.*, pp. 522–545.

4.3. W. R. Buell and B. A. Bush, "Mesh Generation—A Survey," *JEInd*, Vol. 95, No. 1, 1973, pp. 332–338.

4.4. J. Suhara and J. Fukuda, "Automatic Mesh Generation for Finite Element Analysis," *USJap2 Conf.*, pp. 607–624.

4.5. H. A. Kamel and H. K. Eisenstein, "Automatic Mesh Generation in Two and Three Dimensional Inter-Connected Domains," *IUTAM Sympos.*, pp. 455–475.

4.6. W. J. Batdorf and S. S. Kapur, "FLING—A Fortran Language for Interactive Graphics," *ONR Sympos.*, pp. 513–542.

4.7. H. N. Christiansen, "Displays of Kinematic and Elastic Systems," *WPAFB3 Conf.*

4.8. R. D. Bousquet, D. N. Yates, W. W. Sable, and T. J. Vinson, "The Development of Computer Graphics for Large Scale Finite Element Codes," *WPAFB3 Conf.*

4.9. R. Douty and S. Shore, "Technique for Interactive Computer Graphics in Design," *JSTDiv*, Vol. 97, No. ST1, 1971, pp. 273–288.

4.10. W. F. Bates and W. M. Cox, "Structural Analysis by Computer Graphics," *JSTDiv*, Vol. 95, No. ST11, 1969, pp. 2433–2448.

4.11. C. A. Felippa, "An Alphanumeric Finite Element Mesh Plotter," *IJNME*, Vol. 5, No. 2, 1972, pp. 217–236.

4.12. Panel Discussion, "The Large General Purpose Code," *CompSh Conf.*, pp. 1063–1106.

(Section 18.5)

5.1. *NASTRAN: User's Experiences*, NASA Langley Design Center, Langley Station, Va., September 11–12, 1972 (N72-32867, N.T.I.S.).

5.2. P. V. Marcal, ed., *On General Purpose Finite Element Computer Programs*, papers presented at Winter Annual Meeting, ASME, New York, November 30, 1970.

5.3. C. W. McCormick, "Shell Analysis with Large General Purpose Computer Programs," *CompSt*, Vol. 1, Nos. 1/2, 1971, pp. 323–332.

5.4. D. N. Yates, W. W. Sable, and T. J. Vinson, "The DAISY Code," *ONR Sympos.*, pp. 175–210.

5.5. R. D. Logcher, "The Development of ICES STRUDL," *USJap2 Conf.*, pp. 591–606.

5.6. S. D. Hansen, G. L. Anderton, N. E. Connacher, and C. S. Dougherty, "Analysis of the 747 Aircraft Wing-Body Intersection," *WPAFB2 Conf.*, pp. 743–788.

5.7. R. H. Mallett and R. W. Braun, "Structural Analysis Practices for Large Scale Systems," *WPAFB2 Conf.*, pp. 789–838.

5.8. A. F. Grisham, "The Boeing SST Prototype Internal Loads Analysis System and Procedures," *WPAFB3 Conf.*

5.9. H. S. Iyengar, N. Amin, and L. Carpenter, "Computerized Design of World's Tallest Building," *CompSt*, Vol. 2, Nos. 5/6, 1972, pp. 771–783.

5.10. R. K. Henrywood, "The Design, Development, Documentation and Support of a Major Finite Element System," *CompAD*, Vol. 5, No. 3, 1973, pp. 160–165.

(Section 18.6)

6.1. N. Perrone, "Compendium of Structural Mechanics Computer Programs," *CompSt*, Vol. 2, No. 3, 1972, pp. 305–438 (also available as N72-32026, N.T.I.S.).

6.2. See Ref. 34 of Chapter Twelve.

6.3. J. E. Akin, D. L. Fenton, and W. C. T. Stoddart, *The Finite Element Method: A Bibliography of Its Theory and Applications*, Report EM-72-1, University of Tennessee, Knoxville, February 1972.

6.4. W. Allen, "Shared Software in the Production Environment," *CompSt*, Vol. 3, No. 1, 1973, pp. 59–76.

Index

in various elements, 43, 146-147, 182,
185, 187, 200, 201-205
and zero energy modes, 187
Stress intensity factor, 347
Stress-strain relations, coordinate transformation of, 211-214
for creep, 306-307
decomposition of matrix, 106, 108
linearly elastic, 11-14, 16, 83, 144, 148,
179-180, 182, 198
for plasticity, 293-296
temperature dependent, 156, 308
Structure size, and assembly of elements,
31-32, 35, 71-72
Sturm sequence, 250-252, 276
Subparametric elements, 113
Substructures, 227-231, 319, 346
Successive substitution, *see* Direct iteration
Superparametric elements, 113, 177-178
Superposed nodes, and error, 138
Support elements, stiff, 41-42, 288-289,
290, 318, 319
Support reactions, calculation of, 29, 39-42
Surface tractions, 8-9, 60-61, 78, 200
Symbolic generation of matrices, 342
Systems programmer, 48

Tangent stiffness, 277, 284-285, 288, 296,
298-301
Taylor series, 276-277, 280, 315
Thermal stress, *see* Initial stress and strain
Thickness of element, variable, 83, 114,
117, 168, 174, 177, 181, 185, 196,
206, 333
Timoshenko beam, 333
Torsion of shafts, 150
Transformation, *see* Coordinate transformation
Transient problems, *see* Dynamic response
Transverse shear strain and stiffness, accounting for, 175, 177, 194, 204, 331,
333
as error source, 208-209, 226, 253
and stress computation, 185, 201
and "thin" elements, 161, 173, 175, 208
209
Triangle, area of, 82-83, 340
centroid of, 84, 336

Triangular coordinates, *see* Area Coordinates
Truncation error, *see* Computational error
Truss elements, buckling and nonlinearity,
263, 280-283
mass matrix, 239, 256
stiffness matrices, 29-30, 34, 79-80, 232-
233
Two-force element, *see* Truss elements

Underrelaxation, 286, 290, 297
Undetermined coefficients, *see* Generalized
coordinates
Unstable structures, *see* Collapse, indication
of

Variational principles and methods, 61-64,
329-330; *see also* Functional
Vibration problems, beam elements for, 333
"economized" solutions, 244-248
as eigenvalue problems, 241-244
flutter, 255
large amplitude, 254
with membrane forces, 267
rotating structures, 254
in shells of revolution, 170, 254-255
Virtual work, and buckling, 265
for derivation of basic equation, 95
of forces and nodal loads, 65, 78, 96, 213,
221
principle of, 55
Volume, calculation of, 114, 185, 315, 340
and Jacobian matrix, 111, 114, 117
Von Mises criterion, 293

Warped elements, 292
Weighted residual methods, 345
Work, and potential energy, 54-55; *see also*
Virtual work

Yielding, *see* Material nonlinearity

Zero-energy deformation modes, control of,
186-188, 202, 204
described, 115
detection of, 115, 314
occurrences of, 115, 130-131, 140, 186-
187, 202, 204, 331
and stress computation, 187